Michael P. Monahan

STUDY GUIDE AND REVIEW MANUAL OF HUMAN ANATOMY

REGIONAL
SYSTEMIC
APPLIED

KEITH L. MOORE, M.Sc., Ph.D., F.I.A.C.
Professor and Chairman, Department of Anatomy, University of
Toronto, Faculty of Medicine, Toronto, Ontario, Canada; Formerly
Professor and Head of Anatomy in the University of Manitoba

EDWARD J. H. NATHANIEL, M.D., M.S., Ph.D.
Professor, Department of Anatomy, Faculties of Medicine and
Dentistry, University of Manitoba, Winnipeg, Canada

KAZUMASA HOSHINO, M.D., D. Med. Sc., M.I.A.C.
Professor, Department of Anatomy, Faculties of Medicine and
Dentistry, University of Manitoba, Winnipeg, Canada

TRIVEDI V. N. PERSAUD, M.D., Ph.D., D. Sc., M.R.C. (Path.)
Professor, Department of Anatomy, Faculties of Medicine and
Dentistry, University of Manitoba, Winnipeg, Canada

M. H. LINDSAY GIBSON, M.Sc., Ph.D.
Associate Professor, Department of Anatomy, Faculties of Medicine
and Dentistry, University of Manitoba, Winnipeg, Canada

1976

W. B. SAUNDERS COMPANY / PHILADELPHIA / LONDON / TORONTO

W. B. Saunders Company:　　West Washington Square
　　　　　　　　　　　　　　Philadelphia, PA 19105

　　　　　　　　　　　　　　1 St. Anne's Road
　　　　　　　　　　　　　　Eastbourne, East Sussex BN21 3UN, England

　　　　　　　　　　　　　　833 Oxford Street
　　　　　　　　　　　　　　Toronto, Ontario M8Z 5T9, Canada

Study Guide and Review Manual of HUMAN ANATOMY　　　　ISBN 0-7216-6494-6

c 1976 by W. B. Saunders Company. Copyright under the International Copyright Union. All rights reserved. This book is protected by copyright. No part of it may be reproduced, stored in a retrieval system, or transmitted in any form or by any means, electronic, mechanical, photocopying, recording, or otherwise, without written permission from the publisher. Made in the United States of America. Press of W. B. Saunders Company.

Last digit is the print number:　 9　 8　 7　 6　 5　 4　 3　 2　 1

PREFACE

During the last decade there has been much controversy about "new" and "old" anatomy and some uncertainty about how much anatomy should be taught. Because of the curtailment of time available for gross anatomy, there has been a trend to teach the anatomy that will be useful in clinical practice. Students naturally wish to know what the faculty expects them to learn during their undergraduate medical years. We believe much uncertainty can be dispelled by defining the instructional objectives of the undergraduate curriculum, and by emphasizing the foundation of anatomical knowledge upon which more can be added when detailed knowledge becomes necessary.

Mager* has said, "If you give a learner a copy of your objectives, you may not have to do much else." The implications of this statement support our view that medical students must accept responsibility for acquiring and continuing their education.

The objectives listed at the beginning of each chapter in this Study Guide and Review Manual define what students should be able to do in Anatomy when they begin their clinical studies. The objectives are based on the view that it is neither necessary nor desirable to expect students to master the vast amount of anatomical knowledge that exists and was once taught. By restricting the amount of formal instruction to "core anatomy" and by formulating instructional objectives, students are afforded the opportunity of acquiring this knowledge for themselves through lectures, reading, dissection, use of prosected material, models and audiovisual aids, and by interaction with basic science and clinical teachers.

The objectives defined in this guide are based on the stated goal of our faculty "to produce a physician who, at graduation, is able to embark upon any one of the potential pathways of graduate education and practice". Acquisition of anatomical knowledge beyond the stated objectives will depend greatly upon the branch of medicine in which the student wishes to practise.

Because multiple-choice examinations are being used more and more, and are formidable even to the best prepared, commonly used types of these questions have been developed around each region and system of the body, with special emphasis on the stated instructional objectives. Many questions have been formulated to exemplify the clinical applications of gross anatomy and to vitalize the students' interest in clinically oriented anatomy. Questions dealing with clinical problems and requiring anatomical knowledge are used frequently, and emphasis is placed on living and radiological anatomy because these aspects are routinely used by most doctors. The test questions are intended for those wishing to determine the state of their knowledge, and to improve their skills with multiple-choice examinations.

The terminology used is the <u>Nomina Anatomica</u>, third edition, 1966, translated into English where applicable. Names that are widely used clinically appear in parentheses.

---

*Mager, R. F. Preparing Instructional Objectives. Fearon Publishers, Belmont, California, 1962.

In this Study Guide and Review Manual, anatomy is treated both by regions and by systems. Many structures are found in different regions of the body, e.g., the vagus nerves, but it is not feasible to dissect the body system by system for technical reasons; thus, it is dissected region by region. Moreover a good knowledge of the structures occupying any given region is absolutely essential for many clinical purposes, e.g., in cut-wounds of the throat. Hence, a study of regional anatomy is scarcely less important than that of the various systems.

Because we have attempted to establish what aspects of anatomy are important for every medical student to know, we should appreciate receiving constructive criticism of the instructional objectives and questions. We should like the questions and answers to be free of ambiguity and representative not only of important aspects of anatomy, but also of high standards of education.

The radiographs were taken by Dr. Paul Major of the Departments of Radiology and Anatomy; the illustrations were drawn by one of us (Dr. M.H. Lindsay Gibson), and the photographs were taken by Brenda Bell, Medical Photographer, Department of Anatomy. We thank Mr. Brian Decker, Associate Medical Editor of the W.B. Saunders Company, for encouraging us to do this Study Guide and Review Manual, and for his helpful advice during its preparation. In typing the manuscript, we have had the friendly help of Barbara Clune, Roslyn Hoad, Roberta Biedron, and Rosemary Fletcher.

Winnipeg, Canada   The Authors

USER'S GUIDE

This guide is designed to help you study, and later review, human anatomy by providing learning objectives and various types of multiple-choice questions based on these objectives.

The questions are not intended as a substitute for careful study, but to enable you to detect areas of weakness and to afford you the opportunity to correct deficits in your knowledge. Although answers to questions are explained and relevant notes are given, you should consult your textbook for a comprehensive review of difficult concepts and processes. Through discussion of weak areas with your colleagues and instructors, you can test your ability to do the things listed as objectives at the beginning of each chapter.

To use this guide most effectively, the following steps are suggested.

1. Read the objectives listed at the beginning of the chapter you plan to study.

2. Carefully study the appropriate chapter(s) in your textbook, focusing on the objectives.

3. After study, or the lecture, answer the multiple-choice questions. The questions indicate how much knowledge you are expected to acquire to fulfill the objectives. The questions are similar to those used in various board and school multiple-choice examinations, and are designed to be answered at the rate of about one per minute.

    As you complete each set of questions, check your answers. If any of your answers are wrong, read the notes and explanations and study the appropriate material and illustrations in your textbooks before proceeding to the next set of questions.

4. If you get 80 per cent or more of the questions correct on the first trial, or during a subsequent review, you have performed very well and should have no difficulty answering similar questions based on the objectives given in this guide.

5. When you have completed the study of a region, attempt the review examination on the region. If your level of performance is not superior, determine where your knowledge is defective and attempt the examination at a later date.

6. As you complete the study of a system, attempt the review examination on the system. If you do poorly on the examination, revise your knowledge and attempt the examination at a subsequent time. Your level of performance should be no lower than the superior-very superior range.

CONTENTS

PART ONE.   REGIONAL ANATOMY

1. THE UPPER LIMB. 3
    Axilla and Pectoral Region -- Shoulder -- Arm
    and Elbow -- Forearm and Wrist -- Hand.
    GGO:70; B:75; M:298; S:359; WW:318,421,532,639,1063.*

2. THE LOWER LIMB. 38
    Hip and Gluteal Region -- Thigh -- Knee and
    Popliteal Fossa -- Leg -- Ankle and Foot.
    GGO:164; B:309; M:298; S:489; WW:344,422,559,673,1063.

3. THE HEAD AND NECK. 71
    Skull, Mandible, and Hyoid Bone -- Interior of the Cranium
    -- Superficial Structures of the Head, Neck, and Face --
    Anterior Triangle of the Neck -- Submandibular and Parotid
    Regions -- Posterior Triangle of the Neck -- Deep Structures
    of the Neck -- Root of the Neck -- Ear -- Oribital Region --
    Deep Strucutres of the Face -- Nose and Paranasal Sinuses --
    Oral Region -- Pharynx and Larynx.
    GGO:550; B:451; M:136; WW:255,496,686,746.

4. THE THORAX. 129
    Thoracic Wall -- Diaphragm -- Thoracic Cavity -- Pleura
    and Lungs -- Pericardium and Heart.
    GGO:258; B:391; M:125,167,239; S:41; WW:419,516,599,1190.

5. THE ABDOMEN. 166
    Abdominal Walls -- Peritoneal Cavity -- Hollow Abdominal
    Viscera -- Solid Abdominal Viscera -- Blood Vessels,
    Nerves, and Lymphatics.
    GGO:350; B:175; M:175; S:175; WW:660,778,1270.

6. THE PELVIS AND PERINEUM. 221
    Pelvic Walls -- Pelvic Peritoneum and Viscera --
    Perineum.
    GGO:438; B:257; M:198; S:263; WW:351,442,527,668,1254.

7. THE BACK. 249
    Vertebral Column and Joints -- Muscles, Nerves and
    Blood Supply -- Spinal Cord and Meninges.
    GGO:508; B:327; M:281; S:811; WW:510.

---
*The key to these reference books is on page viii.

PART TWO.   REVIEW EXAMINATIONS ON THE SYSTEMS

8. THE ARTICULAR, MUSCULAR, AND SKELETAL SYSTEMS. 269
9. THE CARDIOVASCULAR SYSTEM. 276
10. THE DIGESTIVE SYSTEM. 283
11. THE REPRODUCTIVE SYSTEM. 290
12. THE RESPIRATORY SYSTEM. 295
13. THE URINARY SYSTEM. 300
14. THE NERVOUS SYSTEM. 303

PART THREE.   REVIEW EXAMINATIONS ON THE REGIONS

15. THE UPPER LIMB. 311
16. THE LOWER LIMB. 317
17. THE HEAD AND NECK. 323
18. THE THORAX. 329
19. THE ABDOMEN. 335
20. THE PELVIS AND PERINEUM. 341
21. THE BACK. 347

KEY TO THE REFERENCE BOOKS

The letter combinations below the list of contents in the first seven chapters refer to the textbooks listed below. The letters refer to the books and the numbers to the page(s) in the books where information related to the material may be found, e.g., GGO:70 refers to page 70 in Gardner, Gray, O'Rahilly.

    B -- Basmajian, J.: Grant's Method of Anatomy. 9th ed., Baltimore, Williams and Wilkins Co., 1975.

    GGO-- Gardner, E., Gray, D.J., and O'Rahilly, R.: Anatomy, A Regional Study of Human Structure. 4th ed., Philadelphia, W.B. Saunders Co., 1975.

    S -- Snell, R.S.: Clinical Anatomy For Medical Students. 2nd ed., Boston, Little, Brown and Company, 1975.

    M -- Moore, K.L.: The Developing Human. Clinically Oriented Embryology. Philadelphia, W.B. Saunders Co., 1973.

    WW -- Warwick, R., and Williams, P.L.: Gray's Anatomy. 35th British ed. Philadelphia, W.B. Saunders Co., 1973.

TO OUR WIVES

Marion, Doris, Shigeko, Gisela, and Trish

PART ONE

REGIONAL ANATOMY

1. THE UPPER LIMB

OBJECTIVES

<u>Axilla and Pectoral Region</u>

Be Able To:

* Define the boundaries of the axilla and list its contents.

* Delineate the extent and relationships of the axillary artery and vein, including their branches.

* Schematically illustrate the formation of the brachial plexus of nerves and its branches.

* Draw and label the location and drainage patterns of the axillary group of lymph nodes and vessels, showing its relationships to the axillary blood vessels.

* Describe the pectoral muscles as to their attachments, functions, innervation, and relationships to the axillary artery.

* Discuss the anatomical location of the female breast on the thoracic wall, illustrating its general structure and lymphatic drainage.

<u>Shoulder</u>

Be Able To:

* Point out the anatomical features of the scapula, the clavicle, and the proximal half of the humerus, using anatomical specimens and radiographs.

* Use simple sketches to illustrate the articulations of the shoulder girdle and its ligamentous and muscular support.

* Describe the muscles responsible for movements of the shoulder girdle with specific reference to movements of the scapula and clavicle.

* Discuss the arterial anastomosis around the scapula.

<u>Arm and Elbow</u>

Be Able To:

* Compare and contrast the attachments, the functions, and the innervation of the muscles of the anterior and posterior compartments of the arm.

* Discuss the clinical significance of the following anatomical relationships

in the arm: proximity of the axillary, radial, and ulnar nerves to the humerus; displacement fractures above and below the insertion of the deltoid muscle; and displacement of supracondylar fractures.

* Use bones or radiographs to describe the articulating surfaces of the elbow and proximal radio-ulnar joints, the movements permitted, and the supporting ligaments.

* Illustrate the arterial anastomoses around the elbow.

* Make a sketch of the cubital fossa showing its boundaries and contents and write a note on its clinical significance.

Forearm and Wrist

Be Able To:

* Prepare a sketch illustrating the muscles originating from the medial and lateral epicondyles and supracondylar ridges of the humerus, indicating their anatomical position, insertion, functions, and innervation.

* Describe with the aid of a drawing the muscles originating from and acting between the radius and ulna, indicating their general insertion, functions, and innervation.

* Describe the course of the following branches of the brachial artery: ulnar, radial, and common, anterior, and posterior interosseous arteries.

* Give the anatomical basis and clinical significance for locating the brachial and radial pulses.

* Write brief notes on the distal radio-ulnar and wrist (radiocarpal) joints, describing their articulating surfaces, movements, and relations.

* Briefly describe the attachments of the flexor retinaculum at the wrist, listing the tendons, nerves, and blood vessels passing superficial and deep to it.

* Give an account of the clinical significance of the flexor retinaculum at the wrist.

Hand

Be Able To:

* Describe the arrangement of the carpal bones.

* Define the movements of the thumb and fingers correlating them with the innervation of the thenar, hypothenar, and other intrinsic and extrinsic muscles of the hand.

* Define the tendons forming the boundaries of the "anatomical snuff-box", giving its relation to the following: radial artery and nerve, cephalic vein, and scaphoid bone.

* Describe the motor and sensory deficiencies of the hand resulting from injury of: the radial nerve in the midhumeral region; the median and ulnar nerves at the elbow; and the radial nerve at the wrist.

AXILLA AND PECTORAL REGION

FIVE-CHOICE COMPLETION QUESTIONS

DIRECTIONS: Each of the following questions or incomplete statements is followed by five suggested answers or completions. SELECT THE ONE BEST ANSWER in each case and then underline the appropriate letter at the lower right of each question.

1. THE AXILLARY ARTERY EXTENDS FROM THE OUTER BORDER OF THE FIRST RIB TO THE:
   A. Lower border of pectoralis major muscle
   B. Upper border of pectoralis minor muscle
   C. Lower border of teres major muscle
   D. Upper border of subscapularis muscle
   E. Outer border of the second rib   A B C D E

2. THE _____ MUSCLE IS ATTACHED TO THE STERNUM.
   A. Brachialis   D. Pectoralis minor
   B. Deltoid   E. Subclavius
   C. Pectoralis major   A B C D E

3. THE _____ ARTERY IS NOT A BRANCH OF THE AXILLARY ARTERY.
   A. Lateral thoracic   D. Suprascapular
   B. Subscapular   E. Thoracoacromial
   C. Superior thoracic   A B C D E

4. ALL THE FOLLOWING STATEMENTS ABOUT THE FEMALE BREAST ARE TRUE EXCEPT:
   A. Consists of 15-20 lobes
   B. Nipple contains ampullae
   C. Contains fibrous septa
   D. Lies entirely beneath the deep fascia
   E. Has an axillary tail   A B C D E

5. THE CORDS OF THE BRACHIAL PLEXUS ARE NAMED AFTER THEIR RELATIONSHIP TO:
   A. Each other
   B. The first part of the axillary artery
   C. The subclavian artery
   D. The third part of the axillary artery
   E. The second part of the axillary artery   A B C D E

6. WHICH NERVE PASSES DEEP TO THE DELTOID MUSCLE?
   A. Ulnar   D. Axillary
   B. Radial   E. Musculocutaneous
   C. Median   A B C D E

7. WHICH OF THE FOLLOWING STRUCTURES DOES NOT FORM A BOUNDARY OF THE AXILLA?
   A. Humerus   D. Latissimus dorsi muscle
   B. Brachialis muscle   E. Pectoralis major muscle
   C. Intercostal muscle   A B C D E

SELECT THE ONE BEST ANSWER

8. THE _____ VEIN COURSES BETWEEN THE DELTOID AND PECTORALIS MAJOR MUSCLES.
   A. Axillary
   B. Basilic
   C. Brachial
   D. Cephalic
   E. Median cubital

   A B C D E

---------------------- ANSWERS, NOTES AND EXPLANATIONS ----------------------

1. C. The axillary artery, a continuation of the subclavian artery, begins at the lateral border of the first rib. It courses through the axilla and terminates at the lower border of the teres major muscle. The axillary artery continues as the brachial artery.

2. C. The pectoralis major muscle, besides attaching to the sternum, also originates from the clavicle, the upper six costal cartilages, the aponeurosis of the external oblique muscles, and inserts into the crest of the greater tubercle of the humerus.

3. D. The branches of the axillary artery are: the superior thoracic, thoracoacromial, lateral thoracic, subscapular, and the anterior and posterior circumflex humeral arteries. The suprascapular artery is a branch of the thyrocervical trunk from the first part of the subclavian artery. The suprascapular artery crosses the superior margin of the scapula to supply the supraspinatus and infraspinatus muscles, and is involved with the scapular anastomosis.

4. D. The female breast contains 15-20 lobes, each of which drains via a lactiferous duct. Its terminal expanded portion forms the lactiferous sinus which opens at the summit of the nipple. The nipple is surrounded by a pigmented area of skin called the areola. The lobes of the breast are separated from each other by fibrous connective tissue septa. The gland lies within the superficial fascia except for a part that pierces the deep fascia and extends laterally and upward into the axilla. This portion of the gland is known as the axillary tail.

5. E. The cords of the brachial plexus are named according to their relationship to the second part of the axillary artery, which lies deep to the upper portion of the pectoralis minor muscle. All three cords of the plexus lie above and lateral to the first part of the artery. As they course distally, the medial cord crosses behind the artery to reach the medial side of the second segment of the artery. The posterior cord becomes posterior to the second part of the artery, whereas the lateral cord retains its lateral relationship to the artery.

6. D. All these nerves are formed from the three cords of the brachial plexus of nerves, and are originally medial to the deltoid muscle. Only the axillary nerve reaches the deep aspect of the deltoid muscle, after passing to the back of the arm and through the quadrangular space.

7. B. The walls of the axilla are formed laterally by the humerus, biceps brachii and coracobrachialis muscles; posteriorly by the latissimus dorsi, subscapularis and teres major muscles; medially by the upper intercostal and serratus anterior muscles and the ribs; and anteriorly by the pectoralis major, pectoralis minor and subclavius muscles.

8. D. The cephalic vein begins on the dorsum of the hand and ascends on the radial aspect of the anterior forearm and elbow region. It then ascends along

the lateral margin of the biceps brachii muscle and courses between the medial fibers of the deltoid and the upper lateral fibers of the pectoralis major muscles, the deltopectoral triangle. It then pierces the clavipectoral fascia and joins the axillary vein.

MULTI-COMPLETION QUESTIONS

DIRECTIONS: In each of the following questions or incomplete statements, ONE OR MORE of the completions given is correct. At the lower right of each question, underline A if 1, 2 and 3 are correct; B if 1 and 3 are correct; C if 2 and 4 are correct; D if only 4 is correct; and E if all are correct.

1. THE AXILLARY NERVE INNERVATES THE FOLLOWING MUSCLE(S):
   1. Deltoid
   2. Teres major
   3. Teres minor
   4. Serratus anterior   A B C D E

2. THE MEDIAN NERVE:
   1. Is derived from the lateral and medial cords of the brachial plexus
   2. Passes through the cubital fossa
   3. Gives rise to the anterior interosseous nerve
   4. Supplies the interossei muscles   A B C D E

3. WHICH OF THE FOLLOWING NERVES ARISE(S) FROM THE POSTERIOR CORD OF THE BRACHIAL PLEXUS OF NERVES?
   1. Dorsal scapular
   2. Thoracodorsal
   3. Suprascapular
   4. Subscapular   A B C D E

4. THE BRANCH(ES) OF THE SECOND PART OF THE AXILLARY ARTERY IS (ARE) THE:
   1. Supreme thoracic
   2. Subscapular
   3. Anterior circumflex humeral
   4. Thoracoacromial   A B C D E

5. THE BASILIC VEIN:
   1. Arises on the dorsum of the hand
   2. Joins the internal jugular vein
   3. Accompanies the median antebrachial cutaneous nerve
   4. Is a deep vein of the forearm   A B C D E

6. THE PECTORALIS MAJOR MUSCLE TAKES ORIGIN FROM THE:
   1. Anterior aspect of the medial portion of the clavicle
   2. Aponeurosis of the external oblique muscle of the abdomen
   3. Anterior aspect of the sternum
   4. Upper six bony ribs   A B C D E

7. THE ARTERIAL SUPPLY OF THE MAMMARY GLAND IS DERIVED FROM THE _____ ARTERIES.
   1. Internal thoracic
   2. Intercostal
   3. Lateral thoracic
   4. Superior epigastric   A B C D E

8. THE AXILLARY LYMPH NODES DRAIN:
   1. The upper limb
   2. The mammary gland
   3. Part of the trunk
   4. The root of the neck   A B C D E

9. THE POSTERIOR WALL OF THE AXILLA IS FORMED BY THE _____ MUSCLES.
   1. Teres major
   2. Latissimus dorsi
   3. Subscapularis
   4. Subclavius   A B C D E

A	B	C	D	E
1,2,3	1,3	2,4	only 4	all correct

10. THE DEEP VEINS IN THE UPPER EXTREMITY ARE THE:
    1. Axillary   3. Brachial
    2. Subclavian   4. Basilic   A B C D E

11. THE LATERAL PECTORAL NERVE:
    1. Innervates only the pectoralis minor muscle
    2. Arises from the lateral cord of the brachial plexus
    3. Is usually found lateral to the medial pectoral nerve
    4. Pierces the clavipectoral fascia   A B C D E

------------------------ANSWERS, NOTES AND EXPLANATIONS------------------------

1. **B.** <u>1 and 3 are correct</u>. The axillary nerve arises from the posterior cord of the brachial plexus, passes through the quadrangular space accompanied by the posterior humeral circumflex artery, and supplies both the deltoid and teres minor muscles. Teres major and serratus anterior are innervated by the lower subscapular (also from the posterior cord) and the long thoracic nerves, respectively.

2. **A.** <u>1, 2 and 3 are correct</u>. The median nerve receives contributions from the ventral primary rami of spinal nerves C5, 6, and 7, via the lateral cord and C8 and T1, in the arm and enters the cubital fossa, where it gives off branches to some muscles of the flexor compartment of the forearm. The interosseous nerve supplies the flexor pollicis longus, pronator quadratus, and the lateral half of the flexor digitorum profundus muscles. In the hand, the median nerve supplies three thenar muscles, the lateral two lumbricals, the palmar skin of the lateral three and a half digits, and the dorsal skin of the tips of these same digits.

3. **C.** <u>2 and 4 are correct</u>. All these nerves are branches of the brachial plexus, however, only the thoracodorsal and the subscapular nerves are branches of the posterior cord of the brachial plexus. The dorsal scapular nerve arises chiefly from the ventral ramus of the fifth cervical spinal nerve (C5) and the suprascapular nerve (C5,6) is the branch of the upper trunk of the brachial plexus.

4. **D.** <u>4 only is correct</u>. The axillary artery can be divided into three segments according to its relationship to the pectoralis minor muscle. The first part lies between the lateral border of the first rib and the upper border of the pectoralis minor muscle; the second part lies behind the pectoralis minor muscle, and the third part extends from the lower border of pectoralis minor to the lower border of teres major muscle. The thoracoacromial and lateral thoracic arteries are branches of the second part, the supreme (highest) thoracic artery is a branch of the first part, whereas the branches of the third part are the subscapular and the anterior and posterior circumflex humeral arteries.

5. **B.** <u>1 and 3 are correct</u>. The basilic vein arises superficially on the dorsum of the hand and ascends on the posteromedial surface of the forearm. Before reaching the elbow it curves around the forearm to the anteromedial surface. Here it is joined by the median cubital vein and ascends up the arm, piercing the deep fascia about the middle of the arm. It joins the brachial veins to form the axillary vein at the lower border of the teres major muscle.

6. **A.** <u>1, 2, and 3 are correct</u>. The origin of the pectoralis major muscle is broad and includes: (1) the clavicular portion, from the anterior surface of

the medial half of the clavicle; (2) the sternal portion, from the anterior surface of the manubrium and body of the sternum; (3) the costal portion, commonly from the second to sixth <u>costal cartilages</u>, but occasionally from the first and seventh costal cartilages as well; and (4) the aponeurosis of the external oblique muscle of the abdomen.

7. B. <u>1 and 3 are correct</u>. The arterial supply to the mammary gland is derived from branches of the subclavian and axillary arteries. The internal thoracic artery, a branch of the subclavian, supplies the gland directly via its perforating branches of the second, third and fourth intercostal spaces. The lateral thoracic and thoracoacromial branches of the axillary artery supply the lateral and superior aspects of the gland proper, including the axillary tail.

8. E. <u>All are correct</u>. The axillary lymph nodes can be divided into five groups: lateral, posterior, pectoral, central, and apical. The lateral and posterior groups of lymph nodes drain the upper limb and the posterior aspect of the shoulder, respectively. The pectoral group drains the breast. The pectoral, lateral, and posterior groups all drain into the large central group. The apical group of nodes, situated below the clavicle in the clavipectoral fascia, drains the central group and empties eventually into the venous system.

9. A. <u>1, 2, and 3 are correct</u>. The teres major, latissimus dorsi and subscapularis muscles form the posterior wall of the axilla. The teres major and subscapularis muscles originate from the scapula; the subscapularis form the subscapular fossa, and the teres major form the posterior surface of the lower lateral border of the scapula. The latissimus dorsi originates from the posterior part of the iliac crest, the lumbar fascia, the spines of the lower six thoracic vertebrae, and the lower three or four ribs. The three muscles converge towards the humerus, forming the posterior axillary wall. The teres major and latissimus dorsi muscles insert on the medial lip and floor of the bicipital groove, respectively. The subscapularis muscle inserts on the lesser tuberosity of the humerus. The subclavius muscle forms part of the anterior axillary wall.

10. B. <u>1 and 3 are correct</u>. The deep veins follow the arteries and have similar names. The axillary vein is the main vein carrying blood from the upper limb. At the outer border of the first rib, the axillary vein continues as the subclavian vein in the neck. Hence, the subclavian vein is not a deep vein in the upper extremity. The brachial veins are on each side of the brachial artery and join the axillary vein near the lower border of the subscapularis muscle. The basilic vein is a superficial vein of the upper limb.

11. C. <u>2 and 4 are correct</u>. The lateral pectoral nerve arises from the lateral cord of the brachial plexus of nerves; contains nerve fibers from the anterior divisions of the fifth, sixth, and seventh cervical spinal nerves; pierces the clavipectoral fascia medial to the medial pectoral nerve and innervates chiefly the pectoralis major muscle, but contributes fibers to the pectoralis minor muscle as well. The lateral pectoral nerve sends a branch over the first part of the axillary artery to the medial pectoral nerve, and forms a loop through which the lateral pectoral nerve carries motor fibers to the pectoralis minor muscle. The median and lateral designations of the medial and lateral pectoral nerves refer to the cords of their origins, not to their topographical relationships.

FIVE-CHOICE ASSOCIATION QUESTIONS

DIRECTIONS: Each group of questions below consists of a numbered list of descriptive words or phrases accompanied by a diagram with certain parts indicated by letters, or by a list of lettered headings. For each numbered word or phrase, SELECT THE LETTERED PART OR HEADING that matches it correctly. Then insert the letter in the space to the right of the appropriate number. Sometimes more than one numbered word or phrase may be correctly matched to the same lettered part or heading.

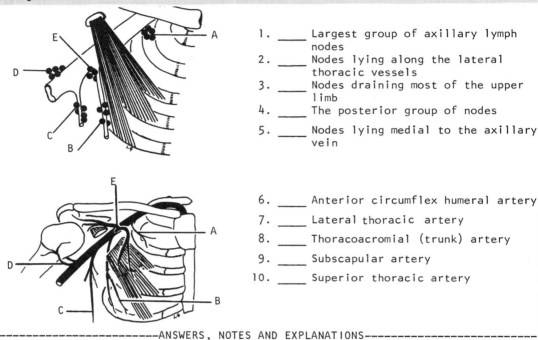

1. ____ Largest group of axillary lymph nodes
2. ____ Nodes lying along the lateral thoracic vessels
3. ____ Nodes draining most of the upper limb
4. ____ The posterior group of nodes
5. ____ Nodes lying medial to the axillary vein

6. ____ Anterior circumflex humeral artery
7. ____ Lateral thoracic artery
8. ____ Thoracoacromial (trunk) artery
9. ____ Subscapular artery
10. ____ Superior thoracic artery

----------------------------ANSWERS, NOTES AND EXPLANATIONS----------------------------

1. **E.** The central nodes lie at the base of the axilla and receive lymph from the lateral, pectoral, and posterior groups of nodes. They form the largest group (10-14 nodes) and are often palpable.

2. **B.** The pectoral nodes surround the lateral thoracic vessels as they follow the lateral margin of the pectoralis minor muscle. This group of nodes receives most of the lymph draining from the breast, however, other groups of nodes also drain the breast.

3. **D.** The lateral (brachial) group of 10-14 lymph nodes, lying posterior to the axillary vein, drains the superficial and deep lymphatics of the upper limb. The deep lymphatics ascend in the arm accompanying the larger vessels.

4. **C.** The subscapular (posterior) nodes accompany the subscapular vein along the lateral border of the scapula. These nodes and their efferent vessels drain the posterior aspect of the shoulder.

5. **A.** The apical (infraclavicular) group of nodes lies medial to the proximal part of the axillary vein, superior to the upper border of the pectoralis minor muscle. These nodes lie deep to the clavipectoral fascia and drain lymph from the other groups of axillary nodes; occasionally they directly drain the breast. The apical group then drains, via two or three efferent

subclavian trunks, to the jugulosubclavian venous junction or to the lower deep cervical group of nodes.

6. **D.** The anterior circumflex humeral is a small artery arising from the third part of the axillary artery. It lies on the anterior surface of the surgical neck of the humerus, and supplies the shoulder joint and the head of the humerus. It anastomoses with the larger posterior circumflex humeral artery.

7. **B.** The lateral thoracic artery, along with the thoracoacromial artery, arises from the second part of the axillary artery. The second part of the axillary artery is defined as the segment located deep to the pectoralis minor muscle. The lateral thoracic artery courses along the lower border of the pectoralis minor muscle and supplies the serratus anterior and pectoral muscles.

8. **E.** The thoracoacromial artery, a short trunk, arises from the second part of the axillary artery. As it passes around the margin of pectoralis minor muscle, it gives rise to four branches: pectoral, acromial, clavicular, and deltoid, supplying the regions indicated by their names.

9. **C.** The subscapular artery is the largest branch of the third part of the axillary artery. It arises at the lower border of the subscapularis muscle and follows it to the inferior angle of the scapula. The subscapular artery supplies adjacent muscles and freely anastomoses with the suprascapular and dorsal scapular arteries.

10. **A.** The superior thoracic artery is a small branch that arises from the first part of the axillary artery near the lower border of the subclavius muscle. It supplies the subclavius and pectoral muscles and a small part of the thoracic wall.

SHOULDER

FIVE-CHOICE COMPLETION QUESTIONS

DIRECTIONS: Each of the following questions or incomplete statements is followed by five suggested answers or completions. SELECT THE ONE BEST ANSWER in each case and then underline the appropriate letter at the lower right of each question.

1. WHICH OF THE FOLLOWING MUSCLES CONTRIBUTES TO THE STABILITY OF THE SHOULDER JOINT?
   A. Subscapularis
   B. Supraspinatus
   C. Infraspinatus
   D. Teres minor
   E. All of the above

   A B C D E

2. ALL THE FOLLOWING MUSCLES WILL ROTATE THE ARM MEDIALLY EXCEPT THE:
   A. Latissimus dorsi
   B. Teres major
   C. Infraspinatus
   D. Subscapularis
   E. Pectoralis major

   A B C D E

3. THE LONG THORACIC NERVE INNERVATES THE _____ MUSCLE.
   A. Trapezius
   B. Rhomboid major
   C. Latissimus dorsi
   D. Pectoralis major
   E. Serratus anterior

   A B C D E

SELECT THE ONE BEST ANSWER

4. CONTRACTION OF THE MIDDLE FIBERS OF THE TRAPEZIUS MUSCLE
   WILL CAUSE:
   A. Shrugging of the shoulder
   B. Extension of the arm
   C. Retraction of the scapula
   D. Rotation of the glenoid cavity upwards
   E. Flexion of the neck towards the same side

   A B C D E

5. WHICH OF THE FOLLOWING MUSCLES IS THE PRINCIPAL PROTRACTOR
   OF THE SCAPULA?
   A. Serratus anterior   D. Trapezius
   B. Pectoralis major   E. None of the above
   C. Pectoralis minor

   A B C D E

6. A YOUNG CHILD PRESENTS WITH A FRACTURE OF THE CLAVICLE AT THE
   JUNCTION OF THE LATERAL AND MIDDLE THIRDS. THE MEDIAL AND
   LATERAL FRAGMENTS ARE TILTED UPWARD AND DOWNWARD, RESPECTIVELY.
   UPWARD DISPLACEMENT OF THE MEDIAL HALF IS CAUSED BY WHICH OF
   THE FOLLOWING MUSCLES?
   A. Deltoid   D. Trapezius
   B. Pectoralis major   E. Subclavius
   C. Sternocleidomastoid

   A B C D E

7. THE CORACOID PROCESS OF THE SCAPULA
   IS INDICATED BY _____.

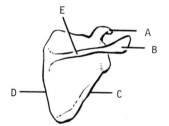

   A B C D E

8. THE SUPRASPINATUS MUSCLE OF THE
   "ROTATOR CUFF" IS INDICATED BY
   _____.

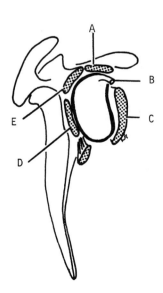

   A B C D E

---------------------- ANSWERS, NOTES AND EXPLANATIONS ----------------------

1. E. The shoulder joint is relatively unstable, due to the shallow glenoid fossa of the scapula compared to the large head of the humerus, and the weak capsule and ligaments attached to the articulating bones. It is the most commonly dislocated large joint and its stability is greatly dependent on the subscapularis, supraspinatus, infraspinatus, and teres minor muscles which attach the proximal end of the humerus to the scapula. The subscapular lies in front of the joint, the supraspinatus above, and the infraspinatus and teres minor muscles are behind. The inferior aspect of the joint is unsupported by muscles and inferior dislocations are not uncommon. The subscapularis muscle is innervated by the upper and lower subscapular nerves, the supraspinatus and infraspinatus by the suprascapular, and the teres minor by the axillary nerve.

2. C. The infraspinatus muscle is principally a lateral rotator of the arm. It is innervated by the suprascapular nerve. Although the subscapularis, pectoralis major, latissimus dorsi, and teres major muscles are all involved with internal rotation of the arm, the muscle primarily involved in this action is the subscapularis.

3. E. The long thoracic nerve is derived from the ventral primary rami of C5, 6, and 7; it innervates the serratus anterior muscle which plays an important role in normal abduction of the humerus by rotating the scapula. Various forms of trauma (crushing, cutting, etc.) interrupt fibers of the nerve, causing winging of the scapula and making elevation of the arm above the horizontal impossible.

4. C. The broad, expansive trapezius muscle originates from the medial third of the superior nuchal line of the occipital bone, the external occipital protuberance, the spines of the last cervical and all thoracic vertebrae, the ligamentum nuchae and the supraspinous ligaments. The trapezius has a continuous insertion into two bones: its uppermost fibers attach to the posterior border of the lateral third of the clavicle, its middle fibers attach to the medial border of the acromial process and the upper border of the spine of the scapulae, and its lower fibers attach to the medial end of the scapular spine. The upper fibers elevate, the middle fibers retract, and the lower fibers pull the scapula downwards.

5. A. The principal protractor of the scapula is the serratus anterior muscle which originates from the outer surface and upper border of the first eight ribs, and inserts into the medial border of the scapula on its costal surface. It is innervated by the long thoracic nerve. In addition to drawing the scapula forward, this muscle, in collaboration with the trapezius, is involved in rotation of the scapula and in raising the arm to 180° of abduction. Other muscles assisting in protraction of the scapula are the pectoralis major and pectoralis minor. In contrast, the middle fibers of the trapezius muscle cause retraction of the scapula.

6. C. The sternocleidomastoid muscle would pull the medial fragment of the clavicle upward. The contractile power of this muscle, attached to the superior medial aspect of the clavicle, overrides the combined contractile power of the subclavius and pectoralis major muscles. The deltoid muscle, attached to the lateral end of the clavicle, would depress the lateral fragment. Moreover, the weight of the arm, due to gravity, would contribute to the downward displacement of the lateral fragment.

7. A. The coracoid process, a beak-like bony projection of the scapula points forward and slightly laterally. Its tip can be palpated below the junction of

the lateral and middle thirds of the clavicle. It serves as a site for muscular and ligamentous attachments. In the diagram, B represents the acromion process; C, the lateral border; D, the medial border; and E, the spine of the scapula.

8. **A.** The supraspinatus muscle, along with the subscapularis (C), teres minor (D) and infraspinatus (E) muscles, form the "rotator cuff" of the shoulder joint which contributes greatly to its stability. The structure B represents the long head of the biceps muscle surrounded by the synovial membrane of the joint. The long head of the biceps is attached to the supraglenoid tubercle of the scapula.

MULTI-COMPLETION QUESTIONS

DIRECTIONS: In each of the following questions or incomplete statements, ONE OR MORE of the completions given is correct. At the lower right of each question, underline A if 1, 2 and 3 are correct; B if 1 and 3 are correct; C if 2 and 4 are correct; D if only 4 is correct; and E if all are correct.

1. WHICH STATEMENT(S) ABOUT THE STERNOCLAVICULAR JOINT IS (ARE) TRUE?
   1. Has an articular disc
   2. Provides the only bony attachment between the appendicular and axial skeletons
   3. Is divided into compartments
   4. Is a fibrous joint   A B C D E

2. MUSCLES ATTACHED TO THE CLAVICLE INCLUDE THE:
   1. Subclavius   3. Pectoralis major
   2. Trapezius   4. Deltoid   A B C D E

3. WHICH OF THE FOLLOWING STRUCTURES BOUNDS THE QUADRANGULAR (QUADRILATERAL) SPACE:
   1. Rhomboid major muscle
   2. The long head of the triceps
   3. Biceps tendon
   4. Neck of the humerus   A B C D E

4. THE MUSCLES ATTACHED TO THE CORACOID PROCESS OF THE SCAPULA ARE:
   1. Deltoid   3. Subclavius
   2. Short head of biceps brachii   4. Pectoralis minor   A B C D E

5. THE CORACOCLAVICULAR LIGAMENT:
   1. Supports the acromioclavicular joint
   2. Consists of two ligamentous bands
   3. Resists downward movement of the scapula
   4. Assists in suspension of the scapula   A B C D E

6. WHICH OF THE FOLLOWING STATEMENTS ABOUT THE CLAVICLE IS (ARE) TRUE?
   1. Develops in membrane
   2. First bone to begin ossification in the embryo
   3. Commonly fractured bone
   4. Its fulcrum of movement is at the costoclavicular ligament   A B C D E

A	B	C	D	E
1,2,3	1,3	2,4	only 4	all correct

7. WHICH STATEMENT(S) ABOUT THE DELTOID MUSCLE IS(ARE) TRUE?
   1. Innervated by C5 and 6 fibers
   2. An important stabilizer of the shoulder joint
   3. A medial rotator of the arm
   4. A lateral rotator of the arm

   A B C D E

8. THE GLENOID FOSSA IS ROTATED UPWARDS BY WHICH OF THE FOLLOWING MUSCLES?
   1. Rhomboids
   2. Trapezius
   3. Levator scapulae
   4. Serratus anterior

   A B C D E

9. THE SUPRASPINATUS AND INFRASPINATUS MUSCLES ARE SUPPLIED BY WHICH OF THE FOLLOWING NERVES?
   1. Lower subscapular
   2. Long thoracic
   3. Upper subscapular
   4. Suprascapular

   A B C D E

10. MUSCLES ATTACHED TO THE SCAPULA ARE:
    1. Serratus anterior
    2. Teres major
    3. Rhomboid major
    4. Deltoid

    A B C D E

---------------------- ANSWERS, NOTES AND EXPLANATIONS ----------------------

1. **A. 1, 2, and 3 are correct.** The sternoclavicular joint is a synovial joint, with two compartments formed by an articular disc. The articulating surfaces are between the medial end of the clavicle, the first costal cartilage, and the manubrium of the sternum. This joint is the only bony attachment between the appendicular and axial divisions of the skeleton. This double plane joint allows anterior and posterior movements in the medial compartment, and elevation and depression of the shoulder in its lateral compartment.

2. **E. All are correct.** The trapezius muscle inserts on the posterior superior border of the lateral third of the clavicle. The subclavius muscle inserts into the inferior surface of the clavicle. The pectoralis major and deltoid muscles take their origin from the superior surface of the clavicle, the deltoid from the lateral and the pectoralis major from the medial half.

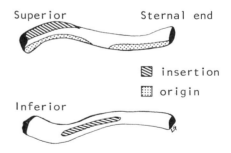

3. **C. 2 and 4 are correct.** The superior boundary of the quadrilateral space is the subscapularis muscle anteriorly and the teres minor muscle posteriorly. It is bounded inferiorly by teres major, medially by the long head of the triceps, and laterally by the surgical neck of the humerus. The axillary nerve and the posterior circumflex humeral vessels pass through this space.

4. **C. 2 and 4 are correct.** The short head of the biceps attaches to the bony beak of the scapula, the coracoid process. The coracobrachialis muscle also takes origin from this process and plays a minor role in shoulder movements. The coracoid process can be palpated deep to a visible depression below the lateral end of the clavicle. The pectoralis minor muscle, attached to the superomedial border of the coracoid process, is involved with pulling the shoulder forward and downward.

5. E. All are correct. The coracoclavicular ligament reinforces the acromioclavicular joint and resists downward movement of the scapula. This strong ligament consists of two ligamentous parts, the conoid and trapezoid ligaments, separated by a bursa. In a fall on an outstretched arm, the acromion is prevented from being driven below the clavicle by the horizontally directed trapezoid ligament. Rupture of this ligament produces dislocation of the acromioclavicular joint. Together with muscles, this ligament assists in suspending the scapula.

6. E. All are correct. All these statements about the clavicle are true. The clavicle serves the important function of a "strut" for the upper limb. Fractures of the clavicle are quite common, particularly in children, in which case the shoulder projects medially, forward, and downward. Membranous ossification of this bone commences as early as the fifth week in the embryo. Surprisingly, the fulcrum of movement of the clavicle is not at the sternoclavicular joint, but at the strong costoclavicular (rhomboid) ligament which anchors the medial end of the clavicle to the first costal cartilage lateral to the joint.

7. E. All are correct. The deltoid muscle originates from the lateral third of the superior surface of the clavicle, the lateral margin and upper surface of the acromion, and the inferior lip of the spine of the scapula. It inserts into the deltoid tuberosity of the humerus. The deltoid muscle is an important stabilizer of the shoulder joint, chiefly during abduction of the arm. It receives its nerve supply from the axillary (circumflex) nerve (C5,6). Its middle fibers abduct, its posterior fibers extend and laterally rotate, and its anterior fibers flex and medially rotate the arm.

8. C. 2 and 4 are correct. The glenoid fossa of the scapula is rotated upwards in abduction of the upper limb. The muscles involved in this movement are the trapezius and the serratus anterior.

9. D. Only 4 is correct. The supraspinatus and infraspinatus muscles are innervated by branches of the suprascapular nerve (C5,6). Although both muscles are innervated by the same nerve, they have different functions. The supraspinatus is an abductor of the humerus, whereas the infraspinatus is a lateral rotator.

10. E. All are correct. The teres major and deltoid muscles originate from the scapula. The other two muscles, the serratus anterior and rhomboid major, are inserted into the scapula.

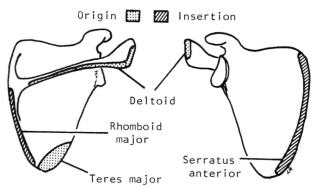

FIVE-CHOICE ASSOCIATION QUESTIONS

DIRECTIONS: Each group of questions below consists of a numbered list of descriptive words or phrases accompanied by a diagram with certain parts indicated by letters, or by a list of lettered headings. For each numbered word or phrase, SELECT THE LETTERED PART OR HEADING that matches it correctly. Then insert the letter in the space to the right of the appropriate number. Sometimes more than one numbered word or phrase may be correctly matched to the same lettered part or heading.

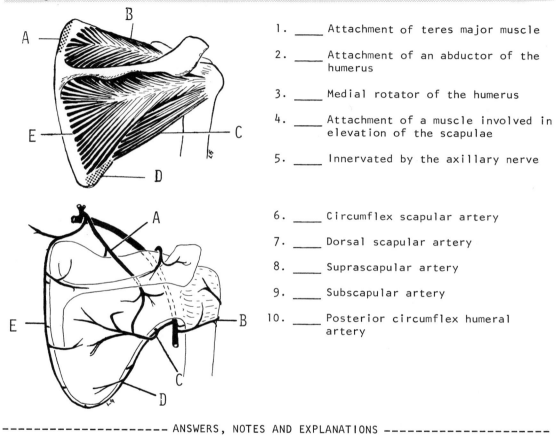

1. ____ Attachment of teres major muscle
2. ____ Attachment of an abductor of the humerus
3. ____ Medial rotator of the humerus
4. ____ Attachment of a muscle involved in elevation of the scapulae
5. ____ Innervated by the axillary nerve

6. ____ Circumflex scapular artery
7. ____ Dorsal scapular artery
8. ____ Suprascapular artery
9. ____ Subscapular artery
10. ____ Posterior circumflex humeral artery

---------------------- ANSWERS, NOTES AND EXPLANATIONS ----------------------

1. **D.** The teres major muscle originates from the dorsal aspect of the inferior angle of the scapula and inserts on the posterior surface of the greater tubercle of the humerus.

2. **B.** The supraspinatus muscle originates from the supraspinous fossa of the scapula and inserts on the superior aspect of the greater tubercle. This muscle, along with the infraspinatus muscle (labelled E in the drawing), is innervated by the suprascapular nerve.

3. **D.** The teres major muscle is a medial rotator of the humerus against resistance. It also aids in adduction and extension of the humerus; this is especially so when tested against resistance.

4. **A.** The levator scapulae muscle inserts on the dorsal aspect of the medial border between the superior angle and the root of the spine of the scapulae.

It originates from the transverse process of the first four cervical vertebrae. This muscle, along with the rhomboid major and minor muscles, rotates the scapula so that the glenoid fossae faces downwards.

5. C. The teres minor is innervated by the axillary nerve and is a lateral rotator of the humerus.

6. C. The circumflex scapular artery, a large branch of the subscapular artery, enters the infraspinous fossa. It supplies the contents of these fossa and anastomoses freely with the suprascapular, subscapular and dorsal scapular arteries. It also supplies branches to the posterior portion of the deltoid and triceps muscles.

7. E. The descending branch of the transverse cervical artery (dorsal scapular), a branch of the subclavian, runs along the medial border of the scapula where it supplies the rhomboids and trapezius muscles. It anastomoses with the subscapular and suprascapular arteries.

8. A. The suprascapular artery, a branch of the thyrocervical trunk, on reaching the superior border of the scapula supplies the contents of the supraspinous and infraspinous fossae. It also supplies the acromioclavicular and shoulder joints.

9. D. The subscapular artery, the largest branch of the third part of the axillary artery, courses down the lateral border of the scapula to the inferior angle and supplies the contents of the subscapular and infraspinous fossae. It anastomoses with the lateral thoracic, intercostal and transverse cervical (colli) arteries. These potential anastomoses enable a collateral circulation to become established in case of ligature of the subclavian artery beyond the origin of the thyrocervical trunk, and the axillary artery proximal to the origin of the subscapular artery.

10. B. The posterior circumflex humeral, a branch of the axillary artery, passes through the quadrangular space winding around the surgical neck of the humerus, supplies the shoulder joint and the deltoid muscle. It anastomoses with the anterior circumflex humeral artery, the smaller of the two.

ARM AND ELBOW

FIVE-CHOICE COMPLETION QUESTIONS

DIRECTIONS: Each of the following questions or incomplete statements is followed by five suggested answers or completions. SELECT THE ONE BEST ANSWER in each case and then underline the appropriate letter at the lower right of each question.

1. WHERE IS THE ULNAR NERVE MOST COMMONLY DAMAGED?
   A. In the axilla
   B. In the midarm
   C. At the elbow
   D. In the forearm
   E. None of the above

   A B C D E

2. THE TRICEPS BRACHII MUSCLE IS SUPPLIED MAINLY BY THE _____ ARTERY.
   A. Supratrochlear
   B. Profunda brachii
   C. Posterior circumflex
   D. Ulnar recurrent
   E. Perforating

   A B C D E

SELECT THE ONE BEST ANSWER

3. THE STRUCTURE INDICATED BY THE ARROW IS THE _____.
   A. Trochlear notch
   B. Coronoid fossa
   C. Trochlea
   D. Radial fossa
   E. Radial notch

   A B C D E

4. THE STRUCTURE INDICATED BY THE ARROW IS THE _____.
   A. Olecranon fossa
   B. Lateral epicondyle
   C. Olecranon
   D. Medial epicondyle
   E. Head of radius

   A B C D E

5. A FUNCTION OF THE BICEPS BRACHII MUSCLE IS _____ OF THE FOREARM.
   A. Extension
   B. Adduction
   C. Supination
   D. Pronation
   E. Circumduction

   A B C D E

6. IN MOVEMENTS OF THE ARM, THE CORACOBRACHIALIS MUSCLE ASSISTS IN:
   A. Abduction
   B. Extension
   C. Rotation
   D. Supination
   E. Flexion

   A B C D E

7. THE SUPERIOR ULNAR COLLATERAL ARTERY IS INDICATED BY _____.

   A B C D E

8. THE RADIAL RECURRENT ARTERY IS INDICATED BY _____.

   A B C D E

---------------------- ANSWERS, NOTES AND EXPLANATIONS ----------------------

1. **C.** Injuries to the ulnar nerve most commonly occur at the elbow. The ulnar nerve lies between the medial epicondyle and the olecranon process at the elbow. Thus, fractures of the medial epicondyle are often associated with damage to the ulnar nerve. The sensation of pain, tingling, and numbness is caused by compression of this nerve against the medial epicondyle (commonly called the "funny bone"). In addition, the nerve lies relatively superficial at the wrist where it also could be injured. Damage to the ulnar nerve in the axilla or in the forearm is not common.

2. **B.** The triceps brachii muscle is the principal component of the posterior osteofacial compartment of the arm. The blood supply to this muscle is via branches of the profunda brachii artery. Other important structures in this compartment are the ulnar and radial nerves; the former passes through the compartment, whereas the latter provides motor fibers to the triceps muscle and continues into the forearm where it provides the motor innervation to the extensor muscles of the forearm.

3. **B.** The coronoid fossa, a small depression, lies on the anterior surface of the humerus, proximal (above) to the trochlea. This depression accommodates the anterior margin of the coronoid process of the ulna during flexion of the elbow.

4. **C.** The arrow indicates the olecranon of the ulna which articulates with the olecranon fossa of the distal end of the humerus.

5. **C.** The biceps brachii muscle, as its name implies, has two heads: a long head originating from the supraglenoid tubercle of the scapula, and a short head from the coracoid process of the scapula. The principal insertion of this muscle is on the radial tuberosity. An expansion from the tendon, known as the bicipital aponeurosis, fuses with deep fascia of the forearm. The biceps is both a strong supinator of the forearm and a strong flexor of the elbow joint.

6. **E.** The coracobrachialis is a flexor and a weak adductor of the arm. It originates from the tip of the coracoid process and inserts into the middle of the medial aspect of the humerus. It, like the brachialis and biceps brachii muscles, is innervated by the musculocutaneous nerve.

7. **B.** The superior ulnar collateral artery and the inferior ulnar collateral artery (C in the diagram) are branches of the brachial artery and are involved in the anastomoses about the elbow. These branches anastomose with the posterior (D) and anterior ulnar recurrent (not labelled) arteries, respectively.

8. **E.** The radial recurrent artery, a branch of the radial artery, anastomoses with the profunda brachii artery (A) and forms the radial contribution to the anastomoses around the elbow. The radial recurrent artery arises immediately below the elbow. It passes between the superficial and deep branches of the radial nerve. It then ascends behind the brachioradialis muscle, in front of the supinator and brachialis muscles. It supplies these muscles and the elbow joint.

MULTI-COMPLETION QUESTIONS

DIRECTIONS: In each of the following questions or incomplete statements, ONE OR MORE of the completions given is correct. At the lower right of each question, underline **A if 1, 2 and 3** are correct; **B if 1 and 3** are correct; **C if 2 and 4** are correct; **D if only 4** is correct; and **E if all** are correct.

1. WHICH OF THE FOLLOWING STATEMENTS ABOUT THE ELBOW JOINT IS(ARE) CORRECT?
   1. Has collateral ligaments
   2. Shares its synovial membrane with the proximal radioulnar joint
   3. Is supplied by the musculocutaneous and ulnar nerves
   4. Is a multiaxial joint   A B C D E

2. THE CONTENTS OF THE CUBITAL FOSSA INCLUDE THE:
   1. Radial artery   3. Median nerve
   2. Radial nerve   4. Ulnar nerve   A B C D E

3. THE NERVE SUPPLY TO THE ELBOW JOINT INCLUDES:
   1. Radial   3. Median
   2. Ulnar   4. Musculocutaneous   A B C D E

4. THE MUSCLE(S) INSERTED ON THE GREATER TUBERCLE OF THE HUMERUS IS(ARE) THE:
   1. Supraspinatus   3. Teres minor
   2. Subscapularis   4. Teres major   A B C D E

5. THE BICIPITAL APONEUROSIS OF THE BICEPS BRACHII MUSCLE PASSES ANTERIOR (SUPERFICIAL) TO THE:
   1. Brachial artery   3. Median nerve
   2. Radial nerve   4. Cephalic vein   A B C D E

6. THE BRACHIAL ARTERY:
   1. Bifurcates opposite the neck of the humerus
   2. Proximally, it lies medial to the humerus
   3. In the cubital fossa is lateral to the biceps tendon
   4. Terminates as the radial and ulnar arteries   A B C D E

7. A PATIENT PRESENTS WITH A TRANSVERSE FRACTURE OF THE HUMERAL SHAFT JUST BELOW THE ATTACHMENT OF THE DELTOID MUSCLE. THE OBVIOUS ELEVATION OF THE DISTAL FRAGMENT IS DUE TO THE ACTION OF WHICH OF THE FOLLOWING MUSCLES?
   1. Brachialis   3. Pectoralis major
   2. Teres major   4. Triceps brachii   A B C D E

8. A SUPRACONDYLAR FRACTURE OF THE HUMERUS, WITH RESULTANT POSTERIOR DISPLACEMENT OF THE DISTAL FRAGMENT, USUALLY INVOLVES WHICH OF THE FOLLOWING STRUCTURES?
   1. Median nerve   3. Brachial artery
   2. Radial nerve   4. Ulnar nerve   A B C D E

9. A HOCKEY PLAYER FELL ON HIS ELBOW WITH HIS ARM ABDUCTED PRODUCING A FRACTURE OF THE SURGICAL NECK OF THE HUMERUS. THIS RESULTED IN ELEVATION AND ADDUCTION OF THE DISTAL FRAGMENT. THE ADDUCTION IS CAUSED BY WHICH OF THE FOLLOWING MUSCLES?
   1. Pectoralis major   3. Teres major
   2. Supraspinatus   4. Coracobrachialis   A B C D E

A	B	C	D	E
1,2,3	1,3	2,4	only 4	all correct

10. THE RADIAL NERVE:
    1. Is a continuation of the lateral cord of the brachial plexus
    2. Passes towards the lateral aspect of the elbow in the spiral groove
    3. Supplies the pronator teres muscle
    4. Is the principal nerve to the extensor muscles of the upper limb

    A B C D E

11. WHICH OF THE FOLLOWING STATEMENTS IS (ARE) TRUE FOR THE ULNAR NERVE?
    1. Largest nerve derived from the medial cord of the brachial plexus
    2. Carries fibers from C6, 7, and 8
    3. Has no branches in the arm
    4. In the arm, it lies lateral to the brachial artery

    A B C D E

-------------------- ANSWERS, NOTES AND EXPLANATIONS --------------------

1. **A. 1, 2, and 3 are correct.** The elbow joint is a uniaxial hinge joint between the distal end of the humerus and the proximal ends of the radius and ulna. Sharing the capsule and the synovial membrane of this articulation is the proximal radioulnar joint, a pivot joint. The elbow joint has a loose capsule which is thickened on its medial and lateral sides into medial (ulnar) and lateral (radial) collateral ligaments. These ligaments reinforce the capsule. The elbow joint is supplied by the musculocutaneous, ulnar, radial, and median nerves.

2. **A. 1, 2, and 3 are correct.** The cubital fossa, a V-shaped space anterior to the elbow, contains a number of important structures. From medial to lateral, they are: the median nerve, the brachial artery and its branches, the biceps tendon, and the radial nerve. The upper limit of the fossa is a line drawn between the epicondyles of the humerus. The converging walls of the fossa are formed by the brachioradialis and pronator teres muscles. The floor consists of the brachialis and supinator muscles; the roof is formed by the deep fascia, reinforced by the bicipital aponeurosis. The roof is crossed obliquely by the median cubital vein, a common site for intravenous injections and blood transfusions.

3. **E. All are correct.** The nerve supply to the elbow joint follows the principle (Hilton's Law): that nerves supplying muscles moving a joint also give rise to its articular branches, containing sensory fibers. Thus the radial, ulnar, median, and musculocutaneous nerves all give articular branches to the elbow joint.

4. **B. 1 and 3 are correct.** The muscles inserting into the greater tubercle of the humerus are the: supraspinatus, infraspinatus, and teres minor. The subscapularis muscle inserts into the lesser tubercle of the humerus, whereas the teres major muscle inserts on the medial lip of the bicipital groove of the humerus.

5. **B. 1 and 3 are correct.** The bicipital aponeurosis, arising from the medial border of the biceps tendon, passes medially and obliquely and blends with the deep fascia of the forearm. The aponeurosis lies on the brachial artery and the median nerve. The cephalic vein and radial nerve lie lateral to the aponeurosis.

6. **C.** <u>2 and 4 are correct</u>. The brachial artery begins at the lower border of the teres major muscle as a continuation of the axillary artery. It descends superficially on the medial aspect of the biceps and coracobrachialis muscles, covered by superficial and deep fascia. It passes anterior to the elbow, medial to the tendon of the biceps, deep to the bicipital aponeurosis, and bifurcates in the lower portion of the cubital fossa, into the radial and ulnar arteries.

7. **D.** <u>Only 4 is correct</u>. As the fracture line is below the insertion of the deltoid muscle, the proximal fragment would be abducted. The distal fragment would be elevated principally by the triceps brachii muscle, but the biceps brachii and coracobrachialis muscles would also contribute to this displacement.

8. **A.** <u>1, 2, and 3 are correct</u>. Posterior displacement of the distal humeral fragment would usually involve the radial and median nerves and the brachial artery. Injury to these structures is caused by the rough fragmented ends of the humerus.

9. **B.** <u>1 and 3 are correct</u>. The surgical neck of the humerus, the region immediately distal to the tubercles, is a common site of fractures. The fracture line lies above the insertions of the pectoralis major, teres major, and latissimus dorsi muscles. The supraspinatus muscle abducts the small proximal fragment, whereas the distal fragment is elevated and adducted. The elevation results from contraction of the deltoid, biceps brachii, and coracobrachialis muscles; the adduction is due to the pectoralis major, teres major, and latissimus dorsi muscles.

10. **C.** <u>2 and 4 are correct</u>. The radial nerve (C5, 6, 7, 8, and T1), a continuation of the posterior cord of the brachial plexus, is the principal nerve to the extensor muscles of the upper limb, and to the skin of the posterior aspect of the arm, forearm, and most of the hand. In the arm it crosses the posterior aspect of the humerus, from medial to lateral, in the spiral groove and then descends towards the elbow region. Here, the nerve passes anterior to the lateral epicondyle of the humerus. The pronator teres muscle is a pro- of the forearm; it lies on the flexor aspect and is innervated by the median nerve.

11. **B.** <u>1 and 3 are correct</u>. The ulnar nerve is the largest branch of the medial cord of the brachial plexus and carries fibers from C7, 8, and T1. It receives no fibers from C6. It courses through the arm, medial to the proximal part of the brachial artery and gives rise to no branches until it reaches the elbow region. Here it lies in a groove on the posterior aspect of the medial epicondyle and gives off a few articular branches to the elbow joint.

FIVE-CHOICE ASSOCIATION QUESTIONS

DIRECTIONS: Each group of questions below consists of a numbered list of descriptive words or phrases accompanied by a diagram with certain parts indicated by letters, or by a list of lettered headings. For each numbered word or phrase, SELECT THE LETTERED PART OR HEADING that matches it correctly. Then insert the letter in the space to the right of the appropriate number. Sometimes more than one numbered word or phrase may be correctly matched to the same lettered part or heading.

ASSOCIATION QUESTIONS

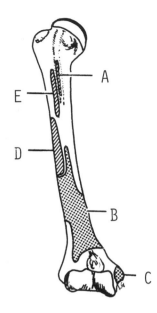

1. ____ The origin of a strong flexor of the forearm.

2. ____ The insertion of an abductor of the arm.

3. ____ The insertion of an extensor and medial rotator of the arm.

4. ____ A common site of origin for the flexor muscles of the forearm.

5. ____ The insertion of an adductor of the arm.

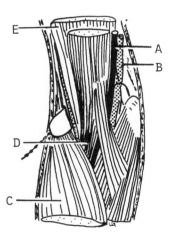

6. ____ The biceps tendon

7. ____ The median nerve

8. ____ The brachialis muscle

9. ____ The brachioradialis muscle

10. ____ The brachial artery

---------------------- ANSWERS, NOTES AND EXPLANATIONS ----------------------

1. **B.** The brachialis muscle, inserted on the anterior aspect of the coronoid process of the ulna, is a strong flexor of the forearm. This muscle, along with the biceps and coracobrachialis muscles, is innervated by the musculocutaneous nerve (C5, 6, 7).

2. **D.** The deltoid muscle inserts into the lateral aspect of the middle of the shaft of the humerus. It originates from the front of the lateral third of the clavicle, the lateral border of the acromion and the inferior margin of the spine of the scapula. This muscle is innervated by the axillary nerve (C5, 6).

3. **A.** The latissimus dorsi muscle originates from the spines of the lower six thoracic vertebrae, all lumbar and upper sacral vertebrae through the lumbo-

dorsal fascia, and the posterior third of the outer lip of the iliac crest. It inserts in the floor of the bicipital groove of the humerus and is innervated by the long thoracic nerve (C6, 7, 8).

4. C. The six flexor muscles of the forearm all originate from the medial epicondyle of the humerus. They are arranged in three layers: superficial, the flexor carpi radialis, flexor carpi ulnaris, and palmaris longus; intermediate, the flexor digitorum superficialis; and deep, the flexor digitorum profundus and flexor pollicis longus muscles. These muscles all insert either into the wrist or the bones of the hand.

5. E. The pectoralis major muscle inserts into the lateral lip of the bicipital groove. It originates from the front of the medial half of the clavicle, the sternum and the costal cartilages of the upper six ribs. This muscle is a powerful adductor and medial rotator of the arm.

6. D. The biceps tendon arises from the fusion of its two bellies and inserts into the posterior part of the tuberosity of the radius. The strong bicipital aponeurosis originates from the biceps tendon, passes medially and obliquely to blend with the deep fascia of the forearm.

7. A. The median nerve is the most medial structure contained in the cubital fossa. From medial to lateral, these structures are: the median nerve, the brachial artery, the biceps tendon, and the radial nerve (not labelled).

8. E. The brachialis muscle, lying deep to the biceps, originates from the lower half of the humeral shaft and inserts into the tuberosity of the ulna, inferior to the coracoid process. This muscle and the supinator muscle form the floor of the cubital fossa.

9. C. The brachioradialis muscle forms the lateral boundary of the cubital fossa, whereas the pronator teres (not labelled) forms the medial boundary. These muscles originate from the lateral and medial supracondylar ridges of the humerus respectively, and approach each other as they descend to insert into the radius.

10. B. The brachial artery enters the cubital fossa on the medial aspect of the biceps muscle. As it lies in the middle of the fossa, it is related to the bicipital aponeurosis superficially, the median nerve medially and the biceps tendon laterally. In the fossa, the brachial artery divides into radial and ulnar arteries.

FOREARM AND WRIST

FIVE-CHOICE COMPLETION QUESTIONS

DIRECTIONS: Each of the following questions or incomplete statements is followed by five suggested answers or completions. SELECT THE ONE BEST ANSWER in each case and then underline the appropriate letter at the lower right of each question.

1. THE EXTENSOR CARPI ULNARIS MUSCLE IS SUPPLIED BY WHICH OF THE FOLLOWING NERVES?
   A. Superficial radial
   B. Ulnar
   C. Median
   D. Deep radial
   E. Anterior interosseous

   A B C D E

SELECT THE ONE BEST ANSWER

2. THE DISTAL END OF THE RADIUS USUALLY ARTICULATES WITH WHICH OF THE FOLLOWING BONES?
   A. Trapezium and trapezoid
   B. Lunate and trapezoid
   C. Capitate and hamate
   D. Trapezoid and scaphoid
   E. Scaphoid and lunate

   A B C D E

3. THE SUPERFICIAL FLEXOR MUSCLES OF THE FOREARM ORIGINATE FROM THE _____ OF THE HUMERUS:
   A. Lateral surface
   B. Posterior surface
   C. Anterior surface
   D. Medial epicondyle
   E. Lateral epicondyle

   A B C D E

4. THE PULSE PALPATED MEDIAL TO THE TENDONS OF THE FLEXOR DIGITORUM SUPERFICIALIS AND PROFUNDUS MUSCLES IS THE PULSE OF THE _____ ARTERY.
   A. Ulnar
   B. Radial
   C. Median
   D. Brachial
   E. Princeps pollicis

5. THE STRUCTURE IMMEDIATELY DEEP TO THE PRONATOR QUADRATUS MUSCLE IS THE:
   A. Ulnar nerve
   B. Radial artery
   C. Flexor pollicis longus muscle
   D. Interosseous membrane
   E. Flexor digitorum profundus muscle

   A B C D E

6. THE PRONATOR TERES AND PRONATOR QUADRATUS MUSCLES ARE INNERVATED BY A BRANCH OF THE _____ NERVE.
   A. Axillary
   B. Median
   C. Musculocutaneous
   D. Radial
   E. Ulnar

   A B C D E

7. THE COMMON INTEROSSEOUS ARTERY IS A BRANCH OF THE _____ ARTERY.
   A. Brachial
   B. Ulnar
   C. Radial
   D. Anterior interosseous
   E. Posterior interosseous

   A B C D E

8. WHICH OF THE FOLLOWING MUSCLES IS INNERVATED BY BOTH THE ULNAR AND MEDIAN NERVES?
   A. Flexor digitorum superficialis
   B. Flexor digitorum profundus
   C. Pronator quadratus
   D. Pronator teres
   E. None of the above

   A B C D E

9. AN ELDERLY PATIENT PRESENTS WITH A FRACTURE OF THE UPPER THIRD OF THE RIGHT RADIUS, WITH THE DISTAL FRAGMENT OF THE RADIUS AND HAND PRONATED. THE PROXIMAL END OF THE FRACTURED RADIUS DEVIATES LATERALLY, PRIMARILY AS A RESULT OF CONTRACTION OF THE _____ MUSCLE.
   A. Pronator teres
   B. Supinator
   C. Pronator quadratus
   D. Brachioradialis
   E. Brachialis

   A B C D E

-------------------- ANSWERS, NOTES AND EXPLANATIONS --------------------

1. **D.** The deep radial nerve arises from the radial nerve in the lateral aspect of the cubital fossa and courses within the supinator muscle lateral to the neck of the radius. It then enters the posterior compartment of the forearm, accompanies the posterior interosseous artery, and terminates as articular

branches to the wrist and carpal joints. As it courses through the posterior compartment, it gives off muscular branches to the extensor carpi ulnaris and to other muscles of the superficial and deep groups of extensor muscles.

2. E. The distal end of the radius and the articular disc form an ellipsoid synovial joint with the proximal row of carpal bones (scaphoid, lunate, and triquetrum), excluding the pisiform. This joint permits movements of flexion, extension, adduction, abduction and circumduction.

3. D. The medial epicondyle of the humerus is the common site of origin for the superficial flexor muscles of the forearm, whereas the lateral epicondyle and supracondyle ridge of the humerus provide the origin for the extensor muscles of the forearm. The flexor muscles of the forearm are eight in number: the pronator teres, flexor carpi radialis, palmaris longus, flexor carpi ulnaris, and flexor digitorum superficialis are the superficial group; the deep group consists of the flexor digitorum profundus, flexor pollicis longus, and pronator quadratus.

4. A. Although the ulnar artery at the wrist may be used for taking the pulse, the radial artery is most commonly used. The pulse of the ulnar artery may be palpated medial to the tendons of the flexor digitorum superficialis and profundus muscles.

5. D. The interosseous membrane lies deep to the pronator quadratus muscle, however, running between them on their way to the wrist are the anterior interosseous artery and nerve. The artery is a branch of the common interosseous artery and the nerve is a branch of the median nerve.

6. B. The pronator teres muscle has two heads of origin: the superficial head arises from the medial epicondyle of the humerus, whereas the deep head arises from the coronoid process of the ulna. This muscle passes downward and laterally, inserting into the middle of the lateral surface (maximal convexity) of the radius. The pronator quadratus arises from the lower quarter of the anterior surface of the ulnar shaft, and inserts into the lower quarter of the anterior surface of the radial shaft. Both muscles are innervated by a branch of the median nerve and pronate the forearm, but the pronator teres is also a flexor of the elbow joint.

7. B. The common interosseous artery is the third named branch of the ulnar artery. It courses posteriorly and inferiorly in the lower region of the cubital fossa, where it divides into the anterior and posterior interosseous arteries on the corresponding sides of the interosseous membrane. The anterior interosseous artery gives branches to the radius and ulna. It also gives rise to the median artery, which accompanies the median nerve, and contributes to the palmar carpal network. The posterior interosseous artery gives rise to the palmar recurrent interosseous artery, which contributes to the anastomoses around the elbow, and a terminal branch which joins the palmar carpal network. The muscles of the forearm receive their vascular supply from numerous unnamed branches of both the anterior and posterior interosseous arteries.

8. B. All these muscles are located in the anterior osteofascial compartment of the forearm. The medial half of the flexor digitorum profundus is supplied by the ulnar nerve; the lateral half by the anterior interosseous branch of the median nerve. The flexor digitorum superficialis, pronator quadratus, and pronator teres muscles are supplied by the median nerve.

9. B. The fracture line of the upper third of the radius lies between the insertions of the supinator and pronator teres muscles. The distal radial fragment and hand are pronated due to the unopposed contraction of pronator teres and

pronator quadratus muscles. The proximal fragment deviates laterally by the unopposed contraction of the supinator muscle. The biceps brachii muscle would also tend to flex the proximal fragment of the radius.

MULTI-COMPLETION QUESTIONS

DIRECTIONS: In each of the following questions or incomplete statements, ONE OR MORE of the completions given is correct. At the lower right of each question, underline A if 1, 2 and 3 are correct; B if 1 and 3 are correct; C if 2 and 4 are correct; D if only 4 is correct; and E if all are correct.

1. THE INTEROSSEOUS MEMBRANE OF THE FOREARM:
   1. Forms the intermediate radioulnar joint
   2. Has fibers running obliquely downward and medially
   3. Is attached to the medial border of the radius
   4. Is pierced by the anterior interosseous vessels   A B C D E

2. THE FOLLOWING FUNCTION(S) MAY BE ATTRIBUTED TO THE COMBINED ACTION OF THE EXTENSOR CARPI RADIALIS LONGUS AND BREVIS AND THE EXTENSOR CARPI ULNARIS MUSCLES:
   1. Radial deviation of the hand
   2. Flexion of the wrist
   3. Ulnar deviation of the hand
   4. Extension of the wrist   A B C D E

3. THE FLEXOR RETINACULUM IS ATTACHED LATERALLY TO WHICH BONE(S)?
   1. Trapezium   3. Scaphoid
   2. Triquetrum   4. Pisiform   A B C D E

4. WHICH OF THE FOLLOWING RELATION(S) OF THE ULNAR ARTERY AT THE WRIST IS (ARE) CORRECT?
   1. Lies superficial to the flexor retinaculum
   2. The pisiform bone lies on its medial side
   3. Lies lateral to the ulnar nerve
   4. Covered by the palmaris brevis muscle   A B C D E

5. THE STRUCTURE(S) CROSSING THE WRIST AND ENTERING THE PALM SUPERFICIAL TO THE FLEXOR RETINACULUM IS (ARE) THE:
   1. Flexor pollicis longus muscle
   2. Ulnar nerve
   3. Median nerve
   4. Superficial palmar branch of the radial artery   A B C D E

6. THE BRACHIORADIALIS MUSCLE:
   1. Inserts into the base of the styloid process of the radius
   2. Acts only at the elbow joint
   3. Is innervated by the radial nerve
   4. Originates from the upper two-thirds of the lateral supracondylar ridge.   A B C D E

7. IN PRONATION AND SUPINATION OF THE FOREARM:
   1. The radius moves across the ulna
   2. The ulna moves across the radius
   3. The radius rotates
   4. The radius and ulna rotate on the humerus   A B C D E

A	B	C	D	E
1,2,3	1,3	2,4	only 4	all correct

8. THE ULNAR NERVE SUPPLIES THE:
   1. Flexor digitorum superficialis muscle
   2. Skin of the dorsum of the hand
   3. Pronator quadratus muscle
   4. Skin of the palm of the hand   A B C D E

9. THE SUPINATOR MUSCLE IS INNERVATED BY BRANCHES OF THE
   _____ NERVE(S).
   1. Ulnar   3. Axillary
   2. Median   4. Radial   A B C D E

10. ASSOCIATED WITH THE "ANATOMICAL SNUFF-BOX" IS (ARE) THE:
    1. Lunate bone   3. Median nerve
    2. Scaphoid bone   4. Radial artery   A B C D E

-------------------- ANSWERS, NOTES AND EXPLANATIONS --------------------

1. E. <u>All are correct</u>. The interosseous membrane is attached to the interosseous borders of the radius and ulna, forming an intermediate joint between these bones. Its strong fibers run obliquely downward and medially so that the force of a fall on an outstretched hand will be transmitted from the radius to the ulna, and to the humerus and scapula. This membrane provides sites for muscle attachments and is pierced in its lower segment by the anterior interosseous vessels.

2. C. <u>2 and 4 are correct</u>. The combined synergistic action of the extensor carpi ulnaris and extensor carpi radialis longus and brevis muscles fixes the wrist during flexion of the hand. The extensor carpi radialis brevis and longus muscles are involved in abduction (radial deviation) of the hand, whereas contraction of the extensor carpi ulnaris adducts the hand (ulnar deviation). Simultaneous contraction of these muscles produces little radial or ulnar deviation of the hand.

3. B. <u>1 and 3 are correct</u>. The flexor retinaculum, a thickening of the deep fascia, forms an osteofascial tunnel on the anterior aspect of the wrist. Laterally this retinaculum is attached to the tubercles of the scaphoid and trapezium bones; medially it is attached to the pisiform and the hook of the hamate. This tunnel contains the median nerve and the flexor tendons of the thumb and fingers. Swelling of the surrounding structures often causes compression symptoms in the hand of median neuritis.

4. E. <u>All are correct</u>. The ulnar artery enters the hand superficial to the flexor retinaculum and lateral to the pisiform bone. The ulnar nerve lies medial to the ulnar artery, deep to the tendon of the flexor carpi ulnaris and the palmaris brevis muscle.

5. C. <u>2 and 4 are correct</u>. From medial to lateral, some notable structures entering the palm superficial to the flexor retinaculum are the: flexor carpi ulnaris tendon, ulnar nerve, ulnar artery, and the palmaris longus tendon.

6. E. <u>All are correct</u>. The brachioradialis muscle, innervated by the radial nerve (C5 - T1), originates from the upper two-thirds of the lateral supracondylar ridge of the humerus, and inserts into the base of the styloid process of the radius. It acts only at the elbow, flexing it.

7. **B.** <u>1 and 3 are correct</u>. In pronation of the forearm, the radius carries the hand with it and is moved obliquely across the anterior aspect of the ulna. The radius rotates so that the proximal end remains lateral and the distal end becomes medial to the ulna. The reversal of this movement, returning the radius to the anatomical position (parallel to the ulna), is called supination.

8. **C.** <u>2 and 4 are correct</u>. In the hand the ulnar nerve divides into a superficial branch, supplying the skin on the medial one and a half fingers, and a deep branch innervating the hypothenar muscles, the medial two lumbrical, all interossei, and the adductor pollicis muscles. The dorsal branch of the ulnar nerve, given off above the wrist joint, passes to the dorsum of the hand and supplies the skin of the medial one and a half fingers. The ulnar nerve provides motor fibers to the flexor carpi ulnaris and the medial half of the flexor digitorum profundus muscles of the forearm.

9. **D.** <u>Only 4 is correct</u>. The supinator muscle, along with the other muscles of the posterior compartment of the forearm, is innervated by the deep radial nerve.

10. **C.** <u>2 and 4 are correct</u>. The floor of the anatomical "snuff-box" is formed by the scaphoid and trapezium bones and is limited proximally by the styloid process of the radius. The radial artery crosses the floor superficially where its pulsation can be felt. The "snuff-box" is bounded medially by the extensor pollicis longus and laterally by the extensor pollicis brevis and abductor pollicis longus tendons. The superficial branch of the radial nerve crosses the tendon of the extensor pollicis longus muscle.

FIVE-CHOICE ASSOCIATION QUESTIONS

DIRECTIONS: Each group of questions below consists of a numbered list of descriptive words or phrases accompanied by a diagram with certain parts indicated by letters, or by a list of lettered headings. For each numbered word or phrase, SELECT THE LETTERED PART OR HEADING that matches it correctly. Then insert the letter in the space to the right of the appropriate number. Sometimes more than one numbered word or phrase may be correctly matched to the same lettered part or heading.

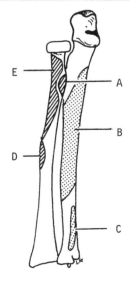

1. ____ Origin of a pronator of the forearm.

2. ____ Insertion of a supinator which originates from the ulna.

3. ____ Insertion of a muscle innervated by the musculocutaneous nerve.

4. ____ Origin of a muscle which flexes the distal phalanx of a finger.

5. ____ Insertion of a forearm pronator.

ASSOCIATION QUESTIONS

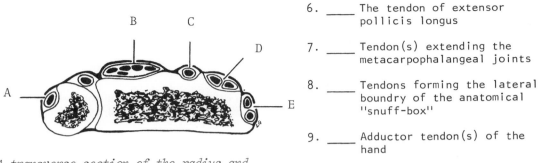

A transverse section of the radius and ulna proximal to the wrist joint.

6. ____ The tendon of extensor pollicis longus

7. ____ Tendon(s) extending the metacarpophalangeal joints

8. ____ Tendons forming the lateral boundry of the anatomical "snuff-box"

9. ____ Adductor tendon(s) of the hand

10. ____ Abductor tendon(s) of the hand

------------------- ANSWERS, NOTES AND EXPLANATIONS -------------------

1. **C.** The pronator quadratus muscle originates from the anterior aspect of the ulnar and inserts on the anterior surface of the radial shaft. This muscle along with the pronator teres pronates the forearm. It is innervated by the anterior interosseous branch of the median nerve.

2. **E.** The supinator muscle originates from the supinator crest and fossa of the ulna as well as from the lateral epicondyle of the humerus, the lateral ligament of the elbow joint and the annular ligament. It inserts into the anterior, lateral and posterior aspects of the neck and shaft of the radius. The supinator muscle, as well as the biceps brachii, is a supinator of the forearm.

3. **A.** The biceps brachii muscle of the arm inserts into the radial tuberosity. The major functions of this muscle are flexion and supination of the forearm; it is innervated by the musculocutaneous nerve (C5, 6, 7).

4. **B.** The flexor digitorum profundus muscle originates from the upper three quarters of the anteromedial surface of the ulna and the neighboring interosseous membrane. It passes beneath the flexor retinaculum, and inserts by four tendons into the base of the distal phalanges of the fingers. The lateral portion of this muscle is innervated by the anterior interosseous branch of the median nerve, and the medial half by the ulnar nerve.

5. **D.** The two heads of the pronator teres muscle unite and insert into the roughened impression in the lateral surface of the middle of the radial shaft. One head of this muscle originates from the medial epicondyle of the humerus and the other from the medial border of the coronoid process of the ulna.

6. **C.** The tendon of the extensor pollicis longus lies in a deep groove lateral to the dorsal tubercle on the posterior surface of the radius. This tendon extends the distal phalanx of the thumb. It forms the medial wall of the anatomical "snuff-box".

7. **B.** The four tendons of the extensor digitorum extend the metacarpophalangeal joints and assist the lumbrical and interossei muscles in extending the proximal and distal interphalangeal joints. The other tendon deep to this group is the extensor indicis which extends the metacarpophalangeal joint of the index finger.

8. E. From lateral to medial the two tendons forming the lateral border of the anatomical "snuff-box" are: the abductor pollicis longus and the extensor pollicis brevis. The former abducts the carpometacarpal joint whereas the latter muscle extends the metacarpophalangeal joint of the thumb.

9. A. The tendon of the extensor carpi ulnaris muscle adducts and extends the hand at the wrist joint. This muscle is innervated by the deep branch of the radial nerve.

10. D. The tendons of the extensor carpi radialis longus and brevis muscles extend and abduct the hand at the wrist. The former is innervated by the radial nerve; the latter by the deep branch of the radial.

H A N D

FIVE-CHOICE COMPLETION QUESTIONS

DIRECTIONS: Each of the following questions or incomplete statements is followed by five suggested answers or completions. SELECT THE ONE BEST ANSWER in each case and then underline the appropriate letter at the lower right of each question.

1. THE CARPAL BONE INDICATED BY THE ARROW IS THE _____.
   A. Trapezoid
   B. Scaphoid
   C. Lunate
   D. Hamate
   E. Capitate

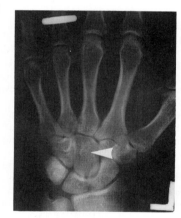

A B C D E

2. THE PULSE OF A PATIENT IS MOST COMMONLY TAKEN BY PALPATING WHICH OF THE FOLLOWING ARTERIES?
   A. Ulnar
   B. Radial
   C. Median
   D. Brachial
   E. Princeps pollicis

A B C D E

3. THE FIRST AND SECOND LUMBRICAL MUSCLES ARE INNERVATED BY THE _____ NERVE.
   A. Deep ulnar
   B. Median
   C. Deep radial
   D. Anterior interosseous
   E. Superficial radial

A B C D E

4. THE DORSAL AND PALMAR INTEROSSEI MUSCLES ARE INNERVATED BY THE _____ NERVE.
   A. Axillary
   B. Median
   C. Musculocutaneous
   D. Radial
   E. None of the above

A B C D E

SELECT THE ONE BEST ANSWER

5. WHICH OF THE FOLLOWING MUSCLES IS AN EXTRINSIC MUSCLE OF THE THUMB?
   A. Opponens pollicis
   B. Extensor pollicis longus
   C. Adductor pollicis
   D. Flexor pollicis brevis
   E. None of the above

6. EXTENSION OF THE THUMB IS WHEN IT IS MOVED:
   A. Laterally from the palm
   B. At right angles to the palmar plane
   C. Alongside the index finger
   D. Straight across the palm
   E. None of the above

7. WHICH OF THE FOLLOWING STRUCTURES LIES OUTSIDE OF THE ANTERIOR (VOLAR) CARPAL TUNNEL?
   A. Median nerve
   B. Palmaris longus
   C. Flexor pollicis longus
   D. Flexor digitorum profundus
   E. Flexor digitorum superficialis

8. THE RADIAL PORTION OF THE FLEXOR RETINACULUM IS ATTACHED TO THE SCAPHOID AND _____ BONES.
   A. Hamate
   B. Lunate
   C. Pisiform
   D. Trapezium
   E. Trapezoid

9. UPON EXAMINATION OF A BUTCHER WHO HAS A DEEP CUT ON HIS PALM, YOU FIND THAT HE IS UNABLE TO ADDUCT HIS THUMB. WHICH OF THE FOLLOWING NERVES HAS MOST LIKELY BEEN SEVERED?
   A. Recurrent motor branch of the median
   B. Superficial ulnar
   C. Deep ulnar
   D. Superficial radial
   E. Deep radial

---------------------- ANSWERS, NOTES AND EXPLANATIONS ----------------------

1. E. The eight carpal bones of the hand are arranged in two rows. The proximal row, medial to lateral, consists of the: pisiform, triquetrum, lunate, and scaphoid; the distal row, medial to lateral, consists of the: hamate, capitate, trapezoid, and trapezium. These bones articulate with each other by synovial plane joints which allow a limited amount of gliding movement.

2. B. The radial artery at the wrist is most commonly used for taking the pulse because it is located superficially at the distal part of the forearm, where it is easily accessible. The pulsation of the artery is strongly felt because it passes immediately anterior to the lower end of the radius. The tendon of the flexor carpi radialis lies medial to the pulsating artery.

3. B. The first and second lumbrical muscles are innervated by the median nerve, whereas the third and fourth lumbricals are innervated by the ulnar nerve.
   Clinically, injury to the median nerve can be recognized by asking the patient to make a fist slowly. If the nerve is damaged, the patient's index and middle fingers will lag behind the ring and little fingers.

4. E. Both the palmar and dorsal interossei muscles are innervated by the deep branch of the ulnar nerve (C7, 8 and T1). In the hand, the middle finger is

used as the reference line when discussing movements of the fingers. The palmar interossei adduct the medial two and the lateral two fingers towards the middle finger, whereas the dorsal interossei abduct the fingers away from the middle finger. The palmar and dorsal interossei flex the fingers at the metacarpophalangeal joints and extend the interphalangeal joints.

5. **B.** The extensor pollicis longus muscle originates from the posterior surface of the ulna and of the interosseous membrane; it inserts on the base of the distal phalanx of the thumb. This muscle is extrinsic because it originates outside the hand, whereas the other muscles are intrinsic, having their origins within the hand.

6. **A.** A thorough knowledge of thumb movements is important because much of the efficiency of the hand depends on the functional state of the thumb. In the anatomical position, the thumb is moved laterally in the same plane as the palm and is known as extension. A medial movement of the thumb in the plane of the palm is flexion, whereas opposition is when the palmar aspect of the thumb touches the palmar aspect of the tip of a finger of the same hand. A forward movement of the thumb at right angles to the palm is called abduction; the reverse movement is adduction.

7. **B.** The only tendon crossing superficial to the flexor retinaculum is the palmaris longus. It is not always present, but when it is, it crosses the upper portion of the retinaculum, inserting into its lower portion and into the palmar aponeurosis. All other long tendons of the anterior forearm entering the hand pass deep to the flexor retinaculum in the anterior carpal tunnel.

8. **D.** The flexor retinaculum, a thickening of deep fascia, bridges the anterior apect of the wrist. It attaches radially (laterally) to the tubercle of the scaphoid and the trapezium, and medially to the pisiform and the hook of the hamate. This retinaculum bridges the median nerve and the extrinsic flexors of the hand.

9. **C.** The adductor pollicis muscle of the thumb is primarily involved in adduction and is innervated by the deep ulnar nerve. The patient would also be unlikely able to adduct, abduct, or flex his fingers at the metacarpophalangeal joints. It should be noted that the three short muscles of the thumb are innervated by the recurrent branch of the median nerve.

MULTI-COMPLETION QUESTIONS

DIRECTIONS: In each of the following questions or incomplete statements, ONE OR MORE of the completions given is correct. At the lower right of each question, underline A if 1, 2 and 3 are correct; B if 1 and 3 are correct; C if 2 and 4 are correct; D if only 4 is correct; and E if all are correct.

1. WHICH OF THE FOLLOWING MUSCLES IS (ARE) INNERVATED BY THE MEDIAN NERVE?
   1. Abductor pollicis brevis
   2. Flexor pollicis brevis
   3. Opponens pollicis
   4. Adductor pollicis   A B C D E

2. THE LUMBRICAL MUSCLES IN THE HAND ARE SUPPLIED BY THE _____ NERVE(S).
   1. Ulnar
   2. Radial
   3. Median
   4. Musculocutaneous   A B C D E

A	B	C	D	E
1,2,3	1,3	2,4	only 4	all correct

3. WHICH OF THE FOLLOWING STATEMENTS ABOUT THE CARPOMETACARPAL JOINT OF THE THUMB IS (ARE) CORRECT?
   1. Has a loose capsule
   2. Trapezium is involved
   3. Has great freedom of movement
   4. Is a gliding joint   A B C D E

4. PARALYSIS OF THE MUSCLES INNERVATED BY THE ULNAR NERVE RESULTS IN THE PHYSICAL DEFORMITY OF THE HAND CALLED "CLAW HAND". WHICH OF THE FOLLOWING IMPAIRMENTS OR LOSSES OF FUNCTION WOULD BE PRESENT WHEN THE NERVE IS INJURED ABOVE THE ELBOW?
   1. Adduction of the hand
   2. Adduction of the thumb
   3. Abduction and adduction of the fingers
   4. Flexion of the medial two distal phalanges   A B C D E

5. THE PROXIMAL ROW OF CARPAL BONES INCLUDES THE:
   1. Scaphoid   3. Pisiform
   2. Trapezium   4. Hamate   A B C D E

6. THE SKIN OF THE THENAR EMINENCE IS INNERVATED BY THE:
   1. Palmar cutaneous branch of the ulnar nerve
   2. Superficial radial nerve
   3. Digital nerve to the thumb
   4. Palmar cutaneous branch of the median nerve   A B C D E

7. A PATIENT, DIAGNOSED TO HAVE NEURITIS OF THE MEDIAN NERVE AT THE WRIST, WOULD PRESENT THE FOLLOWING SYMPTOM(S):
   1. Pain in the index finger
   2. Paresthesia of the little finger
   3. Weakness in abduction of the thumb
   4. Weakness in opposition of the little finger   A B C D E

-------------------- ANSWERS, NOTES AND EXPLANATIONS --------------------

1. **A.** <u>1, 2, and 3 are correct</u>. The four muscles listed represent the short muscles of the thumb. Abductor pollicis brevis, flexor pollicis brevis, and opponens pollicis contribute to the formation of the thenar eminence (the ball of the thumb). The median nerve innervates the abductor pollicis brevis and the opponens pollicis; the flexor pollicis brevis is innervated by the median as well as by the deep branch of the ulnar nerve. The two heads (oblique and transverse) of the adductor pollicis muscle are supplied by a deep branch of the ulnar nerve; this muscle is not considered part of the thenar eminence.

2. **B.** <u>1 and 3 are correct</u>. The four lumbrical muscles in the palm of the hand are numbered (one to four) from the index finger. They arise from tendons of the flexor digitorum profundus. The first two lumbricals are innervated by the digital branch of the median nerve; the third and fourth by the deep branch of the ulnar nerve.

3. **A.** <u>1, 2, and 3 are correct</u>. The carpometacarpal joint of the thumb is classified as a saddle joint and has considerable freedom in its movement. The articulating surfaces of this joint are the trapezium and the proximal end of the first metacarpal bone. This joint is surrounded by a loose capsule supported by ligaments.

4. **E.** <u>All are correct</u>. The interossei and the medial two lumbrical muscles are paralysed, so the fingers cannot be adducted or abducted. The adductor pollicis is similarly affected so adduction of the thumb is also lost. Since the flexor carpi ulnaris is innervated by the ulnar nerve, adduction of the hand will be impaired. The extensor carpi ulnaris, the other adductor of the hand, will still be functional. The proximal phalanges of the medial two fingers cannot be flexed and the distal phalanges cannot be extended. Therefore, the proximal and distal phalanges will be hyperextended and hyperflexed by the unopposed long extensors and flexors, respectively.

5. **B.** <u>1 and 3 are correct</u>. The carpal bones are arranged in two rows of four: the scaphoid, lunate, triquetrum, and pisiform (from lateral to medial) form the proximal row; the trapezium, trapezoid, capitate, and hamate form the distal row. Of particular clinical importance are fractures and dislocations of the carpal bones as they may damage the surrounding muscles, nerves, and blood vessels.

6. **C.** <u>2 and 4 are correct</u>. The thenar eminence is formed by three short muscles of the thumb (abductor pollicis brevis, flexor pollicis brevis, and opponens pollicis). The skin of the thenar eminence is innervated by sensory (cutaneous) nerves: the medial portion by the palmar cutaneous branch of the median nerve, derived from the ventral rami of C6 to C8; the lateral portion by the superficial radial nerve, derived from the ventral rami of C6 to C8; and the hypothenar eminence by the palmar cutaneous branch of the ulnar nerve. The digital nerve to the thumb (more properly called the first common palmar digital nerve of the median nerve) is motor to the short muscles of the thumb, and sensory only distal to the webs of the fingers.

7. **B.** <u>1 and 3 are correct</u>. Neuritis of the median nerve would not produce loss of sensation (paresthesis) or muscle weakness in the little finger, as it is, along with the ring finger, supplied by the ulnar nerve. Crush injuries of the median nerve at the wrist cause pain in the lateral three fingers and muscle weakness in the thenar muscles. The degree of functional loss of the thenar muscles depends on the amount of overlapping by the deep ulnar nerve supplying some of these muscles. The wasting of the thenar muscles resulting from these injuries gives a characteristic flattening of the thenar eminence and a permanent extension of the thumb, often called "ape-hand".

FIVE-CHOICE ASSOCIATION QUESTIONS

DIRECTIONS: Each group of questions below consists of a numbered list of descriptive words or phrases accompanied by a diagram with certain parts indicated by letters, or by a list of lettered headings. For each numbered word or phrase, SELECT THE LETTERED PART OR HEADING that matches it correctly. Then insert the letter in the space to the right of the appropriate number. Sometimes more than one numbered word or phrase may be correctly matched to the same lettered part or heading.

ASSOCIATION QUESTIONS

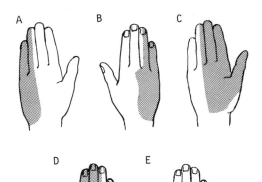

1. ____ Cutaneous distribution of the median nerve on the palmar surface of the hand.

2. ____ Cutaneous distribution of the ulnar nerve on the dorsum of the hand.

3. ____ Cutaneous distribution of the radial nerve on the dorsum of the hand.

4. ____ Cutaneous distribution of the ulnar nerve on the palmar surface of the hand.

5. ____ Cutaneous distribution of the median nerve on the dorsum of the hand.

---------------------- ANSWERS, NOTES AND EXPLANATIONS ----------------------

1. **C.** The cutaneous distribution of the median nerve on the palmar surface of the hand is extensive. It supplies the lateral three and a half fingers, and the lateral aspect of the palm. A small area, the most lateral area of the thenar eminence, is supplied by the radial nerve.

2. **B.** The cutaneous distribution of the ulnar nerve on the dorsum of the hand is to the medial one and a half fingers and the medial half of the dorsal surface of the hand.

3. **E.** The cutaneous distribution of the radial nerve on the dorsum of the hand is to the lateral aspect of the hand and the dorsal surface of the thumb.

4. **A.** The cutaneous distribution of the ulnar nerve on the palmar surface of the hand is to the medial one and a half fingers and the medial portion of the portion of the palm.

5. **D.** The cutaneous distribution of the median nerve on the dorsum of the hand is to the index, middle and lateral half of the ring fingers.

2. THE LOWER LIMB

O B J E C T I V E S

Hip and Gluteal Region

Be Able To:

* Point out the anatomical features of the hip bone and the proximal half of the femur using bones and radiographs.

* Discuss the hip joint with special reference to the role of the ligaments in restricting movements.

* Make simple line drawings illustrating the attachments of the iliotibial tract, the tensor fasciae latae, and the gluteal muscles indicating their innervation and functions.

* Describe the attachments and innervation of the six small lateral rotators of the hip.

* Draw the lumbar and sacral plexuses of nerves showing the roots of origin for the femoral, obturator, and sciatic (tibial and common peroneal) nerves, discussing the muscle compartments and cutaneous areas innervated by them.

Thigh

Be Able To:

* Define the boundaries, floor, and contents of the femoral triangle, and discuss the clinical significance of the femoral sheath.

* Delineate the course of the femoral artery and vein, giving their branches and tributaries, respectively.

* Compare and contrast the anterior, medial, and posterior compartments of the thigh with respect to the attachments, functions, and innervation of the muscles contained in each compartment.

* Describe the superficial and deep groups of inguinal lymph nodes and their drainage patterns in the thigh.

Knee and Popliteal Fossa

Be Able To:

* Use bones or radiographs to point out the features of the distal half of the femur, the condylar region of the tibia, the fibular head, and the patella.

* Discuss the knee joint with special emphasis on the: menisci, cruciate ligaments, collateral ligaments, popliteus tendon, suprapatellar, semimembranous, and prepatellar bursae, giving their clinical significance.

* Schematically draw and label the boundaries and contents of the popliteal fossa.

* Use a simple sketch to illustrate the arterial anastomosis about the knee.

Leg

Be Able To:

* Discuss the attachments, functions, and innervation of the muscles of the anterior, lateral, and posterior compartments of the leg.

* Give the branches of the popliteal artery and the anatomical locations of the pulses of the anterior and posterior tibial arteries in the foot.

Ankle and Foot

Be Able To:

* Use radiographs or bones to demonstrate the articulating surfaces of the ankle (talocrural) joint.

* Make simple labelled sketches to illustrate the ligaments which strengthen the ankle joint.

* Describe the attachments of the flexor, superior and inferior extensor, and peroneal retinacula, and list the structures lying deep to them.

* Compare the medial, lateral, and transverse arches of the sole of the foot, naming the bones involved, indicating the "key-stone" of the longitudinal arches.

* Discuss the importance of the extrinsic muscle tendons, the intrinsic muscles and ligaments in maintaining the above arches.

HIP AND GLUTEAL REGION

FIVE-CHOICE COMPLETION QUESTIONS

DIRECTIONS: Each of the following questions or incomplete statements is followed by five suggested answers or completions. SELECT THE ONE BEST ANSWER in each case and then underline the appropriate letter at the lower right of each question.

1. THE MAJOR MOTOR SIGN IN PARALYSIS OF THE QUADRICEPS FEMORIS MUSCLE IS LOSS OF _____.
    A. Adduction of the thigh
    B. Extension of the thigh
    C. Extension of the leg
    D. Flexion of the leg
    E. Lateral rotation of the leg   A B C D E

SELECT THE ONE BEST ANSWER

2.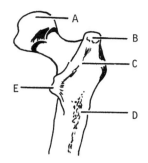

IN THIS POSTERIOR VIEW OF THE PROXIMAL END OF THE FEMUR, THE INTERTROCHANTERIC CREST IS INDICATED BY _____.

A B C D E

3. THE SACRAL PLEXUS OF NERVES IS DERIVED FROM THE ANTERIOR PRIMARY RAMI OF SPINAL NERVES:
   A. L2-5, S1,2
   B. L3,4, S1,2
   C. L4,5, S1,2
   D. L3,4, S1,2,3
   E. None of the above

A B C D E

4. THE LIGAMENT OF THE HIP JOINT RESISTING EXTENSION IS THE:
   A. Transverse
   B. Pubofemoral
   C. Ischiofemoral
   D. Ligament of the head
   E. Iliofemoral

A B C D E

5.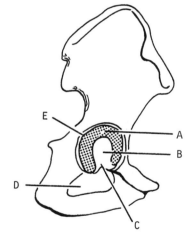

THE ACETABULAR NOTCH IS INDICATED BY _____.

A B C D E

6. THE SUPERIOR GLUTEAL NERVE INNERVATES ALL THE FOLLOWING EXCEPT THE:
   A. Gluteus maximus muscle
   B. Gluteus medius muscle
   C. Gluteus minimus muscle
   D. Tensor fasciae latae muscle
   E. Hip joint

A B C D E

7. THE LUMBOSACRAL PLEXUS OF NERVES IS FORMED BY THE ANTERIOR PRIMARY RAMI OF SPINAL NERVES:
   A. L2-5, S1,2
   B. L2-5, S1-3
   C. L2-5, S1-4
   D. L4,5, S1,2
   E. L4,5, S1-4

A B C D E

8.

THE ACETABULUM IS INDICATED BY _____.

A B C D E

9. THE SCIATIC NERVE LIES BETWEEN THE GREATER TROCHANTER OF THE FEMUR AND THE ISCHIAL TUBEROSITY OF THE PELVIC BONE AFTER IT EMERGES BELOW THE _____ MUSCLE.
   A. Gluteus medius
   B. Obturator internus
   C. Piriformis
   D. Quadratus femoris
   E. Gemelli

A B C D E

10. THE NERVE TO THE OBTURATOR INTERNUS INNERVATES THE MUSCLE OF THE SAME NAME, AS WELL AS THE _____ MUSCLE.
    A. Quadratus femoris
    B. Piriformis
    C. Inferior gemellus
    D. Superior gemellus
    E. Obturator externus

A B C D E

------------------- ANSWERS, NOTES AND EXPLANATIONS --------------------

1. C. The quadriceps femoris muscle, the very powerful knee extensor, consists of the rectus femoris, vastus lateralis, medialis, and intermedius muscles. They are innervated by the femoral nerve (L2, 3, 4). As a group these muscles extend the leg and because the rectus femoris crosses the hip joint, it is also involved in flexion of the thigh.

2. C. The intertrochanteric crest is a ridge of bone connecting the posterior aspects of the greater (B) and lesser (E) trochanters. The gluteal tuberosity (D), along with the iliotibial tract of the fascia lata, serves as an insertion for the gluteus maximus muscle. The articular head of the femur is indicated by A.

3. E. The sacral plexus of nerves is derived from the anterior primary rami of spinal nerves L4, 5 and S1-4. The lumbosacral trunk (L4, 5) joins the sacral nerves to form the sciatic nerve, the largest branch of the plexus. This nerve supplies the lower limb proper, whereas the remaining contributions of the sacral nerves supply pelvic structures and cutaneous branches to the posterior thigh region.

4. E. During extension of the hip joint the iliofemoral ligament becomes taut and permits extension only slightly beyond the vertical. Although the pubofemoral and ischiofemoral ligaments support the fibrous capsule of the hip joint, they play only a minor role in restricting its movements. The transverse ligament and the ligament of the femoral head (ligamentum acetabulare)

have relatively little involvement in hip movements.

5. **C.** The acetabular notch is the inferior incomplete region of the acetabulum (C), leading superiorly into the acetabular fossa (B). The smooth articulating (lunate) surface (A) surrounds the fossa and articulates with the head of the femur. The obturator foramen (D) lies below and medial to the acetabular notch.

6. **A.** The superior gluteal nerve enters the gluteal region through the greater sciatic foramen, above the piriformis muscle. Its upper and lower branches supply the gluteus medius, gluteus minimus, and tensor fasciae latae muscles, as well as the hip joint. The gluteus maximus muscle is innervated by the inferior gluteal nerve.

7. **C.** The lumbosacral plexus of nerves is formed by the anterior primary rami of spinal nerves L2-5 and S1-4. This pattern of plexus formation is found in about 75 per cent of cases; in some cases, contributions are received from L1 or S5, or both. The fourth lumbar nerve divides into two parts: the upper part contributes to the lumbar plexus, and the lower to the sacral plexus.

8. **B.** In this radiograph (PA view), the acetabulum of the hip bone is indicated by B; the anterior superior iliac spine of the ilium is also visible (A). The greater trochanter (E), the intertrochanteric line (D) and the lesser trochanter (C) of the femur may also be distinguished.

9. **C.** The sciatic nerve leaves the pelvis via the greater sciatic notch. It emerges below the piriformis muscle, lying superficial to the obturator internus, the gemelli, and the quadratus femoris muscles. In its course, it lies between the greater trochanter and the ischial tuberosity of the femur and the hip bone, respectively.

10. **D.** The nerve to the obturator internus also supplies the superior gemellus muscle. Likewise, the nerve to the quadratus femoris also supplies the inferior gemellus muscle. The piriformis muscle is supplied by a branch of the sacral plexus, whereas the obturator externus muscle is supplied by the posterior division of the obturator nerve.

MULTI-COMPLETION QUESTIONS

DIRECTIONS: In each of the following questions or incomplete statements, ONE OR MORE of the completions given is correct. At the lower right of each question, underline A if 1, 2 and 3 are correct; B if 1 and 3 are correct; C if 2 and 4 are correct; D if only 4 is correct; and E if all are correct.

1. AN ELDERLY FEMALE PATIENT PRESENTS WITH AN EXTRACAPSULAR INTERTROCHANTERIC FRACTURE OF THE FEMUR. WHICH OF THE FOLLOWING ARTERIES SUPPLY(IES) THE PROXIMAL FRAGMENT OF THE FEMUR?
   1. Medial femoral circumflex
   2. Deep circumflex iliac
   3. Branch of the obturator
   4. Profunda femoris

   A B C D E

2. WHICH STATEMENT(S) IS (ARE) CORRECT CONCERNING THE LATERAL ROTATORS OF THE HIP?
   1. Lie posterior to the hip joint
   2. Each is innervated by a branch of the sacral plexus
   3. Are covered by the gluteus maximus muscle
   4. Are inserted into the lesser trochanter of the femur

   A B C D E

A	B	C	C	E
1,2,3	1,3	2,4	only 4	all correct

3. THE ACETABULAR LABRUM:
   1. Adds stability to the hip joint
   2. Deepens the socket
   3. Consists of fibrocartilage
   4. Is an incomplete ring   A B C D E

4. THE FIBROUS CAPSULE OF THE HIP JOINT:
   1. Is thickened anteriorly by the iliofemoral ligament
   2. Attaches to the base of the femoral head
   3. Is weak inferiorly
   4. Is thicker posterior to the femoral head   A B C D E

5. THE ILIOTIBIAL TRACT:
   1. Is distinct from the tensor fasciae latae
   2. Is a continuation of the gluteal fascia
   3. Inserts into the head of the fibula
   4. Receives muscle fibers from the gluteus maximus   A B C D E

6. WHICH STATEMENT(S) IS (ARE) NOT CORRECT FOR THE GLUTEUS
   MAXIMUS MUSCLE?
   1. Innervated by the inferior gluteal nerve
   2. Rotates the thigh medially
   3. A powerful extensor of the thigh
   4. A powerful adductor of the thigh   A B C D E

7. THE TENSOR FASCIAE LATAE FUNCTIONS IN:
   1. Abduction of the thigh
   2. Flexion of the thigh
   3. Medial rotation of the thigh
   4. Stabilization of the hip joint   A B C D E

8. WHICH OF THE FOLLOWING IS (ARE) TRUE FOR THE GLUTEUS MEDIUS
   MUSCLE?
   1. Originates from the posterior iliac crest
   2. Inserts into the iliotibial tract
   3. Is innervated by the inferior gluteal nerve
   4. Paralysis of it leads to a waddling gait   A B C D E

9. WHICH OF THE FOLLOWING STATEMENTS ABOUT THE SCIATIC NERVE IS
   (ARE) TRUE?
   1. Enters the gluteal region below the piriformis muscle
   2. Divides to form tibial and peroneal nerves
   3. Is the largest nerve in the body
   4. Supplies the gluteus maximus muscle   A B C D E

10. A PATIENT HAS SEVERED THE TIBIAL NERVE ABOVE THE ANKLE ON THE
    POSTERIOR ASPECT OF THE TIBIA. WHICH OF THE FOLLOWING SIGNS
    WOULD BE PRESENT?
    1. Sensory loss on the dorsum of the foot
    2. Sensory loss on the sole of the foot
    3. Foot drop
    4. Clawing of the toes   A B C D E

11. THE OBTURATOR NERVE SUPPLIES THE:
    1. Adductor magnus muscle   3. Skin of the medial thigh
    2. Obturator externus muscle   4. Skin of the anterior thigh A B C D E

A	B	C	D	E
1,2,3	1,3	2,4	only 4	all correct

12. THE DEEP PERONEAL NERVE:
    1. Accompanies the posterior tibial artery
    2. Innervates the anterior compartment muscles of the leg
    3. Supplies skin on the dorsum of the first and second toes
    4. Supplies the tarsal joints   A B C D E

13. WHICH OF THE FOLLOWING IS (ARE) CORRECT FOR THE LUMBAR PLEXUS OF NERVES?
    1. Forms within the psoas major muscle
    2. Has a medial femoral cutaneous branch
    3. A major branch is formed from contributions of L2, 3, and 4
    4. The femoral nerve lies medial to the psoas major muscle   A B C D E

---------------------- ANSWERS, NOTES AND EXPLANATIONS ----------------------

1. **B.** <u>1 and 3 are correct</u>. The proximal fragment of the femur would include the head and neck. It would receive blood from the medial and lateral femoral circumflex arteries and a branch of the obturator artery. This fracture is fairly common in elderly persons, and success of healing depends upon an adequate blood supply. The closer the fracture is to the head of the femur, the poorer the prognosis, because the arterial supply in the head is inadequate without additional blood supply from the other arteries mentioned above.

2. **B.** <u>1 and 3 are correct</u>. The six lateral rotator muscles of the hip: piriformis, obturator externus and internus, quadratus femoris and gemellus superior and inferior are covered by the gluteus maximus muscle. They all lie posterior to the hip joint and with the exception of the obturator externus, which is innervated by the obturator nerve, are supplied by the sacral plexus. These muscles originate from the hip bone and insert onto the greater trochanter, except for the quadratus femoris, which inserts into the middle of the intertrochanteric crest.

3. **A.** <u>1, 2, and 3 are correct</u>. The acetabular labrum, a complete fibrocartilage ring, is located on the margins of the acetabulum of the hip bone. This ring deepens the articular socket for the head of the femur and subsequently stabilizes the hip joint.

4. **B.** <u>1 and 3 are correct</u>. The fibrous capsule of the hip joint attaches superiorly to the hip bone beyond the acetabular circumference, inferiorly to the intertrochanteric line on the anterior aspect. Posteriorly, the line of capsular attachment falls short of the intertrochanteric crest. The head and part of the neck of the femur are enclosed by the capsule. Anteriorly, the fibrous capsule is thickened by the iliofemoral ligament, whereas the remainder of the capsule is supported by the thinner ischiofemoral and pubofemoral ligaments. The capsule is weakest inferiorly.

5. **C.** <u>2 and 4 are correct</u>. The iliotibial tract is a continuation of the gluteal fascia. It receives muscular contributions in the lateral hip region from the tensor fasciae latae and the gluteus maximus muscles. This tract inserts into the lateral lip of the linea aspera of the femur and the lateral condyle of the tibia.

6. **B.** <u>1 and 3 are correct</u>. The gluteus maximus is innervated by the inferior gluteal nerve. It functions as a powerful extensor, a medial rotator, and a weak abductor of the thigh. This muscle is important in running, climbing, and rising from a sitting position. When the lower limbs are fixed it is an extensor of the pelvis or trunk.

7. **E.** <u>All are correct</u>. The tensor fasciae latae acts with the iliopsoas muscle in flexing the thigh and with the gluteus minimus in rotating the thigh medially. It contracts during abduction of the thigh, providing a certain degree of stability to the hip joint.

8. **D.** <u>Only 4 is correct</u>. The gluteus medius muscle originates from the area between the anterior and posterior gluteal lines of the ilium. It inserts into the lateral surface of the greater trochanter and is innervated by the superior gluteal nerve. The gluteus medius abducts and the gluteus minimus medially rotates the thigh. During walking the gluteus medius maintains the pelvis horizontal to the ground, preventing the pelvis from sagging on the free limb side. Paralysis of this muscle gives a characteristic waddling gait.

9. **A.** <u>1, 2, and 3 are correct</u>. The sciatic nerve (L3, 4, S1-3) is the largest nerve in the body. It is the major branch of the sacral plexus and enters the gluteal region by passing through the greater sciatic foramen, deep to the piriformis muscle. The terminal branches of the sciatic nerve are the tibial and peroneal nerves. Although the sciatic nerve is covered by the gluteus maximus muscle, this muscle is innervated by the inferior gluteal nerve (L5, S1, 2) which enters the gluteal region through the greater sciatic foramen, below the piriformis muscle.

10. **C.** <u>2 and 4 are correct</u>. The tibial nerve (L4, 5, S1-3), sectioned in the lower posterior aspect of the leg above the ankle, would result in sensory loss on the sole of the foot. This loss affects posture and locomotion of the lower limb. The intrinsic muscles of the foot would be paralysed. Because the long flexors of the toes are still intact, there is characteristic clawing of the toes. Foot drop and loss of sensation on the dorsum of the foot would result from damage to the common peroneal nerve.

11. **A.** <u>1, 2, and 3 are correct</u>. The obturator nerve (L2-4) enters the thigh through the obturator foramen and supplies the skin on the medial aspect of the thigh. It innervates the obturator externus, gracilis, adductor longus, brevis, and part of the adductor magnus, and occasionally the pectineus. The remaining part of the adductor magnus muscle receives motor fibers from the sciatic nerve; alternatively, the pectineus muscle may be innervated by the femoral nerve.

12. **E.** <u>All are correct</u>. The deep peroneal nerve, a branch of the sciatic, innervates the following muscles of the anterior compartment of the leg: tibialis anterior, extensor digitorum longus, extensor hallucis longus, and peroneus tertius. It also supplies the extensor digitorum brevis of the foot. The deep peroneal nerve supplies the ankle and tarsal joints and an area of skin on the dorsum of the first and second toes.

13. **B.** <u>1 and 3 are correct</u>. The lumbar plexus forms from the primary rami of spinal nerves L2-4 within the substance of the psoas major muscle. It has two major branches, the femoral and obturator nerves, and two minor branches, the genitofemoral and lateral femoral cutaneous nerves. The femoral and lateral cutaneous nerves emerge from the lateral surface of the psoas major muscle, whereas the genitofemoral and obturator nerves emerge from the anterior and medial surfaces. The fourth lumbar nerve also contributes to the sacral plexus.

FIVE-CHOICE ASSOCIATION QUESTIONS

DIRECTIONS: Each group of questions below consists of a numbered list of descriptive words or phrases accompanied by a diagram with certain parts indicated by letters, or by a list of lettered headings. For each numbered word or phrase, SELECT THE LETTERED PART OR HEADING that matches it correctly. Then insert the letter in the space to the right of the appropriate number. Sometimes more than one numbered word or phrase may be correctly matched to the same lettered part or heading.

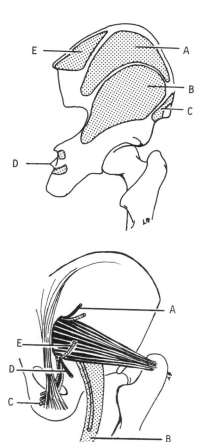

1. ___ Origin of a small muscle inserted into the iliotibial tract

2. ___ Attachment of a muscle innervated by the inferior gluteal nerve

3. ___ Origin of the gluteus minimus muscle

4. ___ Origin of a muscle inserting in part into the gluteal tuberosity

5. ___ Nerve to the obturator internus

6. ___ Nerve to the gluteus maximus

7. ___ Largest arterial branch of the internal iliac artery

8. ___ Nerve to the superior gemellus muscle

9. ___ Inferior gluteal artery

---------------------- ANSWERS, NOTES AND EXPLANATIONS ----------------------

1. **C.** The tensor fasciae latae muscle originates from the outer lip of the iliac crest behind the anterior superior iliac spine, inserting into the iliotibial tract.

2. **E.** The gluteus maximus is innervated by the inferior gluteal nerve. This muscle originates from behind the posterior gluteal line, the dorsal surfaces of the sacrum, coccyx and the sacrotuberous ligament, and the lumbar dorsal fascia. It inserts into the iliotibial tract and the gluteal tuberosity of the femur.

46

3. B. The gluteus minimus muscle originates from the ilium between the anterior and inferior gluteal lines, whereas the gluteus medius (A) muscle originates from the ilium between the anterior and posterior gluteal lines. Both muscles are innervated by the superior gluteal nerve and insert into the greater trochanter of the femur.

4. E. The gluteus maximus inserts into the iliotibial tract and the gluteal tuberosity of the femur. D represents the origins of the gemelli muscles, which are lateral rotators of the thigh.

5. C. The nerve to the obturator internus muscle (L5, S1, 2) passes deep to the sacrotuberous ligament to supply the muscle of the same name. It also gives a branch to the superior gemellus muscle.

6. E. The nerve to the gluteus maximus is the inferior gluteal (L5, S1,2).

7. A. The superior gluteal artery is the largest branch of the internal iliac artery. It leaves the pelvis via the greater sciatic foramen, enters the gluteal region above the piriformis muscle, and supplies the adjacent muscles and the hip joint.

8. C. The nerve to the obturator internus muscle also supplies the superior gemellus muscle, a lateral rotator of the thigh.

9. D. The inferior gluteal artery, a branch of the internal iliac, exits from the pelvis through the greater sciatic foramen. It descends medial to the sciatic nerve (B) accompanying the small posterior femoral cutaneous nerve below the piriformis muscle. This artery supplies the gluteal and hamstring muscles, the hip joint and overlying skin.

THIGH

FIVE-CHOICE COMPLETION QUESTIONS

DIRECTIONS: Each of the following questions or incomplete statements is followed by five suggested answers or completions. SELECT THE ONE BEST ANSWER in each case and then underline the appropriate letter at the lower right of each question.

1. THE BASE OF THE FEMORAL TRIANGLE IS BOUNDED BY THE:
   A. Adductor longus muscle
   B. Inguinal ligament
   C. Sartorius muscle
   D. Femoral sheath
   E. None of the above

   A B C D E

2. THE MEDIAL COMPARTMENT OF THE FEMORAL SHEATH CONTAINS THE:
   A. Femoral canal
   B. Femoral artery
   C. Femoral vein
   D. Femoral nerve
   E. None of the above

   A B C D E

3. IN A FRACTURE AT THE JUNCTION OF THE MIDDLE AND LOWER THIRDS OF THE FEMUR, THE POPLITEAL VESSELS AND NERVE ARE LIKELY TO BE INJURED BECAUSE THE DISTAL FRAGMENT IS PULLED POSTERIORLY BY THE _____ MUSCLE.
   A. Soleus
   B. Gastrocnemius
   C. Semitendinosus
   D. Gracilis
   E. None of the above

   A B C D E

SELECT THE ONE BEST ANSWER

4. WHICH OF THE FOLLOWING IS NOT A BRANCH OF THE FEMORAL ARTERY?
   A. Superficial circumflex iliac   D. Superficial epigastric
   B. Inferior epigastric   E. External pudendal
   C. Profunda femoris

   A B C D E

5. THE SARTORIUS MUSCLE INSERTS ON THE:
   A. Tibial tuberosity
   B. Proximal part of the lateral surface of the tibia
   C. Proximal part of the medial surface of the tibia
   D. Head of fibula
   E. Medial condyle of the tibia

   A B C D E

6. WHICH MUSCLE IS NOT INVOLVED IN FLEXION OF THE THIGH?
   A. Iliacus   D. Rectus femoris
   B. Sartorius   E. Obturator externus
   C. Psoas major

   A B C D E

7. THE GREAT SAPHENOUS VEIN IS LOCATED:
   A. In front of the medial malleolus
   B. Behind the medial malleolus
   C. In front of the lateral malleolus
   D. Behind the lateral malleolus
   E. In the popliteal fossa

   A B C D E

8. THE FOLLOWING MUSCLES ARE EXTENSORS OF THE THIGH EXCEPT THE:
   A. Biceps femoris   D. Gluteus maximus
   B. Adductor magnus   E. Adductor longus
   C. Semitendinosus

   A B C D E

------------------- ANSWERS, NOTES AND EXPLANATIONS --------------------

1. B. The femoral triangle is located on the anterior aspect of the thigh. Its base (superior boundary) is formed by the inguinal ligament; it is bounded laterally by the medial edge of the sartorius, and medially by the lateral margin of the adductor longus muscle. This triangle contains the femoral nerve and vessels.

2. A. The femoral sheath is a funnel-shaped extension of the transversalis and iliopsoas fasciae into the thigh, deep to the inguinal ligament. From lateral to medial, the femoral artery, femoral vein, and femoral canal, separated from each other by septa, occupy the femoral sheath. The femoral nerve, together with the iliopsoas muscle, lies outside the femoral sheath, lateral to the femoral artery. The medially placed femoral canal is occupied by fat, areolar connective tissue, and lymph nodes. It is through this potentially weak medial area that a (femoral) hernia may occur; this hernia is more common in females than males, and is always acquired.

3. B. The distal fragment is pulled posteriorly by contraction of the medial and lateral heads of the gastrocnemius muscle. The fragment would also be elevated by the hamstring and quadriceps muscles. As the popliteal artery lies next to the bone, it is the most likely structure to be injured; the popliteal vein and nerve along with the tibial and common peroneal nerves may also be damaged. The proximal fragment of the femur would be slightly flexed and adducted because of contraction of the iliopsoas and adductor muscles.

4. **B.** Branches of the femoral artery can be arranged into proximal and distal groups. The proximal group includes the superficial epigastric, superficial circumflex iliac, and the external pudendal (superficial and deep) arteries. The distal group includes the profunda femoris and descending genicular arteries. The inferior epigastric artery is a branch of the external iliac artery which arises before the inguinal ligament crosses the major arterial trunk.

5. **C.** The sartorius muscle, an important landmark on the front of the thigh, originates from the anterior superior iliac spine and inserts into the proximal part of the tibia on its medial surface. It flexes both the thigh and the leg and rotates the thigh laterally and the leg medially. The sartorius muscle is innervated by the femoral nerve.

6. **E.** The obturator externus muscle is not a flexor of the thigh; it is a lateral rotator. The other four muscles are involved in flexion of the thigh. The iliacus and psoas major muscles arise from the iliac fossa and lumbar vertebrae, respectively, forming the iliopsoas muscle which inserts by a common tendon on the lesser trochanter of the femur. The rectus femoris muscle, one of four making up the quadriceps femoris, originates from the anterior inferior iliac spine and the posterosuperior aspect of the acetabular rim. It inserts by a tendon into both the base of the patella and the tuberosity of the tibia. The sartorius muscle originates from the anterior superior iliac spine and inserts into the proximal part of the tibia. The iliacus, psoas major, and rectus femoris muscles all cross the hip joint, whereas the sartorius muscle crosses both the hip and knee joints.

7. **A.** The great saphenous vein begins at the medial end of the dorsal venous arch of the foot. It ascends in front of the medial malleolus, crosses the medial surface of the tibia, and then lies in the region of the knee. The great saphenous vein passes through the cribriform fascia to drain into the femoral vein. The short saphenous vein begins at the lateral end of the dorsal venous arch and passes behind the lateral malleolus. It terminates in the popliteal fossa, draining into the popliteal vein. Both these veins may become dilated, sacculated, and tortuous as a result of increased hydrostatic pressure within the vessels, a condition known as varicosis.

8. **E.** The adductor longus muscle is not an extensor, but rather an adductor and a medial rotator of the thigh. The hamstring group of muscles, which includes the biceps femoris, semitendinosus, and semimembranosus, are powerful extensors of the thigh. The gluteus maximus is an important extensor of the hip along with the extensor component of the adductor magnus muscle. The hamstrings and the extensor component of the adductor magnus muscle are innervated by the tibial portion of the sciatic nerve, except for the short head of the biceps femoris which is innervated by the peroneal portion of the sciatic nerve. The gluteus maximus is innervated by the inferior gluteal nerve.

MULTI-COMPLETION QUESTIONS

DIRECTIONS: In each of the following questions or incomplete statements, ONE OR MORE of the completions given is correct. At the lower right of each question, underline A if 1, 2 and 3 are correct; B if 1 and 3 are correct; C if 2 and 4 are correct; D if only 4 is correct; and E if all are correct.

1. THE QUADRICEPS FEMORIS MUSCLE CONSISTS OF WHICH OF THE FOLLOWING MUSCLES?
    1. Iliacus
    2. Rectus femoris
    3. Biceps femoris
    4. Vastus lateralis

    A B C D E

A	B	C	D	E
1,2,3	1,3	2,4	only 4	all correct

2. WHICH STATEMENT(S) IS(ARE) CORRECT FOR THE ADDUCTOR CANAL?
   1. The vastus medialis muscle lies lateral to it
   2. Contains the saphenous vein
   3. Located on the medial aspect of the thigh
   4. Contains the femoral nerve   A B C D E

3. DEEP TO THE INGUINAL LIGAMENT AND WITHIN THE FEMORAL TRIANGLE, THE FEMORAL ARTERY LIES:
   1. Medial to the femoral vein
   2. Medial to the femoral canal
   3. Adjacent to the femoral canal
   4. Medial to the femoral nerve   A B C D E

4. THE INGUINAL GROUP OF LYMPH NODES DRAINS WHICH OF THE FOLLOWING AREAS?
   1. The lower limb
   2. The abdominal wall
   3. The gluteal region
   4. The external genitalia   A B C D E

5. THE VENOUS DRAINAGE OF THE LOWER LIMB:
   1. Is aided by numerous valves
   2. Consists of two isolated systems
   3. Is via deep veins during exercise
   4. Depends on smooth muscle in the venous walls   A B C D E

6. WHICH OF THE FOLLOWING IS(ARE) TRUE FOR THE INGUINAL LYMPH NODES?
   1. Are often palpable
   2. Consist of superficial and deep groups
   3. Are located near the saphenofemoral junction
   4. The deep nodes lie lateral to the femoral vein   A B C D E

7. WHICH STATEMENT(S) ABOUT A FEMORAL HERNIA IS(ARE) CORRECT?
   1. The neck of the hernia lies deep to the inguinal ligament
   2. Lies below and lateral to the pubic tubercle
   3. Is more common in women than in men
   4. Passes through the femoral canal   A B C D E

8. THE LATERAL ROTATOR(S) OF THE THIGH IS(ARE) THE _____ MUSCLES.
   1. Adductor longus   3. Adductor magnus
   2. Adductor brevis   4. Pectineus   A B C D E

9. THE HAMSTRING MUSCLES ARE:
   1. Flexors of the leg
   2. Extensors of the trunk
   3. Extensors of the thigh
   4. Attached to the ischial tuberosity   A B C D E

10. THE ILIOPSOAS MUSCLE:
    1. Inserts into the lesser trochanter of the femur
    2. Acts as a weak medial rotator of the hip
    3. Has a bursa associated with its tendon
    4. Includes fibers from the psoas minor muscle   A B C D E

A	B	C	D	E
1,2,3	1,3	2,4	only 4	all correct

11. IN THE POSTERIOR THIGH, WHICH OF THE FOLLOWING IS (ARE) CORRECT?
    1. The sciatic nerve lies posterior to the adductor component of the adductor magnus muscle
    2. Muscles receive branches from the profunda femoris artery
    3. The semitendinosus tendon forms part of the medial boundary of the popliteal fossa
    4. The inferior gluteal artery descends with the sciatic nerve   A B C D E

12. AN INDUSTRIAL ACCIDENT PRODUCED A FRACTURE OF THE UPPER THIRD OF THE SHAFT OF THE FEMUR. THE DISTAL FRAGMENT WAS ADDUCTED UPWARD AND THE PROXIMAL FRAGMENT WAS ABDUCTED AND LATERALLY ROTATED. CONTRACTION OF WHICH OF THE FOLLOWING MUSCLES PRODUCED THE ABDUCTION AND LATERAL ROTATION?
    1. Obturator internus   3. Gluteal
    2. Quadriceps femoris   4. Iliopsoas   A B C D E

-------------------- ANSWERS, NOTES AND EXPLANATIONS --------------------

1. **C.** <u>2 and 4 are correct</u>. The quadriceps femoris muscle includes the rectus femoris and the vastus lateralis, medialis, and intermedius muscles. The rectus femoris originates from the anterior inferior iliac spine and the posterosuperior aspect of the rim of the acetabulum. It inserts by a common tendon into the base of the patella and the tuberosity of the tibia (ligamentum patella). The vasti muscles arise from the anterior surface of the femur: the vastus lateralis originates from the intertrochanteric line, the greater trochanter, and the gluteal tuberosity; the vastus intermedius originates from the anterolateral upper two-thirds of the shaft of the femur; and the vastus medius originates from the intertrochanteric line, the spiral line, and the medial intermuscular septum. Although the vastus medialis fuses in part with the adductor longus and magnus muscles, it forms, together with the other vasti muscles, the ligamentum patella. The quadriceps femoris extends the leg, and because one of its parts, the rectus femoris, crosses the hip joint, it is also involved in hip flexion. All four muscles of the quadriceps femoris are innervated by the femoral nerve.

2. **B.** <u>1 and 3 are correct</u>. The adductor (subsartorial) canal is located in the middle of the medial aspect of the thigh. The canal is bounded medially by the adductor magnus and laterally by the vastus medialis muscle. This canal is covered by the sartorius muscle and fascia. The femoral vessels, the saphenous nerve, and the nerve to the vastus medialis muscle are the contents of the adductor canal.

3. **D.** <u>Only 4 is correct</u>. From lateral to medial, the structures within the femoral triangle are the: nerve, artery, vein, and canal. The femoral artery, vein, and canal are contained within the femoral sheath, an extension of the fascia transversalis and the iliacus fascia of the thigh.

4. **E.** <u>All are correct</u>. The inguinal lymph nodes drain the superficial thigh region directly and the lower limb indirectly via deep lymphatic channels from the popliteal nodes. The anterolateral abdominal wall below the umbilicus, the gluteal region, part of the uterus, and the anus all drain into the inguinal nodes. The external genitalia, except the glans of the penis or the clitoris, are also drained by the inguinal lymph nodes.

5. **B.** <u>1 and 3 are correct</u>. The lower limb is drained by superficial and deep groups of veins, which have many communications. The major superficial veins

51

are the great (long) and small (short) saphenous veins. The femoral and popliteal veins are the major deep tributaries. The great saphenous vein may become enlarged and tortuous and contain defective valves, producing a varix, e.g., varicose vein. The actions of leg muscles and valves are important in returning blood from the lower limb. During exercise, venous drainage is primarily via deep channels because of changes in pressure.

6. **A.** <u>1, 2, and 3 are correct</u>. The superficial inguinal lymph nodes are located subcutaneously near the saphenofemoral junction, where they are often palpable. There are three to fourteen nodes, of which three to four are deep to the fascia lata of the thigh. These deep nodes lie medial to the femoral vein and one of them is usually found within the femoral ring. The efferent lymphatic channels of these nodes drain to the external nodes of the pelvis.

7. **E.** <u>All are correct</u>. A femoral hernia consists of protrusion of extraperitoneal tissue, with or without abdominal contents, through the femoral ring. It is more common in women than in men. The proximal end of the femoral canal, the femoral ring, lies deep to the inguinal ligament; anterior to the superior ramus of the pubis; lateral to the lacunar ligament; and medial to the femoral vein. The hernial sac pushes the contents of the canal before it and the neck of the sac is found below and lateral to the pubic tubercle. This location of the neck of the sac distinguishes a femoral hernia from an inguinal hernia located medial to the pubic tubercle.

8. **A.** <u>1, 2, and 3 are correct</u>. The adductor longus, brevis, and magnus muscles are lateral rotators of the thigh. These muscles are also powerful adductors of the thigh; they are also aided by the pectineus muscle. All these muscles are innervated by the obturator nerve, except for the pectineus, usually supplied by the femoral nerve, and the extensor part of the adductor muscle, innervated by the sciatic nerve (tibial portion).

9. **E.** <u>All are correct</u>. The hamstring muscles (biceps femoris, semitendinosus, and semimembranosus) all arise from the ischial tuberosity; the biceps femoris also arises from the lateral lip of the linea aspera, the upper part of the lateral supracondylar line (long head), and the lateral intermuscular septum (short head). The biceps femoris inserts into the head of the fibula; the semitendinosus and semimembranosus insert into the medial surface of the tibial shaft and the medial surface of the medial condyle of the tibia, respectively. The hamstring muscles cross the hip and knee joints and are therefore extensors of the thigh and flexors of the leg. These muscles, as a group, can also extend the trunk when the hip and knee joints are fixed.

10. **B.** <u>1 and 3 are correct</u>. The iliopsoas muscle is formed by a broad lateral part from the iliacus muscle and a long medial part from the psoas major. The iliacus originates from the upper part of the iliac fossa and the ala of the sacrum, whereas the psoas major originates from the intervertebral discs and the lumbar vertebrae. The muscle fibers of the iliacus blend with the psoas tendon and insert together into the lesser trochanter of the femur. The tendon of the iliopsoas is usually separated from the hip joint capsule by a bursa. The iliopsoas contains no muscle fibers from the psoas minor muscle, when it is present. The iliopsoas is an important flexor of the thigh and the vertebral column when the hip is fixed; it may also be a weak lateral rotator of the hip. The iliopsoas is innervated in two parts: the psoas major is innervated by the lumbar plexus (L2, 3, and sometimes L1 or 4); the iliacus muscle is innervated by the femoral nerve (L2-4).

11. **A.** <u>1, 2, and 3 are correct</u>. The profunda femoris artery, a branch of the femoral, sends muscular branches to the muscles of the posterior compartment of the thigh. The nerve of this compartment, the sciatic nerve, enters the

thigh below the piriformis muscle with the inferior gluteal artery on its medial aspect. The sciatic nerve, accompanied by a branch from the inferior gluteal artery, passes posterior to the adductor component of the adductor magnus muscle. The semimembranosus and semitendinosus muscles lie medial to the nerve, and the semitendinosus tendon forms part of the medial boundary of the popliteal fossa. The long head of the biceps femoris muscle crosses the nerve obliquely downward, from medial to lateral.

12. **A. 1, 2, and 3 are correct.** The actions of obturator internus, quadriceps femoris, gluteal and gemelli muscles would produce abduction and lateral rotation of the proximal fragment. This fragment would also be flexed because of contraction of the iliopsoas muscle. The distal fragment would be displaced upward and medially by contraction of the adductor and hamstring muscles.

        F I V E - C H O I C E   A S S O C I A T I O N   Q U E S T I O N S

DIRECTIONS: Each group of questions below consists of a numbered list of descriptive words or phrases accompanied by a diagram with certain parts indicated by letters, or by a list of lettered headings. For each numbered word or phrase, SELECT THE LETTERED PART OR HEADING that matches it correctly. Then insert the letter in the space to the right of the appropriate number. Sometimes more than one numbered word or phrase may be correctly matched to the same lettered part or heading.

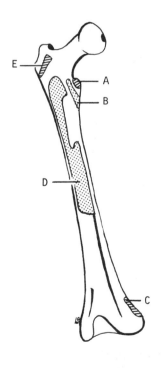

1. ___ The origin of the vastus medialis muscle

2. ___ The attachment of a muscle innervated by the obturator nerve

3. ___ The origin of a muscle joining the deep surface of the quadriceps tendon

4. ___ The insertion of an abductor and medial rotator of the thigh

5. ___ The insertion of a powerful flexor of the thigh

ASSOCIATION QUESTIONS

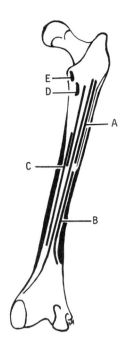

6. ___ The attachment of a gluteal muscle

7. ___ The insertion of the iliacus muscle

8. ___ The origin of a muscle having a dual innervation

------------------- ANSWERS, NOTES AND EXPLANATIONS --------------------

1. **B.** The vastus medialis muscle originates from the intertrochanteric line and the medial lip of the linea aspera; it contributes to the quadriceps tendon and inserts into the tubercle of the tibia. The patella, a sesamoid bone, is found within the quadriceps tendon. The vastus medialis muscle, an extensor of the leg, is innervated by the femoral nerve.

2. **C.** The adductor magnus muscle originates from the outer surface of the inferior ramus of the pubis, the ramus of the ischium, and the ischial tuberosity. It has two functional attachments: the hamstring component attaches to the posterior surface of the femur, along the linea aspera to the medial supracondylar ridge; the adductor component attaches to the adductor tubercle on the medial condyle of the femur. The adductor component, innervated by the obturator nerve, adducts the thigh, whereas the hamstring component, innervated by the sciatic nerve, extends the thigh.

3. **D.** The vastus intermedius muscle originates from the anterolateral surface of the femoral shaft and contributes to the formation of the quadriceps tendon. Like all four muscles of the quadriceps, the vastus intermedius is innervated by the femoral nerve. This muscle is a powerful extensor of the leg.

4. **E.** The gluteus minimus muscle, an adductor and medial rotator of the hip, originates from the outer surface of the ilium between the middle and inferior gluteal lines. It inserts into the anterior surface of the greater trochanter of the femur and is innervated by the superior gluteal nerve.

5. **A.** The psoas major muscle is a powerful flexor and medial rotator of the thigh. It originates from the transverse processes, vertebral bodies, and

discs of the 12th thoracic and the lumbar vertebrae. The tendon of this muscle receives fibers from the ilacus muscle and the common iliopsoas tendon inserts into the lesser trochanter of the femur. The psoas muscle is innervated by branches of the lumbar plexus.

6. A. The gluteus maximus muscle originates from the posterior surface of the ilium, the adjacent surfaces of the sacrum and coccyx, and the sacrotuberous ligament. It inserts into the gluteal tuberosity of the femur and the iliotibial tract.

7. E. The iliacus muscle, forming part of the iliopsoas muscle, inserts into the posterior aspect of the lesser trochanter. This muscle, like the posas major, functions as a flexor and medial rotator of the thigh. It is innervated by a branch of the femoral nerve.

8. B. The short head (B) of the biceps femoris muscle is innervated by the common peroneal nerve, whereas the long head is innervated by the tibial portion of the sciatic nerve. This muscle originates by a long head from the ischial tuberosity and a short head from the linea aspera, lateral supracondylar ridge and lateral intermuscular septum. It inserts by a common tendon into the head of the fibula. The short head of the biceps femoris flexes and laterally rotates the leg; the long head extends the thigh. The insertions of the adductor longus and pectineus muscles are indicated in the diagram by C and D, respectively.

KNEE AND POPLITEAL FOSSA

FIVE-CHOICE COMPLETION QUESTIONS

DIRECTIONS: Each of the following questions or incomplete statements is followed by five suggested answers or completions. SELECT THE ONE BEST ANSWER in each case and then underline the appropriate letter at the lower right of each question.

1. WHICH OF THE FOLLOWING IS NOT A STABILIZER OF THE KNEE JOINT ON ITS LATERAL ASPECT?
   A. Biceps tendon
   B. Popliteus tendon
   C. Iliotibial tract
   D. Transverse ligament
   E. Fibular collateral ligament

   A B C D E

2. THE PATELLA ARTICULATES WITH THE:
   A. Femur
   B. Fibula
   C. Tibia
   D. Femur and fibula
   E. Tibia and fibula

   A B C D E

3. AN INJURY TO THE UPPER LATERAL MARGIN OF THE POPLITEAL FOSSA IS MOST LIKELY TO DAMAGE THE _____ NERVE.
   A. Tibial
   B. Sciatic
   C. Common peroneal
   D. Genicular branch of the obturator
   E. Posterior femoral cutaneous

   A B C D E

4. THE LATERAL (SEMILUNAR CARTILAGE) MENISCUS OF THE KNEE JOINT IS:
   A. More frequently torn than the medial meniscus
   B. Fused with the fibular collateral ligament
   C. Completely attached to the tibia
   D. Grooved by the popliteus tendon
   E. Semicircular in shape

   A B C D E

55

SELECT THE ONE BEST ANSWER

5. THE MAJOR BURSA COMMUNICATING WITH THE SYNOVIAL JOINT SPACE OF THE KNEE IS THE:
   A. Prepatellar
   B. Suprapatellar
   C. Infrapatellar
   D. Semimembranosus
   E. None of the above

   A B C D E

6.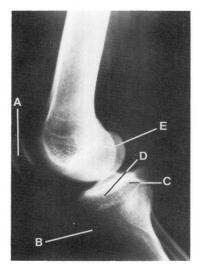

   A CONDYLE OF THE FEMUR IS INDICATED BY _____.

   A B C D E

------------------ ANSWERS, NOTES AND EXPLANATIONS ---------------------

1. **D.** The transverse ligament is not a lateral stabilizer of the knee joint, but it attaches and strengthens the menisci on their anterior aspect. The biceps and popliteus tendons, the iliotibial tract, and the fibular collateral ligament are important lateral stabilizers of the knee joint.

2. **A.** The patella (knee cap) is a large sesamoid bone, about six cm in diameter, which articulates with the patellar surfaces of the femoral condyles. This bone, like all sesamoid bones, is located within a tendon; in this case, the tendon of the quadriceps femoris muscle, called the patellar tendon. It inserts into the tuberosity of the tibia. The function of the patella is to increase the angle of pull of the quadriceps femoris muscle, thereby increasing its power.

3. **C.** The upper lateral boundary of the popliteal fossa is formed by the medial margin of the biceps femoris muscle; the common peroneal nerve lies along its medial margin. Because of its position, close to the neck of the fibula, this nerve is the branch of the sciatic nerve most susceptable to injury. Injury to this nerve produces sensory loss on the lateral surface of the leg and on the dorsum of the foot. Motor loss produces a lack of dorsiflexion and eversion of the foot and extension of the toes, as seen in the condition known as "foot-drop".

4. **D.** The lateral meniscus of the knee is circular in shape and is incompletely attached to the upper aspect of the tibia. Laterally, the meniscus is separated from the fibular collateral ligament by the tendon of the popliteus muscle. The lateral and medial menisci are thick at the periphery of the tibial condyles and become paper thin towards the center. This structure aids in

forming a more stable base for articulation of the femoral condyle. The medial meniscus is more often damaged by tearing than the lateral one.

5. B. The suprapatellar bursa is the major bursa communicating with the knee joint cavity; however, the semimembranosus bursa may also communicate with it. The prepatellar and infrapatellar bursae are located subcutaneously on the anterior aspect of the knee. Repeated, unaccustomed trauma to these bursae may produce an inflammatory response, resulting in an excessive production of synovial fluid and swelling. Such swelling is often painful and is commonly known as "housemaid's knee".

6. E. The femur has two condyles (one of them is indicated by E) which articulate with the patella (A) anteriorly. These condyles also articulate with the plateau-like condyles of the tibia (D). The tibial tuberosity (area of B) is the site of insertion for the ligamentum patellae of the quadriceps femoris muscle. On the lateral surface of the lateral condyle of the tibia, there is an articulating surface for the fibula (C).

MULTI-COMPLETION QUESTIONS

DIRECTIONS: In each of the following questions or incomplete statements, ONE OR MORE of the completions given is correct. At the lower right of each question, underline A if 1, 2 and 3 are correct; B if 1 and 3 are correct; C if 2 and 4 are correct; D if only 4 is correct; and E if all are correct.

1. THE MUSCLE(S) MEDIAL TO THE KNEE JOINT IS (ARE) THE:
   1. Semimembranosus
   2. Semitendinosus
   3. Gracilis
   4. Sartorius

   A B C D E

2. THE POPLITEUS MUSCLE:
   1. Is located posterior to the knee joint
   2. Inserts into the fibula
   3. Unlocks the knee joint at the beginning of flexion
   4. Is innervated by the femoral nerve

   A B C D E

3. THE ANTERIOR CRUCIATE LIGAMENT OF THE KNEE:
   1. Attaches to the lateral surface of the medial condyle of the femur
   2. Prevents backward dislocation of the femur
   3. Is partly extracapsular
   4. Acquires its name from its tibial attachment

   A B C D E

4. THE POPLITEAL FOSSA IS BOUNDED SUPERIORLY BY WHICH OF THE FOLLOWING MUSCLES?
   1. Semitendinosus
   2. Biceps femoris
   3. Semimembranosus
   4. Gracilis

   A B C D E

5. WHICH OF THE FOLLOWING STRUCTURES FORM(S) THE FLOOR OF THE POPLITEAL FOSSA?
   1. Femur
   2. Popliteus muscle
   3. Oblique popliteal ligament
   4. Plantaris muscle

   A B C D E

6. WHICH OF THE FOLLOWING IS (ARE) CORRECT FOR THE KNEE JOINT?
   1. There is a 2-4 degree angle between the axis of the tibia and femur
   2. The capsule of the knee joint is thick and strong
   3. The patella is free of the joint capsule
   4. The knee joint is a condylar type of synovial joint

   A B C D E

A	B	C	D	E
1,2,3	1,3	2,4	only 4	all correct

7. THE TIBIAL COLLATERAL LIGAMENT EXTENDS FROM THE MEDIAL EPICONDYLE
OF THE FEMUR TO THE:
 1. Neck of the fibula
 2. Medial semilunar cartilage
 3. Tuberosity of the tibia
 4. Medial aspect of the tibia
   A B C D E

------------------ ANSWERS, NOTES AND EXPLANATIONS --------------------

1. E. All are correct. The semimembranosus, semitendinosus, gracilis, and sartorius muscles from the thigh all pass medial to the knee joint. They all insert on the anteromedial tibial shaft, except for the semimembranosus which inserts on the posterior surface of the medial tibial condyle. All four muscles are flexors of the knee; the semitendinosus is also a medial rotator of the leg on the femur (thigh). The sartorius muscle is innervated by the femoral nerve, the gracilis by the obturator, and the semimembranosus and semitendinosus by the tibial division of the sciatic nerve.

2. B. 1 and 3 are correct. The popliteus muscle originates from the lateral aspect of the lateral condyle of the femur and the back of the lateral meniscus, and extends obliquely across the back of the knee. It inserts into the posterior aspect of the tibia and is innervated by a branch of the tibial division of the sciatic nerve. The major function of the popliteus muscle is the unlocking of the knee joint at the beginning of flexion.

3. C. 2 and 4 are correct. The anterior cruciate ligament of the knee is entirely intraarticular and is attached to the anterior intercondylar area of the tibia, from which it derived its name. The anterior cruciate ligament passes upwards and backwards to be inserted into the femur on the medial aspect of the lateral femoral condyle. This ligament prevents backward dislocation of the femur and acts as the pivot around which the "screw-home" of the knee movement takes place. This movement locks the knee joint, prohibiting further extension or rotation.

4. A. 1, 2, and 3 are correct. The popliteal fossa, a diamond-shaped space behind the knee, is bounded superiorly by the semitendinosus and semimembranous muscles medially, and the biceps muscle laterally. Inferiorly, the fossa is bounded laterally by the plantaris and the lateral head of the gastrocnemius muscle, and medially by the medial head of the gastrocnemius. The principal contents of the popliteal fossa are the popliteal vessels, the sciatic nerve and its terminal branches, and the common peroneal and tibial nerves.

5. A. 1, 2, and 3 are correct. The floor of the popliteal fossa, from above downwards, is composed of: the popliteal surface of the femur, the oblique popliteal ligament of the knee, and the popliteus muscle. The plantaris muscle forms part of the inferior lateral boundary of the fossa.

6. D. Only 4 is correct. The knee joint is a condylar type of synovial joint permitting flexion, extension, and some rotation. The vertical axis between the tibia and the femur is 10-12 degrees. The fibrous capsule encompassing this joint is rather thin, weak, and often incomplete. The capsule is attached to: the femur above the intercondylar fossa; the margins of the femoral condyles; the margins of the patella and the ligamentum patellae; and the margins of the tibial condyles.

7. C. 2 and 4 are correct. The tibial collateral ligament is attached to the medial semilunar cartilage, and the medial aspects of the articular capsule and the tibia, several cm below the knee joint. This ligament, along with the

cruciate ligaments and the intercondylar eminences, prevents medial displacement of the two long bones articulating at the knee. Following tearing of the cruciate ligaments there is exaggerated movement of the articulating surfaces.

FIVE-CHOICE ASSOCIATION QUESTIONS

DIRECTIONS: Each group of questions below consists of a numbered list of descriptive words or phrases accompanied by a diagram with certain parts indicated by letters, or by a list of lettered headings. For each numbered word or phrase, SELECT THE LETTERED PART OR HEADING that matches it correctly. Then insert the letter in the space to the right of the appropriate number. Sometimes more than one numbered word or phrase may be correctly matched to the same lettered part or heading.

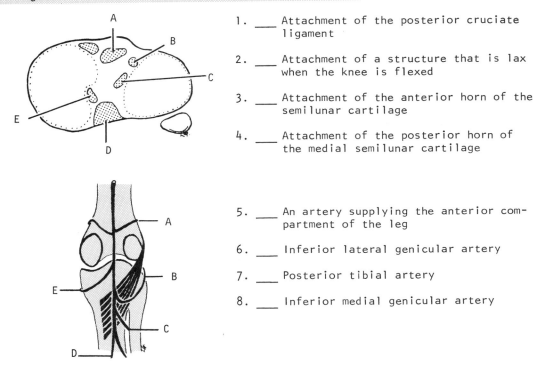

1. ___ Attachment of the posterior cruciate ligament

2. ___ Attachment of a structure that is lax when the knee is flexed

3. ___ Attachment of the anterior horn of the semilunar cartilage

4. ___ Attachment of the posterior horn of the medial semilunar cartilage

5. ___ An artery supplying the anterior compartment of the leg

6. ___ Inferior lateral genicular artery

7. ___ Posterior tibial artery

8. ___ Inferior medial genicular artery

---------------------- ANSWERS, NOTES AND EXPLANATIONS ----------------------

1. D. The posterior cruciate ligament attaches to the posterior impression of the tibia and passes upwards, forwards, and medially, where it attaches to the anterior part of the lateral surface of the medial femoral condyle. This ligament is lax when the knee is extended.

2. A. The anterior cruciate ligament attaches to the second impression of the intercondylar arc and passes upwards, backwards, and laterally, where it attaches to the posterior aspect of the medial surface of the lateral femoral condyle. This ligament is lax when the knee is flexed.

3. B. The attachment of the anterior horn of the lateral semilunar cartilage is indicated by B in the diagram. The transverse ligament binds the anterior horns of both the lateral and medial semilunar cartilages. The attachment of

the posterior horn of the lateral semilunar cartilage is represented by C in the diagram.

4. E. The attachment of the posterior horn of the medial semilunar cartilage is indicated by E in the diagram.

5. C. The anterior tibial artery is a terminal branch of the popliteal artery arising at the level of the lower border of the popliteus muscle. It supplies the anterior compartment of the leg and the dorsum of the foot.

6. B. The inferior lateral genicular artery is the lowest genicular artery to arise from the popliteal artery. The superior lateral genicular artery is labeled A. The popliteus muscle is also illustrated in the diagram.

7. D. The posterior tibial artery supplies the posterior and lateral compartments of the leg and the sole of the foot.

8. E. The inferior medial genicular artery, along with the other genicular arteries around the knee, form an important collateral anastomosis. These arteries compensate for the narrowing of the popliteal artery occurring during extreme flexion of the knee.

L E G

FIVE-CHOICE COMPLETION QUESTIONS

DIRECTIONS: Each of the following questions or incomplete statements is followed by five suggested answers or completions. SELECT THE ONE BEST ANSWER in each case and then underline the appropriate letter at the lower right of each question.

1. THE STRUCTURE ATTACHED TO THE INTERCONDYAR NOTCH OF THE FEMUR IS THE:
   A. Anterior cruciate ligament
   B. Posterior cruciate ligament
   C. Transverse ligament
   D. Popliteus tendon
   E. None of the above

   A B C D E

2. WHICH OF THE FOLLOWING STATEMENTS REGARDING THE GREAT SAPHENOUS VEIN IS CORRECT?
   A. Obliquely crosses the tibia
   B. Lies anterior to the medial malleolus
   C. Originates on the medial side of the big toe
   D. Lies between the medial condyles of the tibia and femur
   E. All of the above

   A B C D E

3. A MUSCLE OF THE POSTERIOR COMPARTMENT OF THE LEG INSERTING INTO THE TENDOCALCANEUS IS THE:
   A. Flexor digitorum longus
   B. Flexor hallucis longus
   C. Tibialis posterior
   D. Popliteus
   E. Soleus

   A B C D E

4. A PLANTAR FLEXOR OF THE FOOT IS THE _____ MUSCLE.
   A. Extensor digitorum longus
   B. Peroneus longus
   C. Peroneus tertius
   E. Popliteus
   D. Tibialis anterior

   A B C D E

SELECT THE ONE BEST ANSWER

5. THE _____ MUSCLE CROSSES BOTH THE KNEE AND ANKLE JOINTS:
   A. Soleus   D. Gastrocnemius
   B. Popliteus   E. Flexor hallucis longus
   C. Peroneus tertius   A B C D E

6. THE PERONEAL ARTERY ARISES FROM THE _____ ARTERY:
   A. Femoral   D. Posterior tibial
   B. Popliteal   E. None of the above
   C. Anterior tibial   A B C D E

------------------- ANSWERS, NOTES AND EXPLANATIONS --------------------

1. E. The intercondylar (fossa) notch, located between the two condyles of the femur, provides no ligamentous or tendinous attachments. The anterior and posterior cruciate ligaments pass through this notch, but attach to the inner surfaces of the lateral and medial femoral condyles, respectively. The transverse ligament binds the anterior horns of the two semilunar cartilages. The popliteus tendon originates from the groove on the lateral surface of the lateral condyle of the femur.

2. E. The great saphenous vein originates superficially on the medial aspect of the big toe and ascends in front of the medial malleolus and crosses the tibia obliquely alongside the saphenous nerve. It passes upwards along the medial border of the tibia, through the area between the medial condyles of the tibia and femur, and ascends further up along the medial side of the thigh. In the thigh, the great saphenous vein passes through the fascia lata, pierces the femoral sheath to join the femoral vein.

3. E. The muscles of the posterior compartment of the leg are innervated by the tibial nerve and are divided into a superficial and a deep group. The superficial group consists of the gastrocnemius, soleus, and plantaris muscles. The first two insert into the tendocalcaneus; the plantaris when present also inserts into the tendocalcaneus. The deep group of muscles forming the posterior compartment are: the flexor digitorum longus, flexor hallucis longus, tibialis posterior, and the popliteus muscles.

4. B. The peroneus longus muscle is both a plantar flexor and an evertor of the foot. The peroneus tertius muscle is an evertor; the extensor digitorum longus and tibialis anterior muscles are dorsiflexors of the foot. The popliteus muscle is involved with rotation at the knee joint.

5. D. The gastrocnemius muscle arises by two heads: a medial head from the popliteal surface of the femur, above the medial condyle near the adductor tubercle; and a lateral head from the lateral aspect of the lateral condyle of the femur. The two heads of this muscle fuse in the middle of the leg and along with the soleus tendon form the tendocalcaneus (Achilles tendon), which inserts into the posterior aspect of the calcaneus. The gastrocnemius thus crosses two joints, the knee and ankle. The ankle jerk, a reflex twitch of the triceps surae (gastrocnemius and soleus muscles), may be produced by tapping the tendocalcaneus. The reflex center for this arc is in the fifth lumbar or first sacral segments of the cord.

6. D. The peroneal artery is a branch of the posterior tibial artery. It descends along the medial crest of the fibula and eventually lies on the interosseous membrane. It supplies the muscles of the posterior compartment of the leg and sends nutrient branches to the tibia, fibula, and calcaneus.

MULTI-COMPLETION QUESTIONS

DIRECTIONS: In each of the following questions or incomplete statements, ONE OR MORE of the completions given is correct. At the lower right of each question, underline A if 1, 2 and 3 are correct; B if 1 and 3 are correct; C if 2 and 4 are correct; D if only 4 is correct; and E if all are correct.

1. THE SUPERIOR TIBIOFIBULAR JOINT IS:
   1. A synovial joint
   2. Affected by ankle movements
   3. Supported by the interosseous membrane
   4. Contains an articular disc

   A B C D E

2. THE ANTERIOR COMPARTMENT OF THE LEG CONTAINS WHICH OF THE FOLLOWING MUSCLES?
   1. Extensor digitorum longus   3. Peroneus tertius
   2. Peroneus longus   4. Sartorius

   A B C D E

3. INVERTOR(S) OF THE FOOT IS (ARE) THE FOLLOWING MUSCLE(S):
   1. Peroneus tertius   3. Plantaris
   2. Tibialis posterior   4. Tibialis anterior

   A B C D E

4. THE SUPERFICIAL PERONEAL NERVE INNERVATES WHICH OF THE FOLLOWING MUSCLES?
   1. Tibialis anterior   3. Peroneus tertius
   2. Peroneus longus   4. Peroneus brevis

   A B C D E

5. WHICH OF THE FOLLOWING MUSCLES IS (ARE) INNERVATED BY THE DEEP PERONEAL NERVE?
   1. Tibialis anterior   3. Peroneus tertius
   2. Extensor digitorum longus   4. Peroneus brevis

   A B C D E

-------------------- ANSWERS, NOTES AND EXPLANATIONS --------------------

1. **A.** 1, 2, and 3 are correct. The superior tibiofibular joint is a synovial joint that contains no disc. Here the head of the fibula articulates with the posterior aspect of the lateral tibial condyle. This joint accommodates movements of the ankle joint. The interosseous membrane connects the interosseous borders of the tibia and fibula.

2. **B.** 1 and 3 are correct. The extensor digitorum longus and peroneus tertius muscles, along with the tibialis anterior and extensor hallucis longus muscles, make up the muscles of the anterior compartment of the leg. All four muscles extend (dorsiflex) the foot at the ankle joint. The tibialis anterior and extensor hallucis longus invert the foot; the peroneus tertius everts the foot. The extensor digitorum longus muscle extends the lateral four toes; the extensor hallucis longus muscle extends the big toe.

3. **C.** 2 and 4 are correct. The tibialis posterior and tibialis anterior muscles are both invertors of the foot. Inversion and eversion of the foot are essential for walking on rough sloping surfaces. Inversion of the foot involves elevation of the medial border of the foot (i.e., turning the sole inwards).

4. **C.** 2 and 4 are correct. The common peroneal nerve divides into two terminal branches, the superficial and deep peroneal nerves, as it curves around the neck and the fibula. Because of its relatively superficial position and direct contact with the lateral side of the neck of the fibula, the common

peroneal nerve could be easily damaged. The peroneus longus and peroneus brevis muscles cause eversion of the foot and are both supplied by the superficial peroneal nerve. The tibialis anterior and the small peroneus tertius muscles are innervated by the deep peroneal branch.

5. **A. 1, 2, and 3 are correct.** The deep peroneal nerve, a branch of the common peroneal, supplies the muscles of the anterior (extensor) compartment of the leg, consisting of the tibialis anterior, extensor hallucis longus, extensor digitorum longus, and peroneus tertius. The peroneus brevis is located with the peroneus longus in the lateral compartment of the leg: both muscles are innervated by the superficial branch of the common peroneal nerve.

FIVE-CHOICE ASSOCIATION QUESTIONS

DIRECTIONS: Each group of questions below consists of a numbered list of descriptive words or phrases accompanied by a diagram with certain parts indicated by letters, or by a list of lettered headings. For each numbered word or phrase, SELECT THE LETTERED PART OR HEADING that matches it correctly. Then insert the letter in the space to the right of the appropriate number. Sometimes more than one numbered word or phrase may be correctly matched to the same lettered part or heading.

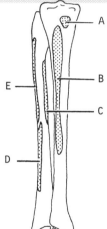

1. ___ Attachment of a muscle innervated by the tibial nerve

2. ___ Origin of the peroneus longus muscle

3. ___ Site of insertion of an extensor muscle of the knee

4. ___ Site of origin of the major dorsiflexor of the foot

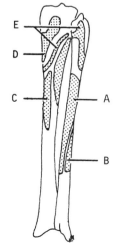

5. ___ Attachment of a flexor muscle of big toe

6. ___ Site of origin of a muscle inserted into the tendocalcaneus

7. ___ Attachment of a muscle inserting into the four lateral toes

8. ___ Site of origin of an evertor of the foot

---------------- ANSWERS, NOTES AND EXPLANATIONS --------------------

1. **C.** The tibialis posterior muscle, innervated by the tibial nerve (L5, S1 and 2), originates from the medial aspect of the fibula. This muscle is located in the posterior compartment of the leg.

2. **E.** The peroneus longus (E) and peroneus brevis (D) muscles originate from the upper two-thirds and lower one-third of the lateral surface of the fibula, respectively. The peroneus longus inserts into the lateral side of the metatarsal of the big toe; the peroneus brevis inserts into the base of the fifth metatarsal. These muscles are both innervated by the superficial peroneal nerve (L4, 5 and S1) and are evertors of the foot. The peroneus longus is also a plantar flexor.

3. **A.** The quadriceps femoris muscle, the strong extensor of the knee, is inserted into the tubercle. This muscle is innervated by the femoral nerve (L2-4).

4. **B.** The tibialis anterior muscle originates from the anterolateral surface of the tibia and inserts into the base of the first metatarsal (big toe) and the medial cuneiform. It is the major dorsiflexor of the foot and also an invertor of the foot. The tibialis anterior is innervated by the deep peroneal nerve (L4, 5, S1), and when injured, results in "foot-drop". Persons with this nerve injury have a characteristic high-stepping gait.

5. **A.** The flexor hallucis longus muscle originates form the lower two-thirds of the posterior aspect of the fibula. It passes through a groove in the talus to reach the sole of the foot and inserts into the distal phalanx of the big toe. The flexor hallucis longus is innervated by the tibial nerve and is responsible for flexing the big toe; it also assists in plantar flexion of the foot.

6. **E.** The soleus muscle originates from the soleal line on the posterior aspect of the tibia and upper fourth of the posterior surface of the fibula. These origins form an inverted horseshoe shaped curve. The soleus, gastrocnemius, and plantaris muscles form a common tendon called the tendocalcaneus (Achilles tendon), inserting into the back of the calcaneus. These three muscles are innervated by the tibial nerve and are responsible for plantar flexion of the foot. The gastrocnemius muscle also flexes the knee and, because of the shortness of its fibers, it can only plantar flex the foot or flex the knee at any one time.

7. **C.** The flexor digitorum longus, a flexor of the lateral four toes, originates from the middle third of the posterior aspect of the tibia and inserts by four small tendons into the distal phalanges of the toes. Like the other muscles of the posterior compartment of the leg, it is innervated by the tibial nerve.

8. **B.** The peroneus brevis muscle, an evertor of the foot, originates from the lower third of the lateral surface of the fibula, and inserts into the base of the fifth metatarsal. It is a muscle of the lateral compartment of the leg. In this diagram, D represents the insertion of the popliteus muscle, a rotator of the tibia upon the femur during the initiation of knee flexion. The peroneus brevis lies deep to the peroneus longus muscle, another evertor of the foot. The peroneus longus is also a plantar flexor of the foot; it acts on the medial side of the foot by depressing the first metatarsal, thereby enabling the inverted foot to remain plantigrade.

ANKLE AND FOOT

FIVE-CHOICE COMPLETION QUESTIONS

DIRECTIONS: Each of the following questions or incomplete statements is followed by five suggested answers or completions. SELECT THE ONE BEST ANSWER in each case and then underline the appropriate letter at the lower right of each question.

1. WHICH OF THE FOLLOWING IS NOT A BONY COMPONENT OF THE MEDIAL LONGITUDINAL ARCH OF THE FOOT?
   A. Talus
   B. Cuboid
   C. Calcaneus
   D. Navicular
   E. First three metatarsals

   A B C D E

2. THE _____ MUSCLE SPANS THE MEDIAL ARCH.
   A. Flexor hallucis longus
   B. Tibialis posterior
   C. Abductor hallucis
   D. Peroneus brevis
   E. Peroneus longus

   A B C D E

3. THE HEEL OF THE FOOT RECEIVES ITS SENSORY INNERVATION FROM THE _____ NERVE.
   A. Sural
   B. Tibial
   C. Saphenous
   D. Medial plantar
   E. Lateral plantar

   A B C D E

4. STRUCTURES LOCATED BEHIND THE LATERAL MALLEOLUS ARE THE:
   A. Dorsalis pedis artery and flexor hallicus longus muscle
   B. Peroneus brevis and peroneus tertius muscles
   C. Peroneus longus and peroneus tertius muscles
   D. Peroneus longus and peroneus brevis muscles
   E. Posterior tibial artery and tibial nerve

   A B C D E

5. WHICH OF THE FOLLOWING STATEMENTS IS CORRECT FOR THE FLEXOR RETINACULUM? IT PASSES BETWEEN THE:
   A. Lateral malleolus and the medial surface of the calcaneus
   B. Medial malleolus and the medial surface of the calcaneus
   C. Lateral malleolus and the lateral surface of the calcaneus
   D. Medial malleolus and the lateral surface of the calcaneus
   E. Tibia and fibula above the ankle joint

   A B C D E

6. THE PULSE OF THE DORSALIS PEDIS ARTERY MAY BE PALPATED:
   A. Behind the medial malleolus
   B. Between the fourth and fifth toes
   C. Posterior to the lateral malleolus
   D. Between the first and second toes
   E. Distal to the ankle joint on the dorsum of the foot

   A B C D E

7. THE TRANSVERSE ARCH IS MAINTAINED BY THE _____ MUSCLE.
   A. Abductor digiti minimi
   B. Abductor hallucis longus
   C. Flexor digitorum brevis
   D. Adductor hallucis
   E. Peroneus brevis

   A B C D E

SELECT THE ONE BEST ANSWER

8.

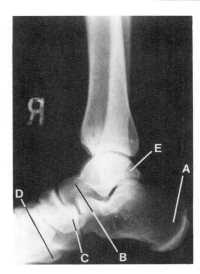

THE TALUS IS INDICATED BY _____.

A B C D E

------------------ ANSWERS, NOTES AND EXPLANATIONS ---------------------

1. **B.** The medial longitudinal arch of the foot is formed by the: calcaneus, talus, navicular, cuneiforms, and first three metatarsals. The cuboid bone is a component of the lateral longitudinal arch. The navicular bone is the "keystone" of the medial longitudinal arch.

2. **C.** The abductor hallucis muscle, part of the first layer of muscles of the sole, spans the extent of the medial arch. It originates from the tuber calcanei and inserts into the base of the proximal phalanx of the big toe. This muscle functions as a flexor and an abductor of the big toe and is innervated by the medial plantar nerve.

3.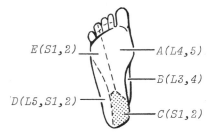

B. The heel of the foot receives its sensory innervation from the tibial nerve (C). The sole of the foot receives its sensory innervation from the sural (D), saphenous (B), medial (A), and lateral (E) plantar nerves as indicated in the diagram.

4. **D.** The peroneus longus and brevis tendons lie behind the lateral malleolus, curving forwards within a common synovial sheath deep to the superior peroneal retinaculum. On the lateral side of the calcaneus, the peroneus longus tendon passes below the peroneal trochlea, deep to the inferior peroneal retinaculum, and enters the foot obliquely where it inserts into the lateral side of the medial cuneiform and the adjacent surface of the first metatarsal bone. The tendon of the peroneus brevis muscle lies above the peroneal trochlea and passes forward to insert into the tuberosity of the fifth metatarsal bone.

5. **B.** The flexor retinaculum, a deep fascial band, passes between the medial malleolus and the medial surface of the calcaneus. The three tendons held in place beneath it, from superior to inferior, are: tibialis posterior, flexor

digitorum longus, and flexor hallucis longus. The tibial nerve and the posterior tibial artery also pass beneath the flexor retinaculum.

6. **E.** The dorsalis pedis artery, a continuation of the anterior tibial, begins distal to the ankle joint on the dorsum of the foot, midway between the malleoli. It lies deep to the extensor hallucis brevis muscle between the tendons of the extensor hallucis longus and the extensor digitorum longus muscles on the navicular bone. Here the pulse is often palpated for assessing the arterial supply to the foot, but weakness of this pulse does not necessarily imply vascular insufficiency.

7. **D.** The transverse (metatarsal) arch, formed by the navicular, cuneiforms, cuboid, and the five metatarsals, is maintained anteriorly by the transverse head of the adductor hallucis muscle. One of the muscles of the third layer of the sole of the foot, it is innervated by the lateral plantar nerve. The transverse head arises from the deep transverse metatarsal ligament and passes medially to insert into the fibrous sheath of the flexor hallucis longus muscle. This transverse head binds the heads of the metatarsal bones. The oblique head originates from the sheath of the peroneus longus, acts as a flexor of the large toe, and inserts into the proximal phalanx of the large toe along with the flexor hallucis brevis. The oblique head of the adductor hallucis muscle acts as a flexor of the large toe.

8. **E.** The talus (E), a tarsal bone of the foot, articulates superiorly with the medial and lateral malleoli of the tibia and fibula to form the ankle joint. Inferiorly, the talus articulates with calcaneus (A) posteriorly, and the navicular bone (B) anteromedially. Together, these three tarsal bones form the talocalcaneonavicular joint. The medial cuneiform (C) and the tuberosity of the fifth metatarsal (D) are also visible.

MULTI-COMPLETION QUESTIONS

DIRECTIONS: In each of the following questions or incomplete statements, ONE OR MORE of the completions given is correct. At the lower right of each question, underline A if 1, 2 and 3 are correct; B if 1 and 3 are correct; C if 2 and 4 are correct; D if only 4 is correct; and E if all are correct.

1. WHICH OF THE FOLLOWING MUSCLES IS (ARE) DORSIFLEXORS OF THE FOOT?
   1. Tibialis anterior
   2. Extensor digitorum brevis
   3. Peroneus tertius
   4. Peroneus brevis   A B C D E

2. THE CALCANEUS OF THE FOOT ARTICULATES WITH WHICH OF THE FOLLOWING BONES?
   1. Cuboid
   2. Fibula
   3. Talus
   4. First cuneiform   A B C D E

3. WHICH OF THE FOLLOWING STATEMENTS IS (ARE) TRUE FOR THE EXTENSOR DIGITORUM BREVIS MUSCLE?
   1. Innervated by the lateral plantar nerve
   2. Divides into four tendons
   3. Adducts the four medial toes
   4. The only muscle on the dorsum of the foot   A B C D E

4. THE LATERAL LONGITUDINAL ARCH OF THE FOOT IS FORMED BY THE:
   1. Calcaneus
   2. Navicular
   3. Cuboid
   4. Talus   A B C D E

A	B	C	D	E
1,2,3,	1,3	2,4	only 4	all correct

5. WHICH OF THE FOLLOWING STATEMENTS IS (ARE) CORRECT FOR THE ANKLE JOINT?
   1. Calcaneus articulates with the fibula
   2. Supported laterally by the deltoid ligament
   3. Has three supporting ligaments medially
   4. Is a hinge joint   A B C D E

6. THE MEDIAL LONGITUDINAL ARCH OF THE FOOT IS SUPPORTED BY THE:
   1. Flexor hallucis longus tendon
   2. Calcaneonavicular ligament
   3. Tibialis posterior tendon
   4. Peroneus longus tendon   A B C D E

7. LIGAMENTS ATTACHING THE TIBIA TO THE FIBULA ARE:
   1. Anterior tibiofibular   3. Posterior tibiofibular
   2. Interosseous   4. Deltoid   A B C D E

8. WHICH OF THE FOLLOWING STATEMENTS IS (ARE) CORRECT REGARDING MUSCLES OF THE SOLE OF THE FOOT?
   1. Strongly support the arches
   2. Functionally arranged in three groups
   3. Innervated by the medial and lateral plantar nerves
   4. Supplied by branches of the posterior tibial artery A B C D E

9. THE DORSALIS PEDIS ARTERY:
   1. Enters the sole of the foot
   2. Is a branch of the peroneal artery
   3. Lies on the extensor hallucis brevis muscle
   4. Begins at a point midway between the malleoli   A B C D E

10. THE FOLLOWING MOVEMENTS OCCUR AT THE ANKLE JOINT:
    1. Dorsiflexion   3. Plantar flexion
    2. Inversion   4. Eversion   A B C D E

------------------ ANSWERS, NOTES AND EXPLANATIONS ---------------------

1. B. 1 and 3 are correct. The tibialis anterior and peroneus tertius muscles are dorsiflexors of the foot, whereas the tibialis anterior is an invertor and the peroneus tertius an evertor. Other dorsiflexors are the extensor digitorum longus and extensor hallucis longus muscles. The extensor digitorum brevis muscle extends the medial four toes, whereas the peroneus brevis muscle is an evertor of the foot. All the dorsiflexors of the foot are innervated by the deep peroneal nerve.

2. B. 1 and 3 are correct. The calcaneus of the foot articulates with three bones: the cuboid, talus, and navicular. The calcaneus, together with the navicular and talus, form the talocalcaneonavicular joint. It also articulates with the talus and cuboid bones to form the subtalar and the calcaneocuboid joints, respectively.

3. C. 2 and 4 are correct. The extensor digitorum brevis, an intrinsic muscle, is the only muscle on the dorsum of the foot. It originates from the sinus tarsi of the calcaneus, the latter dividing into four tendons which join the long extensor tendons of the toes. The extensor digitorum brevis is innervated by the deep peroneal nerve and aids in extending the medial four toes at the metatarsophalangeal and interphalangeal joints.

68

4. **B.** <u>1 and 3 are correct.</u> The lateral longitudinal arch of the foot is formed by the calcaneus, cuboid, and lateral two metatarsal bones. The cuboid bone acts as the "key-stone" in this arch.

5. **D.** <u>Only 4 is correct.</u> The ankle (talocrural) joint is formed by the trochlea of the talus and the medial and lateral malleoli of the tibia and fibula, respectively. This articulation is a hinge joint, allowing only plantar and dorsiflexion of the foot. The joint capsule is supported medially by the medial (deltoid) ligament and laterally by three ligaments: the anterior talofibular, the posterior talofibular, and the intervening calcaneofibular ligaments. These strong ligaments prevent anterior and posterior slipping of the tibia and fibula on the talus.

6. **A.** <u>1, 2, and 3 are correct.</u> The foot has three bony arches, two longitudinal and one transverse. The prominent medial longitudinal arch is formed by the following bones: calcaneus, talus, navicular, three cuneiforms, and three medial metatarsals. The three arches are maintained by the configurations of the articulating bones, ligaments and tendons. Contributing to the support of the medial longitudinal arch are the calcaneonavicular ("spring") ligament and the tendons of the tibialis posterior and flexor hallucis muscles. On the other hand, the peroneus longus tendon, producing the reverse effect and tending to pull the medial side of the foot down, is the most important support for the lateral longitudinal arch. Pes planus (flat foot) and pes cavus (high arch) refer to the clinical conditions of depressed and highly elevated longitudinal arches, respectively.

7. **A.** <u>1, 2, and 3 are correct.</u> The distal ends of the fibula and tibia "grip" the talus to form the ankle (talocrural) joint. The strong fibrous union between the tibia and fibula is called a syndesmosis. The interosseous membrane (ligament) connects the interosseous borders of the tibia and fibula and is considered by some to be an intermediate tibiofibular joint. Anteriorly and posteriorly two strong bands, the anterior and posterior tibiofibular ligaments, also support this syndesmosis. The deltoid ligament supports the medial surface of the ankle joint.

8. **E.** <u>All are correct.</u> The intrinsic muscles of the sole of the foot function together in posture, locomotion, and in support of the bony arches of the foot. They are functionally arranged in three groups: a medial for the big toe, a central for the middle toes, and a lateral for the little toe. For description purposes, these muscles are described in four layers; from superficial to deep, they are: (a) abductor hallucis, flexor digitorum brevis, and abductor digiti minimi; (b) quadratus plantae, the four lumbricals, and tendons of flexor hallucis longus and flexor digitorum longus; (c) flexor hallucis brevis, adductor hallucis, and flexor digiti minimi brevis; and (d) interossei and tendons of the tibialis posterior and peroneus longus. All these muscles are innervated by the medial and lateral plantar nerves and receive their blood supply from branches of the posterior tibial artery.

9. **D.** <u>Only 4 is correct.</u> The dorsalis pedis artery, a continuation of the anterior tibial, begins midway between the malleoli. It lies beneath the inferior extensor retinaculum of the extensor hallucis brevis muscle, between the tendons of the extensor hallucis longus and extensor digitorum longus muscles. At this site the pulse is often palpated for assessing the adequacy of the arterial supply to the foot, but weakness of this pulse does not necessarily imply vascular insufficiency. The deep plantar artery, a terminal branch of the dorsalis pedis, enters the sole of the foot between the heads of the first dorsal interosseus muscle.

10. **B. 1 and 3 are correct.** The ankle joint is a synovial joint of the hinge variety. The articulating bones are the distal ends of the tibia and fibula; together they form a deep socket that accommodates the trochlea of the talus. Movements at this joint are restricted to dorsiflexion (extension) and plantar flexion. Inversion and eversion of the foot occur at the intertarsal joints, in particular at the subtalar (talocalcanean) and transverse tarsal (talocalcaneonavicular and calcaneocuboid) joints.

FIVE-CHOICE ASSOCIATION QUESTIONS

DIRECTIONS: Each group of questions below consists of a numbered list of descriptive words or phrases accompanied by a diagram with certain parts indicated by letters, or by a list of lettered headings. For each numbered word or phrase, SELECT THE LETTERED PART OR HEADING that matches it correctly. Then insert the letter in the space to the right of the appropriate number. Sometimes more than one numbered word or phrase may be correctly matched to the same lettered part or heading.

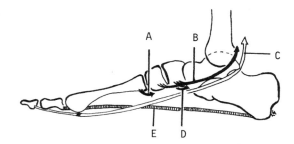

1. ___ Tendon lying in a groove in the talus
2. ___ A powerful evertor of the foot
3. ___ Abductor hallucis muscle
4. ___ Tendon of the tibialis posterior muscle

------------------- ANSWERS, NOTES AND EXPLANATIONS ---------------------

1. **C.** The flexor hallucis longus muscle originates from the posterior aspect of the fibula. Its long tendon passes beneath the flexor retinaculum, lies in a groove in the talus, and runs obliquely across the foot to insert into the inferior aspect of the base of the distal phalanx of the big toe.

2. **A.** The tendon of the peroneus longus muscle crosses the foot obliquely from lateral to medial and inserts into the lateral side of the medial cuneiform and the base of the first metatarsal. This muscle originates from the lateral surface of the fibula and is responsible for eversion of the foot.

3. **E.** The abductor hallucis is an intrinsic muscle on the medial aspect of the foot. It arises from the medial process of the tuber calcanei, extends below the medial arch of the foot, inserting into the base of the proximal phalanx of the big toe. The abductor hallucis muscle abducts and flexes the big toe.

4. **B.** The tendon of the tibialis posterior muscle inserts primarily into the tuberosity of the navicular bone, the medial cuneiform and all other tarsal bones except the talus. It originates from the interosseous membrane, the tibia, and the fibula. The tibialis posterior muscle is the principal invertor of the foot. D in the diagram represents the plantar calcaneonavicular ligament.

3. THE HEAD AND NECK

OBJECTIVES

Skull, Mandible, and Hyoid Bone

Be Able To:

* Use a skull to explain how the bones fit together to form the face and calvaria, and to name the sutures and former sites of the fontanelles.

* Identify the foramina and fissures of the skull through which the cranial nerves, the internal carotid, vertebral and middle meningeal arteries and the internal jugular vein pass.

* Illustrate with simple diagrams the osteology of the mandible and the hyoid bone, and the location of muscle attachments on these bones.

* Recognize the major bones, sinuses, hypophyseal fossa, and other prominent features of the skull in a radiograph of the head.

Interior of the Cranium

Be Able To:

* Define the cranial fossae and point out the important landmarks in them.

* Construct labelled diagrams showing the formation of the circulus arteriosus cerebri (circle of Willis), and discuss its clinical significance.

* Identify the location of the dural venous sinuses using a skull as a guide.

* Discuss the formation of the dural venous sinuses, their major communications with the extracranial venous system, and their role in the circulation of the cerebrospinal fluid (CSF).

* Discuss the cranial meninges, commenting on their clinical significance.

Superficial Structures of the Head, Neck, and Face

Be Able To:

* Prepare simple diagrams illustrating the cutaneous nerves and the superficial blood vessels in the regions of the head, neck, and face.

* Discuss the muscles of facial expression including the epicranius.

Anterior Triangle of the Neck

Be Able To:

   * Define the anterior triangle of the neck and its subdivisions, listing the major structures in each of them.

   * Describe the cervical fascia and its subdivisions.

   * List the contents of the carotid sheath, indicating their topographical relations.

   * Make a diagram showing the branches of the common and external carotid arteries, indicating the territories supplied by them.

   * Discuss the location and functions of the carotid sinus and the carotid body.

   * Describe the formation of the ansa cervicalis and list its effectors, indicating their attachments and functions.

Submandibular and Parotid Regions

Be Able To:

   * Define the submandibular (digastric) and submental triangles, listing their major contents.

   * Illustrate the suprahyoid muscles using a simple diagram to indicate their innervation, blood supply, and venous drainage.

   * Discuss the location, innervation, blood supply, and venous drainage of the submandibular, sublingual, and parotid salivary glands.

   * Describe the clinical significance of the relationship between the parotid gland and its duct and the extracranial distribution of the facial nerve.

   * Illustrate the topographical relationships of the hyoglossus muscle to the following: hypoglossal and lingual nerves; lingual vessels; submandibular and sublingual glands and their ducts.

Posterior Triangle of the Neck

Be Able To:

   * Define the posterior triangle of the neck and list its major contents.

   * Describe the attachments, function, innervation and blood supply of the sternocleidomastoid and trapezius muscles.

   * Illustrate in simple diagrams the origin and distribution of the sensory nerves of the cervical plexus of nerves.

Deep Structures of the Neck

* Sketch the thyroid and parathyroid glands showing their blood supply, venous drainage, and relations.

* Discuss the origin and clinical significance of thyroglossal duct cysts, sinuses and fistulas.

* Give a brief account of the prevertebral fascia and scalene muscles, describing their relations to the phrenic nerve.

* Illustrate the cervical sympathetic trunk and its ganglia.

Root of the Neck

* List the branches of the subclavian artery and state their peripheral distribution.

* Make a simple diagram to illustrate the venous return from the facial and cervical regions to the large veins at the root of the neck.

* Summarize in a simple diagram the superficial and deep lymphatic drainage from the head and neck regions into the veins at the root of the neck.

* Discuss the cause and clinical significance of the neurovascular syndromes related to a cervical rib and the anterior scalene muscle.

Ear

* Describe the external acoustic meatus and state how to inspect the tympanic membrane.

* Illustrate the internal and external surfaces of the tympanic membrane.

* Define the boundaries of the middle ear cavity, identifying major landmarks and related structures.

* Comment on the clinical anatomy of the middle ear in relation to infection.

* Discuss how sound impulses are conducted from the external acoustic meatus to the vestibulocochlear nerve.

* List the structures involved in the maintenance of equilibrium.

* Discuss the clinical significance of the auditory (pharyngotympanic) tube.

* Briefly describe the intracranial course and distribution of the facial nerve, emphasizing its relations to the middle ear.

Orbital Region

* Describe the orbital cavity and list its major contents.

* Make a simple sketch of a sagittal section of the eyelid (palpebra), indicating its contents and innervation.

* List the extrinsic muscles of the eye, giving their attachments, functions, blood supply, venous drainage, and innervation.

   * Explain diagrammatically the action of the eye muscles in bringing about movements of the pupils.

   * Trace the pathway a tear takes from the lacrimal gland to the inferior nasal meatus.

   * Draw a median section and an anterior view of the eyeball, labelling the components and indicating their functions.

Deep Structures of the Face

   * Define the temporal and infratemporal fossae, listing the major structures in them.

   * List the branches of the maxillary artery, indicating the general territories supplied by them.

   * Describe the attachments, innervation, and functions of the muscles of mastication related to the temporomandibular joint.

   * Discuss the distribution of the mandibular nerve and its branches.

   * List the principal contents of the pterygopalatine fossa.

   * State the contents of the greater and lesser palatine canals, the pterygoid canal, and the foramen rotundum, giving their functions.

   * Discuss the function of the pterygopalatine ganglion.

Nose and Paranasal Sinuses

   * Discuss the construction of the nose and its septum, indicating their innervation and blood supply.

   * Describe the lateral wall of the nasal cavity and its relations.

   * State where the paranasal sinuses and the nasolacrimal ducts drain.

   * Describe the location of the olfactory mucosa.

Oral Region

   * Define the boundaries of the oral cavity, indicating major landmarks.

   * Make sketches of the dental arches showing deciduous and permanent teeth.

   * Compare the components, innervation, and blood supply of the hard palate and the soft palate.

   * List the intrinsic and extrinsic muscles of the tongue, indicating their innervation, blood supply, and venous and lymphatic drainage.

Illustrate the distribution of general sensory and taste innervation of the tongue, indicating the cranial nerves involved.

Pharynx and Larynx

* Define the subdivisions of the pharynx.

* Illustrate with simple diagrams the muscles forming the pharyngeal wall, including their attachments, innervation, and blood supply.

* List the distribution of the lymphoid tissues in the nasopharynx and oropharynx, and discuss the clinical significance of the tonsils and adenoids.

* Sketch the skeleton of the larynx and discuss its intrinsic muscles, ligaments, membranes, and joints.

* Discuss the innervation and blood supply of the larynx, and the role of its intrinsic and extrinsic muscles in phonation and respiration.

* Discuss the mechanism of deglutition.

SKULL, MANDIBLE, AND HYOID BONE

FIVE-CHOICE COMPLETION QUESTIONS

DIRECTION: Each of the following questions or incomplete statements is followed by five suggested answers or completions. SELECT THE ONE BEST ANSWER in each case and then underline the appropriate letter at the lower right of each question.

1. ON ITS WAY TO THE BRAIN, THE VERTEBRAL ARTERY PASSES THROUGH THE:

    A. Foramen magnum
    B. Foramen lacerum
    C. Jugular foramen
    D. Foramen spinosum
    E. Foramen ovale

    A B C D E

2. THE SUPRAORBITAL NOTCH TRANSMITS THE _____ VESSELS.

    A. Supratrochlear
    B. Anterior ethmoidal
    C. Supraorbital
    D. Dorsal nasal
    E. Angular

    A B C D E

3. THE MAXILLARY DIVISION OF THE TRIGEMINAL NERVE PASSES THROUGH WHICH FORAMEN?

    A. Spinosum
    B. Ovale
    C. Rotundum
    D. Lacerum
    E. Stylomastoid

    A B C D E

4. THE SELLA TURCICA LIES DIRECTLY ABOVE THE:

    A. Pons
    B. Foramen ovale
    C. Frontal sinus
    D. Sphenoid sinus
    E. Maxillary sinus

    A B C D E

---------------------------ANSWERS, NOTES AND EXPLANATIONS---------------------------

1. **A.** After arising from the subclavian artery at the root of the neck, the vertebral artery ascends between the anterior scalene and longus colli muscles; then it passes through the foramina in the transverse processes of the upper six cervical vertebrae. The vertebral artery curves posteriorly behind the lateral mass of the atlas, passes medially, pierces the dura mater into the vertebral canal, and then enters the cranial cavity through the foramen magnum.

2. **C.** The supraorbital notch, often appearing as the supraorbital foramen, transmits the supraorbital artery and vein, a branch and tributary of the ophthalmic artery and vein, respectively. Other branches of the ophthalmic artery include the anterior ethmoidal, supratrochlear, and dorsal nasal artery. The dorsal nasal artery forms an important anastomosis with the angular artery of the facial artery. The supraorbital nerve is a branch of the frontal nerve from the ophthalmic division of the trigeminal nerve.

3. **C.** The maxillary division of the trigeminal nerve (fifth cranial nerve) passes from the middle cranial fossa through the foramen rotundum into the pterygopalatine fossa. The foramen rotundum is situated immediately inferior to the inferomedial end of the superior orbital fissure in the greater wing of the sphenoid bone.

4. **D.** The sella turcica lies directly above the sphenoid sinus located within the body of the sphenoid bone. The sella turcica is an important landmark in a lateral radiograph of the skull for locating the hypophysis (pituitary gland).

MULTI-COMPLETION QUESTIONS

DIRECTIONS: In each of the following questions or incomplete statements, ONE OR MORE of the completions given is correct. At the lower right of each question, underline A if 1, 2 and 3 are correct; B if 1 and 3 are correct; C if 2 and 4 are correct; D if only 4 is correct; and E if all are correct.

1. WHICH OF THE FOLLOWING STRUCTURES PASS(ES) THROUGH THE FORAMEN OVALE OF THE CRANIUM?
   1. Middle meningeal artery
   2. Greater petrosal nerve
   3. Maxillary nerve
   4. Mandibular nerve   A B C D E

2. WHICH OF THE FOLLOWING IS (ARE) FOUND IN THE SPHENOID BONE?
   1. Pterygoid canal
   2. Foramen spinosum
   3. Foramen rotundum
   4. Stylomastoid foramen   A B C D E

3. WHICH OF THE FOLLOWING STATEMENTS ABOUT THE JUGULAR FORAMEN IS (ARE) FALSE?
   1. Lodges the jugular bulb
   2. Contains the inferior petrosal sinus
   3. Is in close relation to the sigmoid sinus
   4. Transmits cranial nerves VIII, IX, and X A B C D E

A	B	C	D	E
1,2,3	1,3	2,4	only 4	all correct

4. THE ZYMOGATIC ARCH IS FORMED BY THE:
   1. Zygomatic process of the maxilla
   2. Temporal process of the zygomatic bone
   3. Zygomatic process of the frontal bone
   4. Zygomatic process of the temporal bone   A B C D E

5. WHICH OF THE FOLLOWING STATEMENTS ABOUT THE ANTERIOR FONTANELLE IS (ARE) CORRECT?
   1. Located at the junction of the sagittal and coronal sutures
   2. Called the bregma when it is closed
   3. Closed by two years of age
   4. Called the lambda when it is closed   A B C D E

---------------------ANSWERS, NOTES AND EXPLANATIONS---------------------

1. **D.** <u>Only 4 is correct</u>. The structures which normally pass through the foramen ovale are: (1) the mandibular nerve, consisting of a large sensory and a small motor root of the trigeminal nerve; (2) the accessory meningeal artery of the maxillary artery; (3) the emissary vein(s); and sometimes (4) the lesser petrosal nerve on its way to the otic ganglion.

2. **A.** <u>1, 2, and 3 are correct</u>. The sphenoid bone, situated near the central portion of the base of the skull, has a number of important foramina and fissures. The pterygoid canal passes through the pterygoid process. The foramina spinosum, rotundum, and ovale are openings in the greater wing of the sphenoid bone. The stylomastoid foramen is the extracranial aperture at the inferior end of the bony facial canal; it opens between the styloid and mastoid processes.

3. **D.** <u>Only 4 is false</u>. The jugular foramen is a large foramen at the base of the skull, formed by the jugular notch of the jugular process of the occipital bone and the petrous portion of the temporal bone. The anterior part of this foramen transmits the inferior petrosal sinus; the middle part, the cranial nerves IX, X, and XI; and the posterior part, the jugular bulb of the internal jugular vein. The internal jugular vein is the direct continuation of the sigmoid dural venous sinus at the jugular foramen. It should be noted that cranial nerve VIII passes through the internal acoustic meatus.

4. **C.** <u>2 and 4 are correct</u>. The zygomatic arch is formed by the zygomatic process of the temporal bone and the temporal process of the zygomatic bone. The zygomatic processes of the maxilla and frontal bone join the zygomatic bone, without contributing to the formation of the zygomatic arch.

5. **A.** <u>1, 2, and 3 are correct</u>. The anterior fontanelle is the gap or soft spot between the coronal and sagittal sutures in the fetal and infantile skulls and is covered by a fibrous membrane. It is usually obliterated by two years of age and then it is called the bregma. The lambda indicates the site of closure of the posterior fontanelle. The anterior fontanelle is the larger of the two main fontanelles and useful clinically. During parturition, palpation of it is used to determine the position of the fetal head in a vertex presentation. During infancy it may be used to estimate the degree of intracranial pressure, to assess the degree of development of the skull, and also to withdraw blood from the underlying superior sagittal dural sinus.

FIVE-CHOICE ASSOCIATION QUESTIONS

DIRECTIONS: Each group of questions below consists of a numbered list of descriptive words or phrases accompanied by a diagram with certain parts indicated by letters, or by a list of lettered headings. For each numbered word or phrase, SELECT THE LETTERED PART OR HEADING that matches it correctly. Then insert the letter in the space to the right of the appropriate number. Sometimes more than one numbered word or phrase may be correctly matched to the same lettered part or heading.

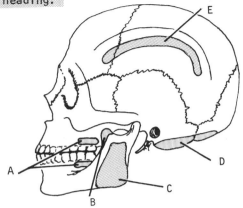

1. ____ Origin of the temporalis muscle
2. ____ Insertion of a muscle innervated by the accessory nerve
3. ____ Insertion of the strongest elevator of the jaw
4. ____ Origin of the accessory muscle of mastication innervated by the facial nerve

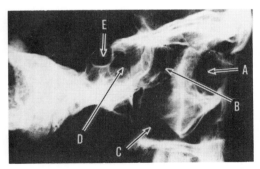

5. ____ Location of the hypophysis (pituitary gland)
6. ____ Sinus draining mucus into the inferoposterior part of the hiatus semilunalis
7. ____ Air sinus immediately inferior to the hypophysis
8. ____ Air sinus usually present at birth

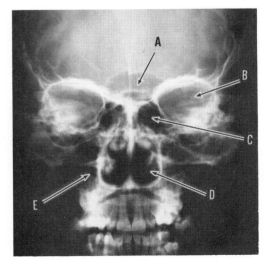

9. ____ Space lateral to ethmoidal air cells
10. ____ Space draining mucus into the anterosuperior portion of the hiatus semilunaris
11. ____ Space draining mucus from the summit of the bulla ethmoidalis
12. ____ A nasal cavity

ASSOCIATION QUESTIONS

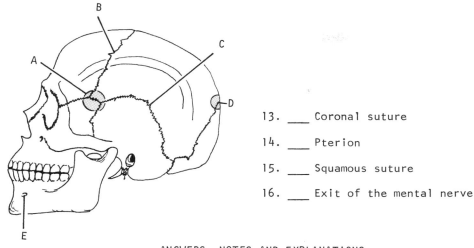

13. ___ Coronal suture

14. ___ Pterion

15. ___ Squamous suture

16. ___ Exit of the mental nerve

---------------------- ANSWERS, NOTES AND EXPLANATIONS ----------------------

1. **E.** The temporalis muscle arises from the periosteum of the temporal fossa below the inferior temporal line and from the deep surface of the temporal fascia.

2. **D.** The accessory nerve (the eleventh cranial nerve) innervates the sternocleidomastoid and trapezius muscles. (The site of attachment of the trapezius is not illustrated in the diagram). The sternocleidomastoid muscle inserts by a strong tendon onto the lateral surface of the mastoid process of the temporal bone and into the lateral half to two-thirds of the superior nuchal line on the occipital bone. It originates by the two heads, clavicular and sternal.

3. **B.** The strongest elevator of the jaw is the temporalis muscle. This fan-shaped muscle arises from the temporal fossa and the deep surface of the temporal fascia. The thick tendon of insertion passes deep to the zygomatic arch and inserts into the apex, anterior border and medial surface of the coronoid process, and the anterior border of the ramus.

4. **A.** All muscles of facial expression are innervated by the facial nerve. The buccinator muscle, the origin of which is illustrated, has the important function in mastication of holding food between the occlusal surfaces of the teeth by: compressing the cheek, drawing the corners of the mouth laterally, and forcing the lips against the teeth. Thus, the buccinator muscle is an accessory muscle of mastication. It arises from the alveolar processes of the maxilla and mandible, lateral to the molar teeth, and from the pterygomandibular raphe. Note that all muscles of mastication are innervated by the trigeminal nerve.

5. **E.** The hypophysis is a very important gland located in the hypophyseal fossa of the sella turcica of the sphenoid bone, posterior to the optic chiasma in the middle cranial fossa.

6. **C.** The maxillary air sinus, the largest of the paranasal air sinuses, is located in the maxilla on each side, lateral to the lateral wall of the nasal cavity. It drains into the opening at the inferoposterior part in the hiatus semilunaris of the middle nasal meatus. As this opening is on the upper

medial wall of the sinus, drainage of it is very poor in the erect posture.

7. D. The sphenoidal air sinus is located immediately inferior to the hypophysis in the body of the sphenoid bone. This sinus is located on each side of the median line, but the intervening septum is often irregular in shape and position, making these cavities asymmetrical. Each sinus opens into the sphenoethmoidal recess.

8. C. The maxillary air sinus is the only paranasal air sinus that may be present at birth. The frontal and sphenoid air sinuses develop at six to seven years of age. The paranasal air sinuses grow continuously until death, but reach their adult proportions during the twenties. These air sinuses are pneumatic areas in the frontal, ethmoid, sphenoid and maxillary bones. These sinuses develop as evaginations of the nasal mucosa by invading the bones surrounding the nasal cavity. The sites of evagination remain as the openings of the air sinuses.

9. B. The large space located lateral to the ethmoidal air cells is the orbital cavity.

10. A. The frontal air sinus drains via the frontonasal duct into the middle nasal meatus at the anterosuperior portion of the hiatus semilunaris. Note the asymmetry between the right and left sinuses. The sinuses are rarely symmetrical because the septum between them frequently deviates from the median plane.

11. C. The middle ethmoidal air cells drain through the opening at the summit of the bulla ethmoidalis in the middle nasal meatus. The anterior ethmoidal air cells open into the hiatus semilunaris inferoposterior to the opening of the frontonasal duct; the posterior ethmoidal air cells drain into the superior nasal meatus.

12. D. The nasal cavity on each side is located medial to the maxillary air sinus (labelled E in the radiograph), inferior to the ethmoidal air cells and superior to the oral cavity.

13. B. The coronal suture is a fibrous joint between the frontal and parietal bones.

14. A. The pterion is the area where the frontal, parietal, greater wing of the sphenoid, and the squamous part of the temporal bones articulate. This area is the former site of the anterolateral fontanelle which closed within two to three months after birth by growth of the surrounding bones. The pterion overlies the anterior branch of the middle meningeal artery on the internal surface of the skull. Therefore, fractures and injuries (e.g., a severe blow) at the pterion may rupture the middle meningeal artery.

15. C. The squamous suture is a fibrous joint between the squamous part of the temporal bone and the parietal bone.

16. E. The mental foramen, located in the mandible, transmits the mental branches of the inferior alveolar nerve and vessels. The mental nerve is a branch of the trigeminal nerve (cranial nerve V). It emerges at the mental foramen and divides into three branches. One branch descends to the skin of the chin and two ascend to the skin and mucous membrane of the lower lip.

INTERIOR OF THE CRANIUM

FIVE-CHOICE COMPLETION QUESTIONS

DIRECTIONS: Each of the following questions or incomplete statements is followed by five suggested answers or completions. SELECT THE ONE BEST ANSWER in each case and then underline the appropriate letter at the lower right of each question.

1. THE CRISTA GALLI SERVES AS AN ATTACHMENT FOR THE:
   A. Falx cerebri
   B. Diaphragma sellae
   C. Falx cerebelli
   D. Tentorium cerebelli
   E. None of the above

   A B C D E

2. THE HYPOPHYSEAL FOSSA IS LOCATED IN WHICH OF THE FOLLOWING BONES?
   A. Ethmoid
   B. Sphenoid
   C. Frontal
   D. Palatine
   E. Maxilla

   A B C D E

3. WHICH OF THE FOLLOWING STATEMENTS ABOUT THE CAVERNOUS SINUS IS FALSE?
   A. The internal carotid artery lies within the cavernous sinus
   B. The hypophysis (pituitary gland) is located medially
   C. The temporal lobe of the cerebral hemisphere is a lateral relation
   D. The trochlear nerve passes through the cavernous sinus
   E. The superior ophthalmic vein drains into the cavernous sinus

   A B C D E

4. CEREBROSPINAL FLUID RETURNS TO THE BLOODSTREAM PRINCIPALLY THROUGH THE:
   A. Interventricular foramen
   B. Lymphatic vessels
   C. Arachnoid villi
   D. Choroid plexuses
   E. Veins in the subarachnoid space

   A B C D E

-------------------- ANSWERS, NOTES AND EXPLANATIONS --------------------

1. A. The crista galli is a median process of the ethmoid bone, shaped like a cockscomb, that extends upward within the anterior cranial fossa. It gives attachment to the falx cerebri, a dural reflection separating the cerebral hemispheres. The diaphragma sellae forms a dural roof for the sella turcica of the sphenoid bone. The falx cerebelli is a sickle-shaped dural process attached superiorly to the tentorium cerebelli and separating the cerebellar hemispheres. The tentorium cerebelli is also the dural process supporting the occipital lobes of the cerebral hemispheres and covering the cerebellum.

2. B. The hypophyseal fossa, containing the hypophysis (pituitary gland), is located in the sella turcica of the sphenoid bone. It is bounded anteriorly by the middle clinoid processes and posteriorly by the dorsum sellae. Inferiorly its floor is formed by the roof of the sphenoid sinus, located in the body of the sphenoid bone. Laterally it is bounded by the cavernous dural venous sinus.

3. D. The paired, cavernous dural venous sinuses are located on each side of the body of the sphenoid bone. They are between the meningeal and periosteal layers of the dura mater. The internal carotid artery and the abducens nerve on each side pass through these sinuses. In their lateral walls, from superior to inferior, the oculomotor, trochlear, ophthalmic, and maxillary nerves pass forward. Thus, the trochlear nerve does not pass through the cavernous

sinus. The temporal lobe of the cerebral hemisphere lies lateral to the sinus. The superior ophthalmic vein passes through the superior orbital fissure as an emissary vein between the cavernous sinus, and the tributaries of the facial vein at the root of the nose.

4. **C.** The cerebrospinal fluid fills the subarachnoid space between the arachnoid and the pia mater. In many areas of the dural venous sinuses, the arachnoid projects into the venous sinuses to form microscopic arachnoid villi. Aggregations of the arachnoid villi, called arachnoid granulations during old age, are most numerous along the superior sagittal sinus. Cerebrospinal fluid diffuses into the blood stream through the membranes of the arachnoid villi and arachnoid granulations. The choroid plexuses within the lateral, third, and fourth ventricles of the brain are believed to be the sites of production of cerebrospinal fluid.

MULTI-COMPLETION QUESTIONS

DIRECTIONS: In each of the following questions or incomplete statements, ONE OR MORE of the completions given is correct. At the lower right of each question, underline A if 1, 2 and 3 are correct; B if 1 and 3 are correct; C if 2 and 4 are correct; D if only 4 is correct; and E if all are correct.

1. CONCERNING THE DURAL VENOUS SINUSES:
   1. All are formed by the meningeal and periosteal layers of the dura mater
   2. Some communicate directly with the subarachnoid space
   3. The sphenoparietal sinus communicates directly with the superior petrosal sinus
   4. The cavernous sinus communicates with the pterygoid plexus of veins via emissary veins   A B C D E

2. WHICH STATEMENT(S) ABOUT THE MIDDLE MENINGEAL ARTERY IS (ARE) CORRECT?
   1. Is extradural in position
   2. Enters the skull through the foramen spinosum
   3. Is accompanied by a sympathetic plexus of nerves
   4. Supplies the parietal lobe of the cerebral cortex   A B C D E

3. WHICH OF THE FOLLOWING ARTERIES CONTRIBUTE(S) TO THE FORMATION OF THE CIRCULUS ARTERIOSUS CEREBRI (CIRCLE OF WILLIS)?
   1. Posterior communicating   3. Anterior cerebral
   2. Posterior cerebral   4. Middle cerebral   A B D C E

-------------------- ANSWERS, NOTES AND EXPLANATIONS --------------------

1. **C,** <u>2 and 4 are correct</u>. Some of the dural venous sinuses are between the periosteal and meningeal layers; others are enclosed in foldings of the meningeal layer. The former type includes the: superior sagittal, transverse, sigmoid, and cavernous sinuses; the latter type includes the: inferior sagittal, straight, and occipital sinuses. At the parietal region, the arachnoid villi project into the venous lacunae which open into the superior sagittal sinus. This sinus communicates with the subarachnoid space, allowing cerebrospinal fluid to enter the venous system. The sphenoparietal sinus drains into the cavernous sinus, which communicates with the superior petrosal sinus. The cavernous sinus communicates with the pterygoid plexus of veins by emissary veins.

A	B	C	D	E
1,2,3	1,3	2,4	only 4	all correct

2. **A. 1, 2, and 3 are correct.** The middle meningeal artery, a branch of the maxillary artery, enters the cranial cavity through the foramen spinosum from the infratemporal fossa. The artery then runs anteriorly and laterally in a groove on the inner surface of the squamous part of the temporal bone and the greater wing of the sphenoid bone, external to the dura mater. Both the anterior and posterior branches of this artery supply blood to the dura mater in the parietal region, but it does not supply the parietal lobe of the cerebral cortex. The postganglionic fibers, arising from the superior cervical sympathetic ganglion, form the perivascular sympathetic plexus of nerves along the middle meningeal artery and act as vasoconstrictors.

3.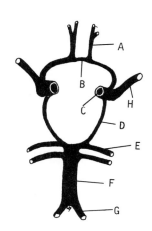

    A. 1, 2, and 3 are correct. The circulus arteriosus cerebri lies in the interpeduncular fossa at the base of the brain, and is the site of anastomosis between the internal carotid and the vertebral arterial systems. This arterial circle is formed: <u>anteriorly</u>, by the right and left anterior cerebral arteries (A), the anterior communicating artery (B) connecting them, and the internal carotid arteries (C); and <u>posteriorly</u>, by the posterior communicating (D) and the posterior cerebral arteries (E). The posterior cerebral arteries are the terminal branches of the basilar artery (F) which is formed by the union of the right and left vertebral arteries (G). Although the middle cerebral artery (H) is the continuation of the internal carotid artery, it does not contribute to the formation of the circle of Willis.

SUPERFICIAL STRUCTURES OF THE HEAD, NECK, AND FACE

FIVE-CHOICE COMPLETION QUESTIONS

DIRECTIONS: Each of the following questions or incomplete statements is followed by five suggested answers or completions. SELECT THE ONE BEST ANSWER in each case and then underline the appropriate letter at the lower right of each question.

1. THE FACIAL NERVE INNERVATES ALL THE FOLLOWING MUSCLES <u>EXCEPT</u>:
    A. Orbicularis oris
    B. Temporalis
    C. Buccinator
    D. Levator labii superioris
    E. Platysma

    A B C D E

2. THE FOLLOWING NERVES SUPPLY THE SKIN OR MUSCLES OF THE EXTERNAL NOSE, <u>EXCEPT</u> THE:
    A. Frontal
    B. Infraorbital
    C. Facial
    D. Nasociliary
    E. Infratrochlear

    A B C D E

SELECT THE ONE BEST ANSWER

3. SENSORY LOSS IN THE SKIN OVERLYING THE PAROTID GLAND SUGGESTS DAMAGE TO THE _____ NERVE.
   A. Great occipital
   B. Great auricular
   C. Zygomaticotemporal
   D. Transverse cervical
   E. None of the above

   A B C D E

4. THE INFRAORBITAL ARTERY IS A BRANCH OF THE:
   A. Transverse facial
   B. Maxillary
   C. Ophthalmic
   D. Internal carotid
   E. Facial

   A B C D E

---------------------- ANSWERS, NOTES AND EXPLANATIONS ----------------------

1. B. The facial nerve innervates all muscles of facial expression including all those listed in this question, except the temporalis. The temporalis, a muscle of mastication, is innervated by the mandibular division of the trigeminal nerve.

2. A. The skin of the external nose is supplied by the lateral nasal branch of the infraorbital nerve, the external nasal branch of the anterior ethmoidal nerve of the nasociliary nerve, and the infratrochlear nerve. The infraorbital nerve is a branch of the maxillary division of the trigeminal nerve and the remaining nerves are branches of the ophthalmic division of this nerve. The nasalis muscle of the external nose is innervated by the facial nerve. Neither branch of the frontal nerve, the supraorbital or the supratrochlear, supplies skin of the external nose.

3. B. The skin overlying the parotid gland is mainly innervated by the great auricular nerve of the cervical plexus derived from C2 and C3, but it receives some fibers from the buccal nerve of the mandibular division of the trigeminal nerve. The greater occipital nerve consists of sensory fibers from the dorsal ramus of the second cervical nerve, and supplies skin of the occipital part of the scalp as far superior as the vertex of the skull. The zygomaticotemporal nerve is a branch of the facial nerve and is motor to some of the muscles of facial expression. The transverse cervical nerve is a branch of the cervical plexus derived from C2 and C3 and supplies the skin overlying the anterior triangle of the neck.

4. B. The infraorbital artery is a branch of the pterygopalatine portion of the maxillary artery. It enters the orbit through the inferior orbital fissure, proceeds forwards in the infraorbital groove in the floor of the orbit, and then passes through the infraorbital canal and emerges via the infraorbital foramen. Branches of this artery supply the maxillary teeth and gingiva. The transverse facial artery is a branch of the superficial temporal artery. The ophthalmic artery is the chief vessel of the orbit, but the infraorbital also contributes to the supply of this region.

M U L T I - C O M P L E T I O N   Q U E S T I O N S

DIRECTIONS: In each of the following questions or incomplete statements, ONE OR MORE of the completions given is correct. At the lower right of each question, underline A if 1, 2, and 3 are correct; B if 1 and 3 are correct; C if 2 and 4 are correct; D if only 4 is correct; and E if all are correct.

A	B	C	D	E
1,2,3	1,3	2,4	only 4	all correct

1. WHICH STATEMENT(S) ABOUT THE CERVICAL PLEXUS OF NERVES IS (ARE) CORRECT?
   1. Formed by branches of the ventral rami of the first four cervical nerves
   2. Supplies the thyrohyoid muscle
   3. Gives rise to the great auricular nerve
   4. Is responsible for cutaneous innervation   A B C D E

2. MUSCLES INNERVATED BY MOTOR FIBERS OF THE FACIAL NERVE ARE THE:
   1. Orbicularis oculi
   2. Anterior belly of digastric
   3. Buccinator
   4. Mylohyoid   A B C D E

3. THE SKIN OF THE HEAD, FACE, AND NECK IS INNERVATED BY BRANCHES OF THE:
   1. Cervical plexus
   2. Dorsal rami of cervical nerves
   3. Trigeminal nerve
   4. Facial nerve   A B C D E

------------------- ANSWERS, NOTES AND EXPLANATIONS ---------------------

1. **E.** <u>All are correct</u>. The cervical plexus of the nerves is formed by the ventral rami of the first four cervical spinal nerves. This plexus lies opposite the first four cervical vertebrae, ventrolateral to the levator scapulae and middle scalene muscles, and deep to the sternocleidomastoid muscle. The cervical plexus of nerves gives rise to: cutaneous branches; the ansa cervicalis (motor to the infrahyoid and geniohyoid muscles); the phrenic nerve (mixed nerve to the diaphragm); proprioceptive contributions to the accessory nerve supplying the sternocleidomastoid and trapezius muscles; and some direct branches to the prevertebral muscles. The thyroid muscle is innervated by a branch from C1 which travels with the hypoglossal nerve for some distance. One of the cutaneous branches of the cervical plexus is the great auricular nerve from C2 and C3.

2. **B.** <u>1 and 3 are correct</u>. Motor fibers of the facial nerve innervate all muscles of facial expression, including the orbicularis oculi and buccinator muscles and the posterior belly of the digastric muscle. The nerve to the mylohyoid muscle, derived from the inferior alveolar nerve (a branch of the mandibular division of the trigeminal nerve), also innervates the anterior belly of the digastric muscle.

3. **A.** <u>1, 2, and 3 are correct</u>. The cutaneous nerves of the head, face, and neck regions are branches of the cervical plexus, the dorsal rami of the cervical nerves, and the trigeminal nerve. The facial nerve has no cutaneous branches. As shown in the following diagram, the dorsal rami of all cervical nerves, except the first, supply areas of the skin behind the posterior border of the sternocleidomastoid muscle. The cutaneous branches of the cervical plexus, derived from C2 to C4, are distributed to the lateral and anterior aspects of the neck anterior to the sternocleidomastoid muscle. This innervation extends to the root of the neck and to areas of skin in the parotid region and surrounding the angle of the mandible and the mastoid process. Anterior to the areas supplied by the cervical plexus, the trigeminal nerve supplies the skin of the face through branches from its three divisions (ophthalmic, maxillary, and mandibular).

Detailed schematic diagram of the cutaneous innervation of the scalp, face, and neck.

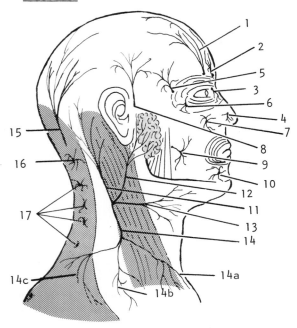

Trigeminal nerve

Ophthalmic division:
1. supraorbital n.
2. supratrochlear n.
3. infratrochlear n.
4. anterior ethmoidal n.

Maxillary division:
5. zygomaticotemporal n.
6. zygomaticofacial n.
7. infraorbital n.

Mandibular division:
8. auriculotemporal n.
9. buccal n.
10. mental n.

Cervical plexus of nerves
11. lesser occipital n.
12. great auricular n.
13. transverse cervical n.
14. supraclavicular n.
    a. medial branch
    b. intermediate branch
    c. lateral branch

Dorsal rami of cervical nerves
15. greater occipital n.
16. third occipital n.
17. cutaneous branches of C4-C7

ANTERIOR TRIANGLE OF THE NECK

FIVE-CHOICE COMPLETION QUESTIONS

DIRECTIONS: Each of the following questions or incomplete statements is followed by five suggested answers or completions. SELECT THE ONE BEST ANSWER in each case and then underline the appropriate letter at the lower right of each question.

1. WHICH OF THE FOLLOWING STATEMENTS IS FALSE? THE SUPERFICIAL LAYER OF THE CERVICAL FASCIA:
   A. Attaches to the hyoid bone
   B. Invests the sternocleidomastoid muscle
   C. Contains the platysma muscle
   D. Forms the suprasternal space
   E. None of the above   A B C D E

2. WHICH OF THE FOLLOWING IS LOCATED OUTSIDE THE CAROTID SHEATH?
   A. Internal jugular vein   D. Vagus nerve
   B. Internal carotid artery   E. None of the above
   C. Cervical sympathetic trunk   A B C D E

86

SELECT THE ONE BEST ANSWER

3. THE INFRAHYOID MUSCLES ARE INNERVATED BY WHICH OF THE FOLLOWING CERVICAL SPINAL NERVES?
   A. C1, C2
   B. C2, C3
   C. C3, C4
   D. C1-C3
   E. None of the above

   A B C D E

4. THE BRANCH OF THE EXTERNAL CAROTID ARTERY ORIGINATING JUST BELOW THE LEVEL OF THE GREATER CORNU OF THE HYOID BONE IS THE _____ ARTERY.
   A. Facial
   B. Occipital
   C. Maxillary
   D. Superior thyroid
   E. Posterior auricular

   A B C D E

5. WHICH OF THE FOLLOWING IS A FALSE STATEMENT?
   A. The anterior triangle of the neck contains cutaneous branches of the cervical plexus of nerves
   B. The digastric triangle contains the submandibular gland
   C. The submental triangle contains the sublingual gland
   D. The muscular triangle contains the infrahyoid muscle
   E. The floor of the posterior triangle of the neck is covered by cervical fascia

   A B C D E

---------------------ANSWERS, NOTES AND EXPLANATIONS---------------------

1. **C,** The platysma muscle lies in the superficial fascia (tela subcutanea). The superficial layer of the cervical fascia encircles the neck and encloses the stenocleidomastoid and trapezius muscles. It forms a roof over the anterior and posterior triangles of the neck, attaches posteriorly to the ligamentum nuchae, anterior to the hyoid bone and superior to the lower border of the mandible. Inferiorly, it attaches to the acromion, clavicle, and manubrium sterni. A slit-like space, formed by the two layers of the superficial layer of the cervical fascia and attached to the anterior and posterior margins of the upper border of the manubrium, is called the suprasternal space. It contains the jugular venous arch connecting the two anterior jugular veins.

2. **C,** The carotid sheath is a tubular neurovascular sheath, formed by a condensation of the cervical fascia. It contains the common and internal carotid arteries, the internal jugular vein, and the vagus nerve. The cervical sympathetic trunk lies posterior to the carotid sheath and anterior to the prevertebral fascia.

3. **D,** The infrahyoid muscles include the: sternohyoid, sternothyroid, omohyoid and thyrohyoid. All these muscles are innervated by the ansa cervicalis (ansa hypoglossi) which is formed from C1, C2, and C3. The superior ramus (descendens hypoglossi; C1, C2) unites with the inferior ramus (descendens cervicalis; C2, C3) from the cervical plexus to form the ansa cervicalis. The thyrohyoid muscle receives its innervation directly from C1 by a branch which accompanies the hypoglossal nerve. The remaining muscles are innervated by fibers from the ansa cervicalis.

4. **D,** The external carotid artery extends from the upper border of the thyroid cartilage to the neck of the mandible, where it terminates by dividing into the maxillary and superficial temporal arteries. It has eight branches. Usually its first branch is the superior thyroid artery arising opposite the thyrohyoid membrane, just below the greater cornu of the hyoid bone. The lingual artery arises opposite the tip of the greater cornu; the facial artery arises immediately above the lingual artery. The occipital and posterior auricular arteries arise from the posterior aspect of the external carotid

artery at the lower and upper borders of the posterior belly of the digastric muscle, respectively.

5. **C.** The submental triangle does not contain the sublingual salivary gland. The neck region is divided into anterior and posterior triangles by the sternocleidomastoid muscle. The anterior triangle may be subdivided into submental (A), digastric (submandibular) (B), muscular (C), and carotid triangles (D). The posterior triangle may be subdivided into subclavian (omoclavicular) (E), and occipital (posterior) (F) triangles. The floor of the posterior triangle is covered by the prevertebral layer of the cervical fascia. The cutaneous branches of the cervical plexus of nerves are located in the roof of the anterior triangle. The digastric triangle contains the submandibular salivary gland. The muscular triangle contains the infrahyoid muscles. The carotid triangle contains the carotid sheath and its contents.

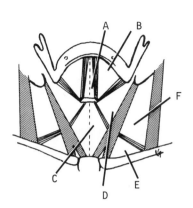

M U L T I - C O M P L E T I O N   Q U E S T I O N S

DIRECTIONS: In each of the following questions or incomplete statements, ONE OR MORE of the completions given is correct. At the lower right of each question, underline A if 1, 2, and 3 are correct; B if 1 and 3 are correct; C if 2 and 4 are correct; D if only 4 is correct; and E if all are correct.

1. WHICH OF THE FOLLOWING STATEMENTS CONCERNING THE THYROHYOID MUSCLE IS (ARE) CORRECT?
    1. Is one of the infrahyoid muscles
    2. Lies lateral to the thyroid notch
    3. Attaches to the hyoid bone lateral to the omohyoid muscle
    4. Is innervated by the ansa cervicalis   A B C D E

2. THE SUPERIOR THYROID ARTERY:
    1. Is the first branch of the external carotid artery
    2. Is the only source of blood supply to the thyroid gland
    3. Has a branch that pierces the thyrohyoid membrane
    4. Runs parallel to the internal laryngeal nerve   A B C D E

3. WHICH STATEMENT(S) ABOUT THE EXTERNAL CAROTID ARTERY IS (ARE) CORRECT?
    1. Divides in the substance of the parotid gland
    2. Extends from the level of the lower border of the thyroid cartilage
    3. At first it is generally anteromedial to the internal carotid artery
    4. Ascends medial to the internal carotid artery   A B C D E

4. WHICH OF THE FOLLOWING MUSCLES IS (ARE) INNERVATED BY THE ANSA CERVICALIS?
    1. Sternohyoid   3. Thyrohyoid
    2. Sternothyroid   4. Omohyoid   A B C D E

A	B	C	D	E
1,2,3	1,3	2,4	only 4	all correct

5. WHICH OF THE FOLLOWING STATEMENTS IS(ARE) CORRECT?
   1. The carotid sinus functions as a pressoreceptor (baroreceptor)
   2. The carotid body functions as a chemoreceptor, sensitive to oxygen tension
   3. The carotid sinus is located near the bifurcation of the common carotid artery
   4. The carotid body is located near the bifurcation of the common carotid artery

   A B C D E

---------------------- ANSWERS, NOTES AND EXPLANATIONS ----------------------

1. E, All are correct. The thyrohyoid muscle is one of the infrahyoid (strap) muscles. It arises from the oblique line of the thyroid cartilage and inserts on the lower border of the greater cornu of the hyoid bone, lateral to the omohyoid muscle. The thyrohyoid muscle is innervated by a branch of the ansa cervicalis, derived from the first cervical nerve via the hypoglossal nerve.

2. B, 1 and 3 are correct. The superior thyroid artery is the first branch of the external carotid artery and descends to the upper pole of each lobe of the thyroid gland, accompanied by the external laryngeal nerve. In addition, the thyroid gland receives a blood supply from the inferior thyroid artery. The internal laryngeal branch of the superior thyroid artery pierces the thyrohyoid membrane along with the internal laryngeal branch of the superior laryngeal nerve.

3. B, 1 and 3 are correct. The external carotid artery extends from the level of the upper border of the lamina of the thyroid cartilage to a point behind the neck of the mandible. As the external carotid artery ascends, it inclines backward and comes to lie lateral to the internal carotid artery.

4. E, All are correct. The superior ramus (descendens hypoglossi; C1, C2) unites with the inferior ramus (descendens cervicalis; C2, C3) from the cervical plexus to form the ansa cervicalis (ansa hypoglossi) which innervates the sternohyoid, sternothyroid, omohyoid and thyrohyoid muscles. The thyrohyoid muscle, however, receives a small branch (travelling with the hypoglossal nerve) directly from C1.

5. E, All are correct. The carotid sinus is located at the terminal portion of the common carotid artery and/or at the root of the internal carotid artery as a spindle-shaped dilation. It is stimulated by changes in blood pressure and serves as a pressoreceptor (baroreceptor). It is innervated by the carotid branch of the glossopharyngeal (IXth cranial) nerve. The carotid body lies at the bifurcation of the common carotid artery; it functions as a chemoreceptor and is sensitive to changes in oxygen tension in the circulating blood. It is innervated by the nerve to the carotid body derived from the pharyngeal branch of the vagus (Xth cranial) nerve. The carotid body is invested by a fibrous capsule from which septa pass into the body, dividing it into lobules. Each lobule consists of masses of large polyhedral glomus cells and flattened supporting cells. Interspersed among these cells are networks of sinusoidal blood vessels.

FIVE-CHOICE ASSOCIATION QUESTIONS

DIRECTIONS: Each group of questions below consists of a numbered list of descriptive words or phrases accompanied by a diagram with certain parts indicated by letters, or by a list of lettered headings. For each numbered word or phrase, SELECT THE LETTERED PART OR HEADING that matches it correctly. Then insert the letter in the space to the right of the appropriate number. Sometimes more than one numbered word or phrase may be correctly matched to the same lettered part or heading.

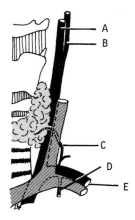

1. ____ An arterial branch to the thyroid gland

2. ____ Gives rise to the internal thoracic artery

3. ____ A branch of the common carotid artery which has no branch in the neck

4. ____ A branch of the thyrocervical trunk

---------------------- ANSWERS, NOTES AND EXPLANATIONS ----------------------

1. **C,** The inferior thyroid artery, a branch of the thyrocervical trunk of the subclavian artery, along with the superior thyroid artery, supplies the thyroid gland.

2. **E,** The subclavian artery gives rise to the internal thoracic artery opposite to the thyrocervical trunk, just medial to the anterior scalene muscle.

3. **B,** The internal carotid artery is one of the two terminal branches of the common carotid artery; it has no branches in the cervical region. The other branch, the external carotid artery, has eight named branches.

4. **C,** The inferior thyroid artery arises from the thyrocervical trunk of the subclavian artery and ascends along the medial border of the anterior scalene muscle deep to the prevertebral fascia, as high as the lower border of the cricoid cartilage. Here, it turns medially, deep to the carotid sheath, and arches downward to the lower pole of the thyroid gland, where it turns upward and divides into terminal branches. Here, the inferior thyroid artery is in close relation to the recurrent laryngeal nerve which may be accidentally injured during operations on the thyroid gland. The closeness of the relationship between the recurrent laryngeal nerve and the inferior thyroid artery makes injury to the nerve possible, especially when ligation of the artery is done close to the thyroid gland.

SUBMANDIBULAR AND PAROTID REGIONS

FIVE-CHOICE COMPLETION QUESTIONS

DIRECTIONS: Each of the following questions or incomplete statements is followed by five suggested answers or completions. SELECT THE ONE BEST ANSWER in each case and then underline the appropriate letter at the lower right of each question.

1. WHICH STATEMENT IS FALSE?
   A. The submandibular ganglion is suspended from the lingual nerve
   B. The stylohyoid ligament is attached to the lesser cornu of the hyoid bone
   C. The sublingual gland makes an impression on the inner surface of the mandible
   D. Most of the ducts of the sublingual gland empty directly into the floor of the mouth
   E. The geniohyoid muscle is supplied by the mylohyoid nerve
   A B C D E

2. A CHILD FALLS ON A NAIL WHICH PENETRATES THE SKIN OVERLYING THE SUBMENTAL TRIANGLE, SLIGHTLY LATERAL TO THE MEDIAN RAPHE OF THE MYLOHYOID MUSCLE. AS THE NAIL PENETRATES THE UNDERLYING TISSUES, WHICH OF THE FOLLOWING MUSCLES WOULD BE THE LAST ONE TO BE PIERCED?
   A. Digastric   D. Genioglossus
   B. Platysma   E. Geniohyoid
   C. Mylohyoid
   A B C D E

3. THE ANTERIOR BELLY OF THE DIGASTRIC MUSCLE IS SUPPLIED BY THE:
   A. Facial nerve   D. Cervical plexus
   B. Mylohyoid nerve   E. Accessory nerve
   C. Ansa cervicalis
   A B C D E

4. THE MEDIAL RELATION OF THE BODY OF THE MANDIBLE IS THE:
   A. Inferior alveolar nerve   D. Stylomandibular ligament
   B. Sphenomandibular ligament   E. Lingual nerve
   C. Buccinator muscle
   A B C D E

5. PREGANGLIONIC SYMPATHETIC FIBERS TO THE SUBMANDIBULAR GLAND SYNAPSE IN WHICH OF THE FOLLOWING GANGLIA?
   A. Otic   D. Pterygopalatine
   B. Ciliary   E. Superior cervical
   C. Submandibular
   A B C D E

6. IN THE SUBMANDIBULAR TRIANGLE, THE HYPOGLOSSAL NERVE PASSES:
   A. Superficial to the mylohyoid muscle
   B. Between the mylohyoid and digastric muscles
   C. Between the hyoglossus and mylohyoid muscles
   D. Deep to the hyoglossus muscle
   E. Between the hyoglossus and middle pharyngeal constrictor muscles
   A B C D E

-------------------- ANSWERS, NOTES AND EXPLANATIONS --------------------

1. **E.** The submandibular ganglion is a parasympathetic ganglion receiving preganglionic secretomotor fibers of the chorda tympani (VIIth cranial nerve), carried by the lingual nerve (Vth cranial nerve). These fibers synapse and

postganlionic fibers proceed to innervate the submandibular and sublingual salivary glands. The submandibular ganglion is suspended from the lingual nerve by two roots. The stylohyoid ligament is a fibrous cord between the styloid process of the temporal bone and the lesser cornu of the hyoid bone; it may be partially ossified. The sublingual gland lodges in the sublingual fossa on the medial surface of the body of the mandible. It drains into the floor of the mouth through several ducts. The geniohyoid muscle is innervated by motor fibers derived from the 1st cervical nerve via the hypoglossal nerve.

2. **D,** The nail would penetrate the skin overlying the submental triangle lateral to the median raphe of the mylohyoid muscle: then it would pass through the platysma, the anterior belly of the digastric, the mylohyoid, the geniohyoid and the genioglossus muscles.

3. **B,** The anterior belly of the digastric muscle is innervated by motor fibers from the mylohyoid nerve which is a branch of the inferior alveolar nerve of the mandibular division (Vth cranial nerve). The posterior belly of this muscle is innervated by the facial (VIIth cranial) nerve.

4. **E,** The lingual nerve runs forward along the medial aspect of the body of the mandible. Sometimes it may be seen through the translucent mucous membrane in the sublingual sulcus at the level of the third mandibular molar tooth. The inferior alveolar nerve and the sphenomandibular ligament are located medial to the ramus of the mandible. The buccinator muscle attaches to the lateral aspect of the body of the mandible; the stylomandibular ligament attaches to the angle of the mandible.

5. **E,** Preganglionic sympathetic fibers derived from the spinal cord traverse the white rami communicantes of the upper three (or four) thoracic spinal nerves, and ascend to the superior cervical sympathetic ganglion where they synapse. The otic, ciliary, submandibular, and pterygopalatine ganglia are parasympathetic ganglia.

6. **C,** The hypoglossal nerve enters the submandibular (digastric) triangle deep to the posterior belly of the digastric muscle. It then passes forwards on the lateral surface of the hyoglossus muscle, deep to the mylohyoid. The hypoglossal nerve innervates all the extrinsic and intrinsic muscles of the tongue.

MULTI-COMPLETION QUESTIONS

DIRECTIONS: In each of the following questions or incomplete statements, ONE OR MORE of the completions given is correct. At the lower right of each question, underline A if 1, 2 and 3 are correct; B if 1 and 3 are correct; C if 2 and 4 are correct; D if only 4 is correct; and E if all are correct.

1. WHICH STATEMENT(S) ABOUT THE FACIAL NERVE IS (ARE) CORRECT?
   1. Motor to the posterior belly of the digastric muscle
   2. Contains taste and secretomotor fibers
   3. Nerve of the hyoid or second branchial arch
   4. Motor to the muscles of mastication   A B C D E

2. WHICH OF THE FOLLOWING ARTERIES SUPPLY(IES) THE PAROTID GLAND?
   1. Maxillary   3. Superficial temporal
   2. Transverse facial   4. External carotid   A B C D E

A	B	C	D	E
1,2,3	1,3	2,4	only 4	all correct

3. CONCERNING THE PAROTID GLAND:
   1. Pus cannot readily spread from the parotid to the submandibular gland
   2. Complete surgical removal of the parotid gland is extremely difficult, if not impossible, without damaging the facial nerve
   3. The parotid gland is innervated by secretomotor fibers of the lesser petrosal nerve
   4. Patients with mumps often complain of severe pain resulting from irritation of the auriculotemporal nerve   A B C D E

4. THE ARTERIAL SUPPLY OF THE SUBMANDIBULAR GLAND IS FROM THE:
   1. Maxillary   3. Inferior alveolar
   2. Lingual   4. Facial   A B C D E

5. THE BLOOD SUPPLY OF THE SUBLINGUAL GLAND IS FROM THE:
   1. Submental   3. Lingual
   2. Maxillary   4. Facial   A B C D E

---------------------- ANSWERS, NOTES AND EXPLANATIONS ----------------------

1. A, 1, 2, and 3 are correct. The facial (VIIth cranial) nerve consists of a larger part which supplies the muscles of facial expression, the posterior belly of the digastric, stylohyoid, and stapedius muscles; and a smaller part (the nervus intermedius) which contains taste fibers from the anterior two-thirds of the tongue, secretomotor fibers for the lacrimal and salivary (submandibular and sublingual) glands, and some pain fibers. The muscles of mastication are supplied by the trigeminal (Vth cranial) nerve.

2. E, All are correct. The parotid gland receives its blood supply from glandular branches of the external carotid, maxillary, superficial temporal, and transverse facial arteries. The maxillary and superficial temporal arteries are the terminal branches of the external carotid, and the transverse facial is a branch of the superficial temporal.

3. E, All are correct. The stylomandibular ligament is the thickening of the deep portion of the parotid fascia passing downward and laterally from the styloid process to the angle of the mandible. This ligament separates the parotid gland from the submandibular gland, therefore pus cannot readily spread between these glands. The facial nerve emerges from the stylomastoid foramen and enters the parotid gland, where it divides into temporofacial and cervicofacial divisions. These further ramify in the substance of the parotid gland and emerge from the gland to innervate the muscles of facial expression. Therefore, the complete surgical removal of the parotid gland (parotidectomy) often results in injury of the facial nerve. The parotid gland is innervated by secretomotor fibers of the glossopharyngeal (IXth cranial) nerve via the lesser petrosal nerve (preganglionic fibers). Its postganglionic fibers, derived from the otic ganglion, reach the gland via the auriculotemporal nerve (mandibular division of the Vth cranial nerve). Because the auriculotemporal nerve is sensory and passes through the gland, irritation of this nerve due to mumps (viral infection) causes pain.

4. D, Only 4 is correct. The submandibular gland receives it blood supply from glandular branches of the facial artery. However, sometimes it may be supplied by the lingual artery.

5. **B,** <u>1 and 3 are correct</u>. The sublingual gland receives its blood supply from the sublingual branch of the lingual artery and from the submental artery which pierces the mylohyoid muscle to reach the gland. The submental artery is a branch of the facial artery.

POSTERIOR TRIANGLE OF THE NECK:

FIVE-CHOICE COMPLETION QUESTIONS

DIRECTIONS: Each of the following questions or incomplete statements is followed by five suggested answers or completions. SELECT THE ONE BEST ANSWER in each case and then underline the appropriate letter at the lower right of each question.

1. THE BOUNDARIES OF THE POSTERIOR TRIANGLE OF THE NECK ARE FORMED BY THE:
   A. Sternocleidomastoid muscle, the posterior midsagittal line of the neck, and the clavicle
   B. Sternocleidomastoid muscle, the lower margin of the mandible, and the clavicle
   C. Sternocleidomastoid muscle, the inferior belly of omohyoid muscle, and the clavicle
   D. Sternocleidomastoid and trapezius muscles, and the clavicle
   E. None of the above

   A B C D E

2. WHICH OF THE FOLLOWING IS NOT FOUND IN THE POSTERIOR TRIANGLE OF THE NECK?
   A. Accessory nerve
   B. Occipital lymph node
   C. Posterior auricular artery
   D. Transverse cervical artery
   E. Brachial plexus of nerves

   A B C D E

3. WHICH STATEMENT ABOUT THE TRAPEZIUS MUSCLE IS <u>FALSE</u>?
   A. Innervated by motor fibers of the accessory nerve
   B. Forms a boundary of the posterior triangle of the neck
   C. Receives its blood supply from the transverse cervical artery
   D. Weakness of this muscle results in drooping of the shoulder
   E. Inserts into the scapula and the length of the clavicle

   A B C D E

------------------------ ANSWERS, NOTES AND EXPLANATIONS ------------------------

1. **D,** The boundaries of the posterior triangle of the neck are formed by the posterior border of the sternocleidomastoid muscle, the anterior border of the trapezius muscle, and the clavicle.

2. **C,** The posterior auricular artery arises from the posterior aspect of the external carotid artery at the level of the upper border of the posterior belly of the digastric muscle. Deep to the parotid gland, it ascends to the notch between the external acoustic meatus and the mastoid process. Therefore, it does not pass through the posterior triangle of the neck. The accessory nerve emerges from the deep aspect of the posterior border of the sternocleidomastoid muscle and runs diagonally and posteroinferiorly to pass beneath the trapezius muscle. It innervates both the sternocleidomastoid and trapezius muscles. In the lower anterior angle of this triangle, the roots and trunks of the brachial plexus of nerves are located adjacent to the subclavian artery. The supraclavicular lymph nodes are located in this region and may be involved in spread (metastasis) of cancer, e.g., from the breast. The occipital lymph nodes are located between the attachments of the

sternocleidomastoid and trapezius muscles overlying the superior nuchal line. The transverse cervical artery, a branch of the thyrocervical trunk, runs laterally across the posterior triangle of the neck to the anterior border of the trapezius muscle which it supplies.

3. E, The trapezius muscle inserts into two bones: its uppermost portion inserts into the lateral third of the clavicle; its middle part on the acromion and upper border of the spine of the scapula; and its lowermost portion into the tubercle of the crest of the scapula.

MULTI-COMPLETION QUESTIONS

DIRECTIONS: In each of the following questions or incomplete statements, ONE OR MORE of the completions given is correct. At the lower right of each question, underline A if 1, 2 and 3 are correct; B if 1 and 3 are correct; C if 2 and 4 are correct; D if only 4 is correct; and E if all are correct.

1. WITHIN THE POSTERIOR TRIANGLE OF THE NECK ARE FOUND THE:
   1. Levator scapulae muscle
   2. Scalenus medius muscle
   3. Accessory nerve
   4. Trapezius muscle   A B C D E

2. PROPRIOCEPTIVE INNERVATION TO THE:
   1. Sternocleidomastoid muscle is by cervical nerves (C2 and C3)
   2. Temporomandibular joint is mainly by the auriculotemporal nerve
   3. Trapezius muscle is by cervical nerves (C3 and C4)
   4. Masseter muscle is by the mandibular nerve   A B C D E

3. THE TRAPEZIUS MUSCLE ORIGINATES FROM THE:
   1. Superior nuchal line
   2. Ligamentum nuchae
   3. Spines of thoracic vertebrae
   4. Spine of the scapula   A B C D E

4. WHICH OF THE FOLLOWING STATEMENTS ABOUT THE STERNOCLEIDOMASTOID MUSCLE IS(ARE) CORRECT?
   1. Is an important landmark bisecting each side of the neck
   2. An accessory muscle of respiration
   3. Arises by two heads
   4. Inserts by two tendons   A B C D E

---------------------- ANSWERS, NOTES AND EXPLANATIONS ---------------------

1. A, 1, 2, and 3 are correct. The trapezius muscle forms the posterior boundary of the posterior triangle of the neck. The floor of the posterior triangle is covered by prevertebral fascia, under which lie splenius capitis, levator scapulae, scalenus medius and posterior muscles. Some authors include the scalenus anterior muscle as a part of the floor. After innervating the sternocleidomastoid muscle, the accessory (XIth cranial) nerve emerges from it and passes through the middle of the posterior triangle to supply the trapezius muscle.

2. E, All are correct. Proprioceptive sensory innervation to the sternocleidomastoid and trapezius muscles is from cervical nerves (C2 and C3) and (C3 and C4), respectively. Proprioceptive fibers to the temporomandibular joint are mainly from the auriculotemporal nerve, a branch of the mandibular division of the trigeminal (Vth cranial) nerve. Those to the masseter muscle are from the masseteric nerve, which is also a branch of the mandibular nerve; some of these proprioceptive fibers also innervate the temporomandibular joint.

3. **A, 1, 2, and 3 are correct.** The trapezius muscle arises from the medial one-third of the superior nuchal line, the external occipital protuberance, the ligamentum nuchae, the spines of all thoracic vertebrae, and the supraspinous ligaments. The trapezius inserts into the scapula, but does not originate from it.

4. **A, 1, 2, and 3 are correct.** The sternocleidomastoid muscle diagonally bisects each side of the neck into the anterior and posterior triangles. By drawing the clavicle and the sternum, and indirectly the ribs, upward, this muscle also serves as an accessory muscle of respiration. It arises by two heads, a sternal head from the anterior surface of the manubrium sterni, and a clavicular head from the upper surface of the medial one-third of the clavicle. This muscle inserts by one strong tendon into the lateral surface of the mastoid process and along the lateral half of the superior nuchal line of the occipital bone.

DEEP STRUCTURES OF THE NECK

FIVE-CHOICE COMPLETION QUESTIONS

DIRECTIONS: Each of the following questions or incomplete statements is followed by five suggested answers or completions. SELECT THE ONE BEST ANSWER in each case and then underline the appropriate letter at the lower right of each question.

1. OF THE FOLLOWING, THE MOST LIKELY CAUSE OF PAINLESS SWELLING IN THE MIDLINE OF THE NECK NEAR THE HYOID BONE IS:
    A. Branchial cyst
    B. Ectopic thyroid gland
    C. Thyroglossal duct cyst
    D. Undescended thymus gland
    E. Branchial vestige

    A B C D E

2. THE ISTHMUS OF THE THYROID GLAND IS USUALLY LOCATED:
    A. Anterior to the cricoid cartilage
    B. At the level of the 3rd to 5th cervical vertebrae
    C. On the cricothyroid membrane
    D. In front of the trachea
    E. In front of the thyroid cartilage

    A B C D E

3. PREGANGLIONIC SYMPATHETIC FIBERS TO THE CERVICAL SYMPATHETIC GANGLIA ARE USUALLY DERIVED FROM WHICH OF THE FOLLOWING SEGMENTS OF THE SPINAL CORD?
    A. C2-C6
    B. C1-T5
    C. T1-L3
    D. C8-T3
    E. T1-T4

    A B C D E

4. WHEN THE NUMBER OF GRAY RAMI COMMUNICANTES (GRC) TO THE CERVICAL SPINAL NERVES IS COMPARED WITH THE NUMBER OF WHITE RAMI COMMUNICANTES (WRC) IN THE REGION ABOVE THE LEVEL OF THE MIDDLE CERVICAL SYMPATHETIC GANGLION:

    A. The number of GRC is greater than the number of WRC when both are present
    B. The number of WRC is greater than the number of GRC when both are present
    C. Only the GRC are present
    D. Only the WRC are present
    E. No difference in number is observed

    A B C D E

SELECT THE ONE BEST ANSWER

5. MOST SYMPATHETIC PREGANGLIONIC FIBERS TO THE HEAD REGION SYNAPSE IN THE:

   A. Inferior cervical ganglion
   B. Middle cervical ganglion
   C. Cervicothoracic ganglion
   D. Superior cervical ganglion
   E. Intermediolateral cell column

   A B C D E

---------------------- ANSWERS, NOTES AND EXPLANATIONS ----------------------

1. C. The most likely cause of a midline swelling in the neck is a thyroglossal duct cyst. These cysts develop from remnants of the thyroglossal duct, the tubular structure that connected the thyroid gland to the tongue in the embryo. Thyroglossal duct cysts may be located anywhere from the base of the tongue near the foramen cecum to the isthmus of the thyroid gland, but they are usually found in the region of the hyoid bone. A midline swelling in the neck could be caused by an ectopic thyroid gland, but this is a very rare condition. In these cases, the thyroid gland fails to descend to its normal position in front of the trachea. Branchial cysts and vestiges are found in the side of the neck, anterior to the sternocleidomastoid muscle.

2. D. The isthmus connects the two lateral lobes of the thyroid gland and usually lies anterior to the second to fourth tracheal rings. This is well below the cricoid cartilage, which is at the level of the sixth cervical vertebra. The thyroid cartilage is located at the level of the third to the fifth cervical vertebrae. It is important to know these relations because of the sites of tracheotomy (tracheostomy), which usually involve the second and third tracheal rings.

3. E. Preganglionic sympathetic fibers to the cervical sympathetic ganglia have their cell bodies in the intermediolateral cell column of the spinal cord T1-T4, occasionally T5.

4. C. Preganglionic sympathetic fibers, derived from the intermediolateral cell column of the spinal cord of segments T1-L2 (or L3), pass via the ventral roots and white rami communicantes of the spinal nerves (T1-L2 or L3) to the paravertebral sympathetic ganglia. Preganglionic fibers to the middle and superior cervical sympathetic ganglia do not synapse in the thoracic ganglia, but ascend in the sympathetic trunk to the cervical ganglia, where they synapse. Postganglionic fibers pass through the gray rami communicantes to join each spinal nerve. From the superior and middle cervical ganglia, postganglionic fibers are contributed to spinal nerves C1-C6. As there is no intermediolateral cell column in the cervical spinal cord, no white rami communicantes from the cervical spinal cord to the sympathetic trunk are present.

5. D. Most preganglionic sympathetic fibers to the head region synapse in the superior cervical ganglion. From here postganglionic fibers, particularly their vascular branches, follow the internal and external carotid arteries and their branches form delicate perivascular plexuses. The superior cervical ganglion is over an inch (3 cm) long and lies in front of the lateral mass of the atlas and axis. The cervicothoracic (stellate) ganglion is irregular in shape, and much larger than the middle cervical ganglion. The middle cervical ganglion is a small, inconstant ganglion lying on the trunk medial to the sixth cervical vertebra. The inferior cervical ganglion lies behind the commencement of the vertebral artery.

MULTI-COMPLETION QUESTIONS

DIRECTIONS: In each of the following questions or incomplete statements, ONE OR MORE of the completions given is correct. At the lower right of each question, underline A if 1, 2 and 3 are correct; B if 1 and 3 are correct; C if 2 and 4 are correct; D if only 4 is correct; and E if all are correct.

1. WHICH OF THE FOLLOWING STATEMENTS ABOUT THE THYROID GLAND IS(ARE) CORRECT?
   1. Is an endocrine gland producing thyroxin
   2. Drains by the superior, middle, and inferior thyroid veins
   3. Has right and left lobes and occasionally a pyramidal lobe
   4. Receives its blood supply from the superior, middle, and inferior thyroid arteries

   A B C D E

2. WHICH STATEMENT(S) ABOUT THE THYROID CARTILAGE IS(ARE) CORRECT?
   1. Comprises two laminae
   2. Derived from the cartilage of the fourth branchial arch
   3. Produces the laryngeal prominence
   4. Has a synovial articulation with the cricoid cartilage

   A B C D E

3. WHICH STATEMENT(S) ABOUT THE JUGULODIGASTRIC LYMPH NODE IS (ARE) CORRECT?
   1. Lies anterolateral to the internal jugular vein
   2. Receives lymph from the tongue and palatine tonsil
   3. Lies below the posterior belly of the digastric muscle
   4. Lies superomedial to the angle of the mandible

   A B C D E

4. THE PHRENIC NERVE:
   1. Runs lateral to the carotid sheath
   2. Arises from cervical nerves 4, 5 and 6
   3. Runs anterior to the prevertebral fascia
   4. Runs vertically downward in front of the scalenus anterior muscle

   A B C D E

-------------------- ANSWERS, NOTES AND EXPLANATIONS --------------------

1. **A,** 1, 2, and 3 are correct. The thyroid gland consists of right and left lobes connected by an isthmus. If present, the pyramidal lobe ascends from the isthmus towards the hyoid bone. The thyroid gland is supplied by the superior thyroid artery (a branch of the external carotid) and the inferior thyroid artery (a branch of the thyrocervical trunk of the subclavian artery). Note there is no middle thyroid artery. Venous drainage of this gland is by the superior and middle thyroid veins to the internal jugular vein, and by the inferior thyroid vein to the brachiocephalic vein.

2. **E,** All are correct. The thyroid cartilage is usually derived from the cartilage of the fourth branchial arch. If the fifth is present, it may contribute to the formation. The anterior borders of the laminae of the thyroid cartilage, fused below, diverge above to form the superior thyroid notch. The thyroid cartilage has a synovial articulation with the cricoid cartilage. The movement of this joint is important in phonation.

3. **A,** 1, 2, and 3 are correct. The jugulodigastric lymph node is one of the superior deep cervical lymph nodes. It is located anterolateral to the internal jugular vein, below the posterior belly of the digastric muscle, and inferoposterior to the angle of the mandible. The jugulodigastric lymph node is mainly concerned with lymphatic drainage of the tongue and palatine tonsil.

In case of tonsillitis, for example, this gland may become palpable and tender.

4. **D, Only 4 is correct.** The phrenic nerve arises from the fourth cervical nerve, with contributions from the third and fifth. The roots of the phrenic nerve unite at the lateral border of the scalenus anterior at the level of the cricoid cartilage; it then descends vertically in front of the scalenus anterior deep to the prevertebral layer of the cervical fascia. This nerve runs posterior to the carotid sheath. The phrenic nerve is the sole motor nerve supply to the diaphragm. It also contains sensory fibers, most of which are pain fibers from the diaphragmatic pleura and peritoneum covering the superior and inferior surfaces of the central portion of the diaphragm, the mediastinal pleura, and the pericardium. This nerve also contains proprioceptive fibers for the diaphragm.

ROOT OF THE NECK

FIVE-CHOICE COMPLETION QUESTIONS

DIRECTIONS: Each of the following questions or incomplete statements is followed by five suggested answers or completions. SELECT THE ONE BEST ANSWER in each case and then underline the appropriate letter at the lower right of each question.

1. WHICH OF THE FOLLOWING STATEMENTS ABOUT THE SUBCLAVIAN ARTERY IS CORRECT?
    A. Its second part gives rise to three major branches
    B. It passes anterior to the anterior scalene muscle
    C. The three cords of the branchial plexus of nerves are designated with reference to this artery
    D. All of the above
    E. None of the above   A B C D E

2. AS IT CROSSES THE FIRST RIB, THE SUBCLAVIAN VEIN LIES:
    A. Between the clavicle and the middle scalene muscle
    B. Posterior to the anterior scalene muscle
    C. Between the middle and posterior scalene muscles
    D. Anterior to the anterior scalene muscle
    E. None of the above   A B C D E

3. WHICH STATEMENT ABOUT THE PHRENIC NERVE IS NOT TRUE?
    A. Consists mainly of the anterior primary ramus of C4
    B. Runs vertically over the middle scalene muscle
    C. Lies deep to the prevertebral fascia
    D. Passes into the mediastinum behind the subclavian vein
    E. None of the above   A B C D E

4. THE RIGHT LYMPHATIC DUCT USUALLY EMPTIES INTO THE VENOUS SYSTEM AT THE JUNCTION OF WHICH OF THE FOLLOWING VEINS?
    A. Right and left brachiocephalic veins
    B. Right and left subclavian veins
    C. Right external jugular and right internal jugular veins
    D. Right external jugular and right subclavian veins
    E. Right internal jugular and right subclavian veins   A B C D E

SELECT THE ONE BEST ANSWER

5. WHICH OF THE FOLLOWING STATEMENTS ABOUT A CERVICAL RIB IS INCORRECT?
   A. The most common type of accessory rib
   B. Attached to the seventh cervical vertebra
   C. May be associated with signs of pressure on the brachial plexus
   D. May be associated with signs of pressure on the subclavian artery
   E. None of the above

   A B C D E

-------------------- ANSWERS, NOTES AND EXPLANATIONS --------------------

1. E. The second part of the subclavian artery usually gives rise to the costocervical artery; occasionally it is a branch of the first part. The descending scapular (dorsal scapular) artery, which is usually a branch of the transverse cervical artery, occasionally arises from the second or third part of the subclavian artery. The second part is narrow and is located posterior to the anterior scalene muscle. The three cords of the brachial plexus of nerves are designated with reference to the second portion of the axillary artery where it is crossed by the pectoralis minor muscle.

2. D. The subclavian vein is the continuation of the axillary vein at the outer border of the first rib. As it crosses the first rib, it lies anterior to the anterior scalene muscle and the phrenic nerve.

3. B. The phrenic nerve consists mainly of the anterior ramus of C4 with some contributions from the anterior primary rami of C3 and C5. It runs deep to the prevertebral fascia over the anterior scalene muscle, and passes into the mediastinum deep to the subclavian vein.

4. E. The right lymphatic duct, about one cm in length, usually empties into the venous system at the junction of the right internal jugular and right subclavian veins after draining the right side of the head and neck, the right upper limb, and the right side of the thorax.

5. E. All the statements about a cervical rib are correct. This type of accessory rib is the most common type and is present in 0.5 to 1 per cent of persons. The presence of a cervical rib may be associated with clinical signs of pressure on the brachial plexus, or the subclavian artery, or both, causing "neurovascular compression syndromes".

MULTI-COMPLETION QUESTIONS

DIRECTIONS: In each of the following questions or incomplete statements, ONE OR MORE of the completions given is correct. At the lower right of each question, underline A if 1, 2 and 3 are correct; B if 1 and 3 are correct; C if 2 and 4 are correct; D if only 4 is correct; and E if all are correct.

1. THE SUBCLAVIAN ARTERIES ARISE FROM THE:
   1. Vertebral
   2. Aortic arch
   3. Common carotid
   4. Brachiocephalic

   A B C D E

2. THE RIGHT RECURRENT LARYNGEAL NERVE USUALLY LOOPS AROUND WHICH OF THE FOLLOWING STRUCTURES:
   1. Aortic arch
   2. Axillary artery
   3. Ligamentum arteriosum
   4. Subclavian artery

   A B C D E

A	B	C	D	E
1,2,3	1,3	2,4	only 4	all correct

3. THE RIGHT LYMPHATIC DUCT RECEIVES LYMPH FROM THE:
   1. Entire head and neck regions
   2. Entire right side of the body
   3. Entire thorax and right side of the upper limb
   4. Right side of the head, neck, thorax and upper limb   A B C D E

4. IN THE NECK REGION, THE BRACHIAL PLEXUS OF NERVES IS RELATED TO WHICH OF THE FOLLOWING MUSCLES?
   1. Anterior scalene   3. Middle scalene
   2. Superior belly of omohyoid 4. Sternohyoid   A B C D E

-------------------- ANSWERS, NOTES AND EXPLANATIONS --------------------

1. C. 2 and 4 are correct. The left subclavian artery arises directly from the arch of the aorta, whereas the right subclavian artery is a branch of the brachiocephalic arterial trunk.

2. D. Only 4 is correct. The right recurrent laryngeal nerve usually loops around the right subclavian artery, whereas the left loops under the arch of the aorta and the ligamentum arteriosum. These nerves then ascend in the neck to supply the larynx.

3. D. Only 4 is correct. The right lymphatic duct receives lymph from the right side of the head, neck, and thorax and the right upper limb via the jugular, subclavian, and bronchomediastinal lymphatic trunks.

4. B. 1 and 3 are correct. The brachial plexus of nerves is formed by the anterior primary rami of the fifth to the eighth cervical nerves and the greater part of the anterior primary ramus of the first thoracic nerve. It emerges between the anterior and middle scalene muscles.

E A R

       F I V E - C H O I C E   C O M P L E T I O N   Q U E S T I O N S

DIRECTIONS: Each of the following questions or incomplete statements is followed by five suggested answers or completions. SELECT THE ONE BEST ANSWER in each case and then underline the appropriate letter at the lower right of each question.

1. WHICH OF THE FOLLOWING STATEMENTS CONCERNING THE MIDDLE EAR IS INCORRECT?
   A. Located in the petrous portion of the temporal bone.
   B. Communicates with the nasopharynx through the auditory tube
   C. Limited inferiorly by the jugular fossa
   D. Bounded medially by the tympanic membrane
   E. Separated from the middle cranial fossa by the tegmen tympani   A B C D E

2. THE FLOOR OF THE MIDDLE EAR (TYMPANIC) CAVITY IS SUPERIOR TO THE:
   A. Carotid canal   D. Jugular fossa
   B. Facial canal   E. Internal acoustic meatus
   C. Auditory tube   A B C D E

SELECT THE ONE BEST ANSWER

3. WHICH OF THE FOLLOWING STATEMENTS ABOUT THE TYMPANIC MEMBRANE IS CORRECT?
   A. Forms a partition between the external and internal acoustic meatus
   B. Has a uniform pearly-white appearance
   C. Is crossed laterally by the chorda tympani nerve
   D. Is attached to the long process of the incus
   E. The umbo is located at the center of its lateral concave surface

   A B C D E

4. WHICH OF THE FOLLOWING STATEMENTS ABOUT THE TYMPANIC CAVITY IS CORRECT?
   A. The epitympanic recess lies superior to its medial wall
   B. Traversed by the greater petrosal nerve
   C. Related inferiorly to the external jugular vein
   D. Traversed by the chorda tympani nerve
   E. The sphenoid bone forms its roof

   A B C D E

5. WHICH OF THE FOLLOWING SEQUENCES IS CORRECT FOR SOUND TRANSMISSION FROM THE EXTERNAL ACOUSTIC MEATUS TO THE EIGHTH CRANIAL NERVE?
   A. Tympanic membrane, malleus, stapes, incus, and cochlea
   B. Tympanic membrane, malleus, incus, stapes, and cochlea
   C. Tympanic membrane, auditory tube, middle ear, and inner ear
   D. Tympanic membrane, incus, malleus, stapes, and cochlea
   E. Tympanic membrane, malleus, incus, stapes, and semicircular canal

   A B C D E

6. THE TWO MUSCLES IN THE MIDDLE EAR ARE INNERVATED BY WHICH OF THE FOLLOWING?
   A. Stapedius and tensor tympani by the trigeminal nerve
   B. Stapedius and tensor tympani by the facial nerve
   C. Stapedius by the trigeminal and tensor tympani by the facial nerve
   D. Stapedius by the facial and tensor tympani by the trigeminal nerves
   E. Stapedius by the facial and tensor tympani by the glossopharyngeal nerves

   A B C D E

7. THE LESSER PETROSAL NERVE:
   A. Contains mainly sympathetic fibers
   B. Contains taste fibers from the soft palate
   C. Contains fibers derived from the glossopharyngeal nerve
   D. Passes through the foramen spinosum into the infratemporal fossa
   E. Contains parasympathetic fibers destined for the pterygopalatine ganglion

   A B C D E

----------------------ANSWERS, NOTES AND EXPLANATIONS----------------------

1. D. The middle ear consists of the tympanic cavity, an air space, and the auditory ossicles, located in the petrous portion of the temporal bone. The tympanic cavity is separated superiorly from the middle cranial fossa by the tegmen tympani; inferiorly from the jugular bulb by the jugular fossa; laterally from the external acoustic meatus by the tympanic membrane; medially from the inner ear by the medial wall; and anteriorly from the internal carotid artery by the carotid canal. The tympanic cavity communicates anteriorly with the nasopharynx via the auditory (pharyngotympanic) tube, and posteriorly

with the mastoid air cells and the mastoid (tympanic) antrum through the aditus ad antrum. The bony layer of the tegmen tympani is very thin particularly during childhood. Thus, middle ear infection may spread through it into the fossa, causing serious meningitis.

2. D. The floor of the middle ear (tympanic) cavity is superior to the jugular fossa. The tympanic cavity communicates anteriorly with the semicanal for the tensor tympani muscle, and below with the nasopharynx through the auditory (pharyngotympanic) tube. Inferior to the tubal opening, the tympanic cavity is related anteriorly to the carotid canal lodging the internal carotid artery. On the medial wall, the bony facial canal forms a slightly curved prominence inferior to the prominence of the lateral semicircular canal. The opening of the internal acoustic meatus is on the posterior surface of the petrous portion of the temporal bone, superior to the jugular foramen. The seventh and eighth cranial nerves enter the internal acoustic meatus and run towards the medial wall of the tympanic cavity.

3. E. The tympanic membrane lies obliquely, forming a 55° angle with the floor of the external acoustic meatus; it separates the external acoustic meatus from the tympanic cavity. The lateral surface of the tympanic membrane is concave. The center of its concavity, called the umbo, is where the handle of the malleus is firmly attached to the medial surface of the tympanic membrane. The chorda tympani nerve is closely related to this portion of the tympanic membrane and crosses the tympanic cavity between the handle of the malleus and the long process of the incus. The surface of the tympanic membrane may be divided into three major parts: the pars flaccida, the pars tensa, and the cone of light. Because of differences in these parts, the tympanic membrane does not have a uniform pearly-white appearance.

4. D. A portion of the tympanic cavity, situated above the level of the tympanic membrane, is called the epitympanic recess; it contains the head of the malleus and the body and short crus of the incus. The tympanic cavity is covered by the tegmen tympani of the petrous portion of the temporal bone; its floor is formed by the jugular fossa, through which it is related to the internal jugular vein and its jugular bulb. The tympanic cavity is traversed by the chorda tympani nerve, medial to the tympanic membrane, and by the lesser petrosal nerve on its way to the otic ganglion.

5. B. For sound perception, air-borne sound impulses reach the tympanic membrane through the external acoustic meatus. Vibrations of the membrane are transmitted through the chain of auditory ossicles (malleus, incus, and stapes), to the fenestra vestibuli, causing vibrations of the perilymph in the internal ear. The internal ear consists of the osseous (bony) and the membranous labyrinths. The bony labyrinth consists of the bony vestibule, three semicircular canals, and the cochlea. They are lined by endosteum and contain perilymph in which is suspended the membranous labyrinth, containing the endolymph. The cochlea is the essential organ for hearing; the semicircular canals function in maintaining equilibrium.

6. D. The stapedius muscle, arising from the pyramidal eminence on the posterior wall of the tympanic cavity, inserts on the neck of the stapes. This muscle is innervated by the facial (seventh cranial) nerve. The tensor tympani muscle arises largely from the cartilaginous portion of the auditory tube and enters a semicanal for this muscle, superior to the auditory tube. Its tendon turns laterally around the processus cochleariformis and inserts on the handle of the malleus. This muscle is innervated by a branch of the nerve to the medial pterygoid muscle, which is a branch of the mandibular nerve (fifth cranial nerve, third division).

7. C. A branch of the glossopharyngeal nerve, the tympanic nerve, reaches the promontory on the medial wall of the tympanic cavity and ramifies to form the tympanic plexus of nerves. The lesser petrosal nerve is formed by a union of fibers from the tympanic plexus and leaves the tympanic cavity to enter the middle cranial fossa. Passing over the petrous portion of the temporal bone toward the foramen lacerum, this nerve leaves the cranial cavity and enters the infratemporal fossa through a foramen for the lesser petrosal nerve, or the foramen ovale, and terminates in the otic ganglion. Parasympathetic fibers in this nerve synapse in this ganglion and postganglionic fibers innervate the parotid gland via the auriculotemporal nerve.

MULTI-COMPLETION QUESTIONS

DIRECTIONS: In each of the following questions or incomplete statements, ONE OR MORE of the completions given is correct. At the lower right of each question, underline A if 1, 2 and 3 are correct; B if 1 and 3 are correct; C if 2 and 4 are correct; D if only 4 is correct; and E if all are correct.

1. WHICH OF THE FOLLOWING STATEMENTS IS (ARE) CORRECT?
   1. The joints between the auditory ossicles are synovial
   2. The tympanic plexus of nerves lies in the floor of the middle ear cavity
   3. The malleus is crossed medially by the chorda tympani nerve
   4. The tensor tympani muscle inserts into the long process of the incus   A B C D E

2. ON THE MEDIAL WALL OF THE MIDDLE EAR CAVITY, WHICH STRUCTURE(S) IS (ARE) RECOGNIZABLE?
   1. Aditus ad antrum
   2. Prominence of the bony facial canal
   3. Pyramidal eminence
   4. Oval window (fenestra vestibuli)   A B C D E

3. THE MIDDLE EAR CAVITY COMMUNICATES WITH WHICH OF THE FOLLOWING?
   1. Internal acoustic meatus   3. External acoustic meatus
   2. Nasopharynx   4. Mastoid air cells   A B C D E

4. THE MIDDLE EAR CAVITY IS RELATED ANTERIORLY TO WHICH OF THE FOLLOWING STRUCTURES?
   1. Auditory tube   3. Internal carotid artery
   2. Internal jugular vein   4. Semicircular canal   A B C D E

5. WHICH OF THE FOLLOWING NERVE COMPONENTS IS (ARE) CONTAINED IN THE FACIAL NERVE?
   1. Special sensory   3. General sensory
   2. Parasympathetic   4. Motor   A B C D E

6. WHICH STATEMENTS ABOUT THE TYMPANIC MEMBRANE ARE CORRECT?
   1. Forms a partition between the internal acoustic meatus and the tympanic cavity
   2. Covered externally by very thin skin
   3. Medially related to the incus
   4. Derived from three embryonic germ layers   A B C D E

A	B	C	D	E
1,2,3	1,3	2,4	only 4	all correct

7. WHICH STATEMENT(S) ABOUT THE TYMPANIC CAVITY IS (ARE) TRUE?
   1. An air space in the temporal bone
   2. Derived from the tubotympanic recess
   3. Communicates with the mastoid antrum
   4. Contains derivatives of the second branchial arch cartilage   A B C D E

-------------------- ANSWERS, NOTES AND EXPLANATIONS --------------------

1. **B.** <u>1 and 3 are correct</u>. The synovial articulation between the malleus and the incus (mallear joint) is a saddle joint, whereas the one between the incus and the stapes is a ball-and-socket type. The chain of the auditory ossicles serves as a lever system. The handle of the malleus attaches to the medial surface of the tympanic membrane. Medial to the malleus, the chorda tympani nerve crosses the tympanic membrane, passes through the petrotympanic fissure and enters the infratemporal fossa. The tympanic plexus of nerves lies under the mucous membrane over the promontory on the medial wall of the middle ear cavity. The tensor tympani muscle inserts into the handle (manubrium) of the malleus, and draws it medially to tense the tympanic membrane.

2. **C.** <u>2 and 4 are correct</u>. On the medial wall of the middle ear cavity, the following structures are recognizable from above downwards: (1) the prominence of the lateral semicircular canal of the inner ear; (2) the prominence of the bony facial canal; (3) the oval window (fenestra vestibuli) to which the base of the stapes is attached; (4) the promontory, formed by the basal turn of the cochlea and covered by the tympanic plexus of nerve; and (5) the round window (fenestra cochleae), closed by the secondary tympanic membrane. On the posterior wall of the middle ear cavity is the aditus ad antrum leading into the mastoid (tympanic) antrum, and below it is the pyramidal eminence containing the stapedius muscle.

3. **C.** <u>2 and 4 are correct</u>. The middle ear cavity communicates anteriorly with the nasopharynx through the auditory (pharyngotympanic) tube, and posteriorly with the mastoid (tympanic) antrum and the mastoid air cells. It communicates neither with the internal nor the external acoustic meatus. The mucous membrane lining the nasal cavity and nasopharynx is continuous with that in the middle ear cavity, the mastoid antrum, and the mastoid air cells. Thus, infections in the nasal cavity or the nasopharynx may spread into the tympanic cavity, causing middle ear infections (otitis media) and mastoiditis.

4. **B.** <u>1 and 3 are correct</u>. The middle ear cavity is related anteriorly to the internal carotid artery lodged in the carotid canal. Superior to this area, the auditory (pharyngotympanic) tube opens. The internal jugular vein is lodged in the jugular fossa, inferior to the floor of the middle ear cavity. The semicircular canals of the inner ear are located medial to this cavity, but the lateral semicircular canal forms a prominence on the medial wall of the tympanic cavity.

5. **E.** <u>All are correct</u>. The facial (seventh cranial) nerve consists of a large motor root and a small nervus intermedius, containing special sensory (taste), secretomotor (parasympathetic) and some pain (general sensory) fibers.

6. **C.** <u>2 and 4 are correct</u>. The oval tympanic membrane (ear drum) is placed obliquely at the inner end of the external acoustic (auditory) meatus. The external epidermal layer is derived from ectoderm of the first branchial groove; its core of connective tissue differentiates from mesenchyme. The tympanic membrane is medially related to the malleus, but not to the incus.

7. **A. 1,2, and 3 are correct.** The middle ear consists largely of an air space, the tympanic cavity, in the temporal bone. The tubotympanic recess of the embryo becomes the tympanic cavity and mastoid antrum. The stapes derived from the second branchial arch (Reichert's) cartilage is in the tympanic cavity, but the other cartilaginous derivative (the styloid process of the temporal bone) is not. The mucous membrane of the middle ear, derived from the tubotympanic recess, covers the structures in the tympanic cavity.

FIVE-CHOICE ASSOCIATION QUESTIONS

DIRECTIONS: Each group of questions below consists of a numbered list of descriptive words or phrases accompanied by a diagram with certain parts indicated by letters, or by a list of lettered headings. For each numbered word or phrase, SELECT THE LETTERED PART OR HEADING that matches it correctly. Then insert the letter in the space to the right of the appropriate number. Sometimes more than one numbered word or phrase may be correctly matched to the same lettered part or heading.

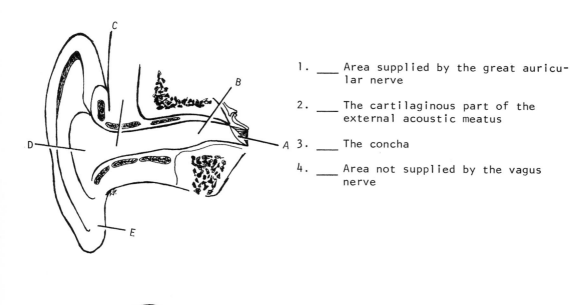

1. ___ Area supplied by the great auricular nerve

2. ___ The cartilaginous part of the external acoustic meatus

3. ___ The concha

4. ___ Area not supplied by the vagus nerve

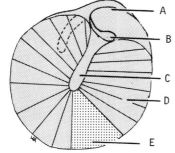

5. ___ The umbo

6. ___ Area attached to the lateral process of the malleus

7. ___ The pars tensa

8. ___ The cone of the light

9. ___ The pars flaccida

---------------------- ANSWERS, NOTES AND EXPLANATIONS ---------------------

1. E. The lobe of the auricle is innervated by the great auricular nerve, a sensory branch of the cervical plexus of nerves.

2. C. The external acoustic meatus is divided into a cartilaginous (outer or lateral third) (C) and an inner bony two thirds (B). It is about 2.5 cm long. For otoscopic examinations of the tympanic membrane, the external acoustic meatus is straightened by pulling the auricle upward and backward.

3. D. The deep concavity of the auricle leading to the external acoustic meatus is called the concha. It is the deepest depression in the auricle.

4. E. The vagus nerve innervates the concha (D), the external acoustic meatus (B and C) and the lateral surface of the tympanic membrane (A), but not the ear lobe (E).

5. C. The lateral surface of the tympanic membrane is concave; the center of its concavity is attached to the handle (manubrium) of the malleus. This area, called the umbo, is an important landmark in otoscopic examination.

6. B. The area attached to the lateral process of the malleus forms the (mallear) prominence.

7. D. The large part of the tympanic membrane, tensely fixed by attachment of the middle fibrous basis to the tympanic plate, is called the pars tensa. The tympanic membrane consists basically of three layers: (1) a middle fibrous basis attached to the tympanic plate of the temporal bone, except anterosuperiorly; (2) a lateral surface covered by epidermis; and (3) a medial surface lined by mucous membrane of the middle ear.

8. E. The anteroinferior quadrant of the tympanic membrane, called the cone of light, reflects light. In middle ear infection, this area, in particular, loses its ability to reflect light.

9. A. The anterosuperior part of the tympanic membrane, called the pars flaccida, is not attached to the tympanic plate of the temporal bone. Its fibrous basis is thinner and is limited by anterior and posterior mallear folds.

ORBITAL REGION

FIVE-CHOICE COMPLETION QUESTIONS

DIRECTIONS: Each of the following questions or incomplete statements is followed by five suggested answers or completions. SELECT THE ONE BEST ANSWER in each case and then underline the appropriate letter at the lower right of each question.

1. WHICH OF THE FOLLOWING IS NOT A COMPONENT OF THE EYELID?
   A. Orbital septum
   B. Muscular layer
   C. Conjunctiva
   D. Tarsal plates
   E. Ciliary muscle

   A B C D E

2. WHICH OF THE FOLLOWING NERVES IS NOT LOCATED WITHIN THE ORBIT?
   A. Infraorbital
   B. Trochlear
   C. Maxillary
   D. Abducens
   E. Oculomotor

   A B C D E

SELECT THE ONE BEST ANSWER

3. IF YOU SEE A PATIENT WITH A DROOPING EYELID, THE FIBERS OF WHICH OF THE FOLLOWING NERVES WOULD YOU SUSPECT ARE INVOLVED?
   A. Optic
   B. Oculomotor
   C. Trigeminal
   D. Sympathetic
   E. Parasympathetic

   A B C D E

4. WHICH OF THE FOLLOWING BONES DOES NOT CONTRIBUTE TO THE BONY ORBIT?
   A. Frontal
   B. Maxilla
   C. Lacrimal
   D. Temporal
   E. Zygomatic

   A B C D E

5. WHICH OF THE FOLLOWING OPENINGS IS NOT PRESENT IN THE ORBITAL CAVITY?
   A. Optic canal
   B. Pterygoid canal
   C. Ethmoidal foramen
   D. Supraorbital foramen
   E. Superior orbital fissure

   A B C D E

6. WHICH OF THE FOLLOWING GANGLIA IS LOCATED IN THE ORBIT?
   A. Pterygopalatine
   B. Geniculate
   C. Ciliary
   D. Semilunar
   E. Optic

   A B C D E

7. WHICH OF THE FOLLOWING STRUCTURES IS NOT FOUND WITHIN THE COMMON TENDINOUS RING OF THE OCULAR MUSCLES?
   A. Oculomotor nerve
   B. Trochlear nerve
   C. Abducens nerve
   D. Nasociliary nerve
   E. Central artery of retina

   A B C D E

8. FIBERS OF WHICH OF THE FOLLOWING NERVES SUPPLY THE DILATOR PUPILLAE MUSCLE?
   A. Motor fibers from the oculomotor
   B. Postganglionic parasympathetic fibers of the otic ganglion
   C. Sympathetic fibers from the superior cervical ganglion
   D. Postganglionic parasympathetic fibers of the ciliary ganglion
   E. None of the above

   A B C D E

9. WHICH OF THE FOLLOWING IS NOT A BRANCH OF THE OPHTHALMIC DIVISION OF THE TRIGEMINAL NERVE?
   A. Posterior ethmoidal
   B. Infratrochlear
   C. Long ciliary
   D. Infraorbital
   E. Lacrimal

   A B C D E

10. WHICH OF THE FOLLOWING MUSCLES IS SUPPLIED BY THE ABDUCENS NERVE?
    A. Medial rectus
    B. Lateral rectus
    C. Inferior rectus
    D. Inferior oblique
    E. Superior oblique

    A B C D E

------------------ ANSWERS, NOTES AND EXPLANATIONS --------------------

1. **E.** The eyelid (palpebra) consists of the following layers: skin, subcutaneous tissue, muscle, tarsofascial layer, and conjunctiva. The muscular layer consists mainly of the palpebral and lacrimal portions of the orbicularis oculi muscle, innervated mainly by the facial nerve, and the levator palpebrae superioris muscle, innervated by the oculomotor nerve and partly by sympathetic fibers. The tarsofascial layer consists of a dense fibrous tarsal plate, the tarsus, a membranous plane, and the orbital septum which is continuous with

the periosteum of the bones of the orbit. The orbital septum constitutes the superior and inferior palpebral fasciae. The tarsal glands are embedded in the posterior surface of the tarsus. The conjunctiva covering the inner surface of each eyelid (the palpebral conjunctiva) is reflected over the anterior portion of the sclera and cornea of the eyeball as the bulbar conjunctiva.

2. C. The maxillary nerve, the second division of the trigeminal, leaves the middle cranial fossa through the foramen rotundum and enters the pterygopalatine fossa, but not the orbit. The oculomotor, trochlear, and abducens (third, fourth and sixth cranial nerves, respectively) pass through the superior orbital fissure. The infraorbital nerve, a branch of the maxillary, enters the orbit through the inferior orbital fissure.

3. D. Drooping of the eyelid results from paralysis of the levator palpebrae superioris muscle. The deep fibers (smooth muscle) of this muscle, called the superior tarsus muscle, insert into the superior tarsus and are innervated by sympathetic fibers from the superior cervical sympathetic ganglion. Other parts (striated muscle) of the levator pulpebrae superioris muscle are innervated by the oculomotor (third cranial) nerve.

4. D. The wall of the bony orbit is formed by the frontal, zygomatic, maxillary lacrimal, ethmoid, and sphenoid bones. The temporal bone does not contribute to the formation of the orbit.

5. B. The following openings are present in the orbital cavity: supraorbital foramen, anterior and posterior ethmoidal foramina, optic canal, superior and inferior orbital fissures, zygomaticoorbital, zygomaticofacial, zygomaticotemporal, infraorbital groove (leading to the infraorbital foramen), and lacrimal fossa (leading to the nasolacrimal duct). The pterygoid canal opens into the pterygopalatine fossa.

6. C. The ciliary ganglion, a parasympathetic ganglion, is located between the optic nerve and the lateral rectus muscle, where preganglionic fibers of the oculomotor nerve synapse. Postganglionic fibers from this ganglion innervate the sphincter pupillae muscle of the iris and the ciliary muscle of accommodation. The optic ganglion, in the infratemporal fossa, and the pterygopalatine ganglion, in the pterygopalatine fossa, are both parasympathetic ganglia. The geniculate and semilunar ganglia are sensory ganglia for the facial and trigeminal nerves, respectively.

7. B. The common tendinous ring is the origin of all four recti muscles: the superior, inferior, medial, and lateral recti. It surrounds the optic canal and part of the superior orbital fissure. The two heads of the lateral rectus muscle originate from each side of the superior orbital fissure. Within this fissure, the superior ophthalmic vein, trochlear nerve, and frontal and lacrimal nerves (from the ophthalmic division of the trigeminal nerve) pass outside the common tendinous ring, whereas the oculomotor, nasociliary, and abducens nerves pass within the ring. The optic nerve and the central artery of the retina also pass within the common tendinous ring.

8. C. The dilator pupillae muscle is innervated by sympathetic fibers from the superior cervical sympathetic ganglion. Contraction of this muscle causes dilatation of the pupil (mydriasis).

9. D. The infraorbital nerve is a branch of the maxillary division of the trigeminal nerve. Both the ophthalmic and maxillary divisions are purely sensory. The ophthalmic division divides into three branches: lacrimal (smallest), frontal (largest), and nasociliary nerves. The frontal nerve divides into two

terminal branches: supraorbital and supratrochlear nerves. The branches of the nasociliary nerve are: the communicating branch of the ciliary ganglion, the long ciliary, the infratrochlear, and the anterior and posterior ethmoidal nerves.

10. **B.** The abducens nerve supplies only the lateral rectus muscle. The superior oblique muscle is supplied by the trochlear nerve; all other muscles of the eyeball (superior, inferior, and medial recti and the inferior oblique muscles) are innervated by the oculomotor nerve.

MULTI-COMPLETION QUESTIONS

DIRECTIONS: In each of the following questions or incomplete statements, ONE OR MORE of the completions given is correct. At the lower right of each question, underline A if 1, 2 and 3 are correct; B if 1 and 3 are correct; C if 2 and 4 are correct; D if only 4 is correct; and E if all are correct.

1. CONCERNING THE LACRIMAL GLAND:
   1. Its excretory ductules open into the lacrimal sac
   2. Receives its secretomotor innervation from a branch of the facial nerve
   3. Parasympathetic fibers to this gland synapse in the ciliary ganglion
   4. Sensory innervation of this gland is by way of a branch of the ophthalmic nerve

   A B C D E

2. THE SUPERIOR OPHTHALMIC VEIN COMMUNICATES DIRECTLY WITH WHICH OF THE FOLLOWING?
   1. Angular vein
   2. Inferior ophthalmic
   3. Cavernous dural venous sinus
   4. Pterygoid venous plexus

   A B C D E

3. THE OPHTHALMIC ARTERY GIVES RISE TO WHICH OF THE FOLLOWING BRANCHES?
   1. Lacrimal
   2. Anterior ethmoidal
   3. Posterior ethmoidal
   4. Central artery of the retina

   A B C D E

4. WHICH STATEMENT(S) ABOUT THE EYE IS (ARE) CORRECT?
   1. The retina is derived from the embryonic forebrain
   2. The lens is a derivative of the surface ectoderm of the embryo
   3. The embryonic optic fissures contain the hyaloid vessels
   4. The eyes develop from all three germ layers

   A B C D E

5. PARALYSIS OF THE SYMPATHETIC NERVE SUPPLY TO THE EYE RESULTS IN:
   1. Constriction of the pupil
   2. Slight drooping of the upper eyelid
   3. Retraction of the eyeball
   4. Vasoconstriction

   A B C D E

6. WHICH OF THE FOLLOWING STATEMENTS ABOUT THE ANTERIOR CHAMBER OF THE EYE IS (ARE) CORRECT?
   1. Located between the iris and lens
   2. Contains vitreous humor
   3. Lies anterior to the lens and ciliary process
   4. Situated anterior to the iris and pupil

   A B C D E

A	B	C	D	E
1,2,3	1,3	2,4	only 4	all correct

7. TEARS PASS THROUGH WHICH OF THE FOLLOWING ROUTES?
   1. Lacrimal papilla, lacrimal sac, inferior nasal meatus
   2. Lacrimal lake, lacrimal caruncle, lacrimal sac, frontonasal duct
   3. Lacrimal punctum, lacrimal canaliculus, lacrimal sac, nasolacrimal duct
   4. Lacrimal lake, lacrimal punctum, lacrimal caruncle, lacrimal sac, nasolacrimal duct

   A B C D E

ANSWERS, NOTES AND EXPLANATIONS

1. **C. 2 and 4 are correct.** The lacrimal gland receives its parasympathetic secretomotor innervation from the greater petrosal nerve. This nerve is derived from the facial (seventh cranial) nerve, the fibers of which synapse in the pterygopalatine ganglion. Its postganglionic fibers travel to the lacrimal gland via the zygomatic, zygomaticotemporal, and communicating branches of the maxillary nerve. They then proceed via the lacrimal nerve of the ophthalmic division of the trigeminal nerve. The lacrimal nerve provides the gland with sensory innervation. Six to twelve excretory ductules of the lacrimal gland open into the superior fornix of the conjunctiva.

2. **A. 1, 2, and 3 are correct.** The superior ophthalmic vein anastomoses anteriorly with the angular vein, a tributary of the facial vein, and ends posteriorly in the cavernous dural venous sinus. In the orbit, it also communicates with the inferior ophthalmic vein which ends in the pterygoid plexus of veins. The facial vein also communicates with the pterygoid plexus of veins via its tributary, the deep facial vein. Because the superior ophthalmic vein has no valves, blood can flow in either direction. In addition, because of its communications with the cavernous dural venous sinus and the pterygoid venous plexus, possible spread of infection via this vein between the inside and the outside of the cranium is of clinical importance. For this reason, the territory of the facial vein, the area around the nose and the upper lip, is sometimes called the "danger area" of the face.

3. **E. All are correct.** In addition to the vessels listed, the ophthalmic artery also gives rise to the long and short posterior ciliary, supraorbital, supratrochlear, and palpebral arteries. The central artery of the retina is of particular clinical importance because it is an end artery, and represents the sole blood supply to the retina. Its occlusion results in complete blindness.

4. **A. 1, 2, and 3 are correct.** The eye develops from three sources: neuroectoderm, surface ectoderm, and mesoderm. Note: only two germ layers are involved. The retina develops from the optic cup, a derivative of the forebrain. The lens develops from the lens vesicle, a derivative of the surface ectoderm. The optic (choroidal) fissures contain the hyaloid vessels, the distal portions of which eventually degenerate, but their proximal portions persist as the central artery and vein of the retina.

5. **A. 1, 2, and 3 are correct.** Interruption of the sympathetic nerve supply to the eye (as a result of injury to the cervical sympathetic trunk or spinal cord segments T1-3) leads to the following: constriction of the pupil (miosis) due to paralysis of the dilator pupillae muscle; slight drooping of the upper eyelid (ptosis) due to paralysis of the smooth muscle component of the levator palpebrae superioris muscle; retraction of the eyeball into the orbit (enophthalmos) also due to paralysis of the tarsal muscle; and vasodilation in the facial and cervical regions (Horner's syndrome).

6. **D.** <u>Only 4 is correct</u>. The anterior chamber of the eye, containing aqueous humor, lies behind the cornea and anterior to the iris and pupil. The posterior chamber of the eye lies posterior to the iris and in front of the lens, the suspensory ligament of the lens, and the ciliary process. It also contains aqueous humor which circulates into the anterior chamber of the eye through the edge of the pupil. The vitreous humor (body), a transparent semi-gelatinous material, is contained in the vitreous chamber posterior to the lens and the ciliary processes. Accumulation of aqueous humor, often due to obstruction of the canal of Schlemm (the scleral venous sinus), results in glaucoma. The loss of transparency in the cornea and the lens results in an opaque cornea and cataract, respectively.

7.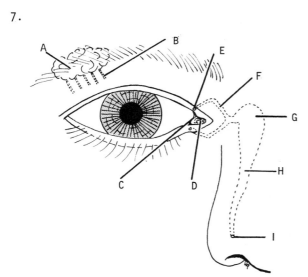

B. <u>1 and 3 are correct</u>. Tears produced by the lacrimal gland (A), located posterior (deep) to the superolateral orbital margin, pass through its excretory ductules (B) into the lateral part of the superior conjunctival fornix. By blinking movements, tears are spread evenly over the eyeball. At the medial corner of the eye, tears accumulate in the area of the lacrimal lake (D) bordered by the semilunar fold of the conjunctiva (C). Opposite the semilunar fold, the lacrimal puncta (E) open at the summits of the lacrimal papillae of the superior and inferior palpebrae. Beginning at the lacrimal punctum, tears pass through the lacrimal canaliculus (F), the lacrimal sac (G), the nasolacrimal duct (H), and enter the inferior nasal meatus (I).

FIVE-CHOICE ASSOCIATION QUESTIONS

DIRECTIONS: Each group of questions below consists of a numbered list of descriptive words or phrases accompanied by a diagram with certain parts indicated by letters, or by a list of lettered headings. For each numbered word or phrase, SELECT THE LETTERED PART OR HEADING that matches it correctly. Then insert the letter in the space to the right of the appropriate number. Sometimes more than one numbered word or phrase may be correctly matched to the same lettered part or heading.

ASSOCIATION QUESTIONS

WHICH GROUP OF MUSCLES IS RESPONSIBLE FOR MOVEMENT OF THE PUPIL TO THE POSITIONS ILLUSTRATED IN THE DIAGRAMS? NOTE: THE ARROWS INDICATE THE MEDIAL SIDE.

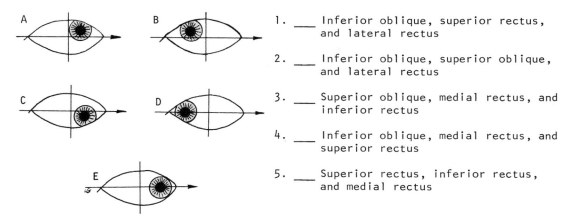

1. ___ Inferior oblique, superior rectus, and lateral rectus

2. ___ Inferior oblique, superior oblique, and lateral rectus

3. ___ Superior oblique, medial rectus, and inferior rectus

4. ___ Inferior oblique, medial rectus, and superior rectus

5. ___ Superior rectus, inferior rectus, and medial rectus

------------------- ANSWERS, NOTES AND EXPLANATIONS ---------------------

1. **B.** Movement of the pupil from the center of the eye to the superolateral corner of the eye is accomplished by contraction of the inferior oblique with the aid of the superior and lateral rectus muscles.

2. **D.** Movement of the pupil from the center of the eye to the lateral angle of the eye is caused principally by contraction of the lateral rectus muscle. Contraction of the inferior and superior oblique muscles maintains the eyeball in a horizontal plane.

3. **C.** Movement of the pupil from the center of the eye to the inferomedial corner is the result of contraction of the superior olbique, aided by the medial and inferior rectus muscles.

4. **A.** Moving the pupil upwards and medially results from contraction of the inferior oblique, medial rectus, and superior rectus muscles.

5. **E.** Movement of the pupil medially (nasalward) is the result of contraction of the medial rectus, aided by the superior and inferior rectus muscles.

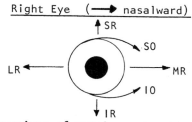

Directions of muscle contractions
SR - superior rectus
LR - lateral rectus
IR - inferior rectus
MR - medial rectus

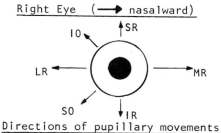

Directions of pupillary movements
SO - superior oblique
IO - inferior oblique

DEEP STRUCTURES OF THE FACE

FIVE-CHOICE COMPLETION QUESTIONS

DIRECTIONS: Each of the following questions or incomplete statements is followed by five suggested answers or completions. SELECT THE ONE BEST ANSWER in each case and then underline the appropriate letter at the lower right of each question.

1. THE BONY ROOF OF THE INFRATEMPORAL FOSSA IS FORMED BY THE:
   A. Tympanic plate of the temporal bone
   B. Zygomatic arch
   C. Tuberosity of the maxilla
   D. Lateral pterygoid plate
   E. Greater wing of the sphenoid bone   A B C D E

2. WHICH OF THE FOLLOWING IS NOT A BRANCH OF THE MAXILLARY ARTERY?
   A. Superficial temporal   D. Deep temporal
   B. Middle meningeal   E. Buccal
   C. Sphenopalatine   A B C D E

3. PARASYMPATHETIC FIBERS TO THE PTERYGOPALATINE GANGLION ARE CARRIED BY THE _____ NERVE.
   A. Glossopharyngeal   D. Deep petrosal
   B. Greater petrosal   E. Oculomotor
   C. Lesser petrosal   A B C D E

4. THE INFRATEMPORAL FOSSA CONTAINS THE FOLLOWING STRUCTURE(S):
   A. Pterygoid plexus of vein D. Chorda tympani
   B. Middle meningeal artery   E. All of the above
   C. Maxillary artery   A B C D E

5. WHICH OF THE FOLLOWING MUSCLES INSERTS ON THE CORONOID PROCESS OF THE MANDIBLE?
   A. Masseter   D. Lateral pterygoid
   B. Temporalis   E. None of the above
   C. Medial pterygoid   A B C D E

6. SYMPATHETIC FIBERS TO THE MIDDLE MENINGEAL ARTERY ARE POST-GANGLIONIC FIBERS WITH THEIR CELL BODIES LOCATED IN THE _____ GANGLION.
   A. Otic   D. Pterygopalatine
   B. Semilunar   E. Superior cervical
   C. Geniculate   A B C D E

7. WHICH OF THE FOLLOWING DOES NOT COMMUNICATE WITH THE PTERYGOID VENOUS PLEXUS?
   A. Sphenoidal emissary vein D. Deep facial vein
   B. Inferior ophthalmic vein E. Maxillary vein
   C. Cavernous venous sinus   A B C D E

-------------------- ANSWERS, NOTES AND EXPLANATIONS --------------------

1. E. The bony roof of the infratemporal fossa is formed by the infratemporal surface of the greater wing of the sphenoid bone. This fossa is bounded anteriorly by the posterior wall of the maxilla (tuberosity), posteriorly by the styloid process, medially by the lateral pterygoid plate of the sphenoid bone,

and laterally by the ramus and coronoid process of the mandible. The zygomatic arch is lateral to the infratemporal fossa.

2. **A.** The maxillary artery, a terminal branch of the external carotid artery, arises in the parotid gland posterior to the neck of the mandible. The other terminal branch is the superficial temporal. The course and branches of the maxillary artery may be described in three parts: mandibular, pterygoid, and pterygopalatine (or terminal). The middle meningeal artery is a clinically important branch arising from the mandibular portion; other branches are the: deep auricular, anterior tympanic, accessory meningeal, and inferior alveolar. Branches of the pteryoid part are the: masseteric, anterior and posterior deep temporal, pterygoid, and buccal arteries. The pterygopalatine part gives rise to the: posterior superior alveolar, infraorbital, descending palatine, pharyngeal, and sphenopalatine arteries, and the artery of the pterygoid canal.

3. **B.** Preganglionic parasympathetic fibers from the nervus intermedius of the facial (seventh cranial) nerve are carried by the greater petrosal nerve and synapse in the pterygopalatine ganglion. Fibers derived from the oculomotor nerve synapse in the ciliary ganglion, whereas those from the glossopharyngeal nerve reach the otic ganglion via the lesser petrosal nerve. The deep petrosal nerve carries postganglionic sympathetic fibers and, together with the greater petrosal nerve, constitutes the nerve of the pterygoid canal.

4. **E.** The infratemporal fossa contains the lower portion of the temporalis muscle, the lateral and medial pterygoid muscles, the maxillary artery and its branches, the pterygoid plexus of veins, the mandibular division of the trigeminal nerve and its branches, the chorda tympani nerve, the otic ganglion, and the buccal fat pads. The middle meningeal is an important branch of the maxillary artery.

5. **B.** The only muscle attached to the coronoid process of the mandible is the temporalis muscle. The masseter and the medial and lateral pterygoid muscles are also muscles of mastication; they attach to the mandible at other sites.

6. **E.** Sympathetic fibers of the perivascular plexus surrounding the middle meningeal artery serve as vasoconstrictors; their cell bodies are located in the superior cervical ganglion. The otic and pterygopalatine ganglia are both parasympathetic, whereas the semilunar and geniculate ganglia are sensory.

7. **C.** The pterygoid venous plexus is located in the infratemporal fossa, partly between the temporalis and lateral pterygoid muscles, and partly between the lateral and medial pterygoid muscles. This plexus communicates with the cavernous venous sinus in the cranial cavity, thus it is of special significance in the spread of infection. However, its communication with the cavernous venous sinus is indirect as follows, via: (1) the emissary veins through the sphenoidal emissary foramina or the foramen ovale; (2) the deep facial, angular, and superior ophthalmic veins; and (3) the inferior ophthalmic vein. There are numerous tributaries of the pterygoid venous plexus which also connect with the tributaries of the maxillary vein and the retromandibular vein.

MULTI-COMPLETION QUESTIONS

DIRECTIONS: In each of the following questions or incomplete statements, ONE OR MORE of the completions given is correct. At the lower right of each question, underline <u>A if 1, 2, and 3</u> are correct; <u>B if 1 and 3</u> are correct; <u>C if 2 and 4</u> are correct; <u>D if only 4</u> is correct; and <u>E if all</u> are correct.

A	B	C	D	E
1,2,3	1,3	2,4	only 4	all correct

1. THE TEMPOROMANDIBULAR JOINT IS WHICH OF THE FOLLOWING TYPES?
   1. Ball and socket
   2. Hinge
   3. Pivot
   4. Gliding

   A B C D E

2. THE MAXILLARY NERVE IS SENSORY AND SUPPLIES THE:
   1. Skin over the upper jaw
   2. Anterior two-thirds of the tongue
   3. Mucous membrane of the soft palate
   4. Mandibular process of the first branchial arch

   A B C D E

3. YOU WOULD HAVE DIFFICULTY PROTRUDING YOUR LOWER JAW IF THE FOLLOWING MUSCLE(S) WAS (WERE) IMPAIRED:
   1. Mylohyoid
   2. Buccinator
   3. Temporalis
   4. Lateral pterygoid

   A B C D E

4. WHICH OF THE FOLLOWING STATEMENTS CONCERNING THE TEMPORAL FOSSA IS (ARE) CORRECT?
   1. Bounded by the temporal line
   2. Contains the pterion in its floor
   3. Its floor is formed by four bones
   4. Contains the temporalis muscle

   A B C D E

5. THE MUSCLES OF MASTICATION INCLUDE THE:
   1. Lateral pterygoid
   2. Buccinator
   3. Temporalis
   4. Mylohyoid

   A B C D E

------------------- ANSWERS, NOTES AND EXPLANATIONS ---------------------

1. **C.** <u>2 and 4 are correct</u>. The temporomandibular joint is the articulation between the temporal bone of the cranium and the mandible. It is a complex joint because its articular cavity is divided by an articular disc into upper and lower compartments, each having a separate synovial cavity. In the upper compartment, the upper surface of the articular disc articulates with the posterior slope of the articular eminence and the mandibular fossa of the temporal bone to form a gliding joint. The gliding movement of the upper joint results from contraction of the lateral pterygoid muscle. In the lower compartment, the inferior surface of the articular disc articulates with the head of the condyle to form a hinge joint.

2. **B.** <u>1 and 3 are correct</u>. The maxillary nerve, the second division of the trigeminal (fifth cranial) nerve, arises from the semilunar ganglion. In addition to supplying the skin over the upper jaw, it supplies the maxillary teeth, the gingiva, the mucous membrane of the roof of the mouth and of the palate. The sensory nerve supply to the mucosa of almost the entire two-thirds of the tongue is from the lingual branch of the mandibular division of the trigeminal nerve. This division also supplies the mandibular process of the first branchial arch in the embryo.

3. **D.** <u>Only 4 is correct</u>. Protraction of the mandible, i.e., moving the lower jaw forward, is produced by the lateral (external) pterygoid muscle. It is attached to the neck of the mandible and the articular disc of the temporomandibular joint. Also assisting in protraction of the mandible are the medial pterygoid and the superficial fibers of the masseter muscles. The temporalis muscle is involved in closing the mouth and retracting the mandible. The mylohyoid muscle, forming the floor of the mouth, assists in opening the mouth.

4. E. All are correct. The temporal fossa is bounded superiorly by the temporal line, to which the temporal fascia is attached, and inferiorly by the zygomatic arch below. This fossa contains the temporalis, a fan-shaped muscle arising from the periosteum of the fossa and the temporal fascia. The temporalis muscle inserts by a thick tendon into the anterior border and medial surface of the coronoid process of the mandible. The floor of the fossa is formed by four bones: parietal, frontal, greater wing of the sphenoid, and squamous part of the temporal. The area where these bones approach each other (about 3 cm behind the external angular process of the orbit) is called the pterion: it overlies the anterior branch of the middle meningeal artery on the internal surface of the skull and the stem of the lateral sulcus of the brain. Thus, skull fractures at the pterion may result in rupture of the middle meningeal artery and extradural hemorrhage.

5. B. 1 and 3 are correct. The muscles of mastication are the temporalis, masseter, lateral and medial pterygoid; they are all innervated by branches of the mandibular nerve. The buccinator muscle is considered to be an accessory muscle of mastication because it keeps the food between the occlusal surfaces of the maxillary and mandibular teeth and out of the vestibule. The buccinator, like all muscles of facial expression, is innervated by the facial nerve. The mylohyoid muscle, a suprahyoid muscle, is innervated by a branch of the mandibular nerve.

NOSE AND PARANASAL SINUSES

FIVE-CHOICE COMPLETION QUESTIONS

DIRECTIONS: Each of the following questions or incomplete statements is followed by five suggested answers or completions. SELECT THE ONE BEST ANSWER in each case and then underline the appropriate letter at the lower right of each question.

1. WHICH OF THE FOLLOWING STATEMENTS IS INCORRECT?
    A. The maxillary air sinus may become infected from an infected tooth
    B. The trigeminal nerve supplies general sensory innervation to the nasal mucosa
    C. The nasolacrimal duct opens in the anterior part of the middle nasal meatus
    D. The facial nerve supplies parasympathetic innervation to the nasal mucosa
    E. The nasal mucosa receives a rich blood supply   A B C D E

2. WHICH OF THE FOLLOWING NERVES DOES NOT SUPPLY GENERAL SENSORY INNERVATION TO THE NASAL CAVITY?
    A. Posterior superior nasal   D. Greater palatine
    B. Posterior ethmoidal   E. Nasopalatine
    C. Anterior ethmoidal   A B C D E

-------------------- ANSWERS, NOTES AND EXPLANATIONS --------------------

1. C. The nasolacrimal duct opens in the anterior part of the inferior nasal meatus. The floor of the maxillary air sinus is formed by the alveolar process of the maxilla and is often marked by small elevations formed by the molar teeth. Particularly when this bony floor is thin, infection of a molar may spread into the maxillary air sinus. The nasal mucosa receives a rich

blood supply from various sources including: the anterior and posterior ethmoidal from the ophthalmic artery; the sphenopalatine and greater palatine from the maxillary artery; and the superior labial from the facial artery. The mucous lining receives its general sensory innervation from the trigeminal nerve and parasympathetic fibers from the greater petrosal of the facial nerve.

2. **B.** The posterior ethmoidal nerve is often absent, but when present, it supplies general sensory innervation to the sphenoidal air sinus and the ethmoidal air cells. General sensory innervation to the nasal cavity is provided by: (1) branches from the ophthalmic division of the trigeminal, the anterior ethmoidal nerve, a branch of the nasociliary, supplying the anterior part of the nasal septum and the lateral wall of the nasal cavity; (2) branches from the maxillary division of the trigeminal, (a) the greater palatine nerve, via its posterior inferior nasal branches, supplies the inferior concha and the inferior and middle nasal meatuses, (b) the nasopalatine and posterior superior nasal nerves innervate the superior and middle nasal conchae, the nasal septum, and the incisive canal, and (c) the anterior superior alveolar nerve supplies the anterior part of the inferior meatus and the floor of the nasal cavity.

MULTI-COMPLETION QUESTIONS

DIRECTIONS: In each of the following questions or incomplete statements, ONE OR MORE of the completions given is correct. At the lower right of each question, underline A if 1, 2, and 3 are correct; B if 1 and 3 are correct; C if 2 and 4 are correct; D if only 4 is correct; and E if all are correct.

1. TO WHICH OF THE FOLLOWING AREAS IS THE OLFACTORY NERVE DISTRIBUTED?
   1. Roof of the nasal cavity
   2. Nasal septum
   3. Superior nasal concha
   4. Nasal vestibule   A B C D E

2. WHICH OF THE FOLLOWING PARANASAL SINUSES OPEN(S) INTO THE MIDDLE NASAL MEATUS?
   1. Maxillary
   2. Middle ethmoidal
   3. Anterior ethmoidal
   4. Posterior ethmoidal   A B C D E

3. WHICH OF THE FOLLOWING ARTERIES SUPPLY(IES) THE NASAL CAVITY?
   1. Anterior ethmoidal
   2. Sphenopalatine
   3. Superior labial
   4. Greater palatine   A B C D E

4. LYMPH FROM THE POSTERIOR PORTION OF THE NASAL CAVITY DRAINS INTO WHICH OF THE FOLLOWING LYMPH NODES?
   1. Superior deep cervical
   2. Retropharyngeal
   3. Jugulodigastric
   4. Submandibular   A B C D E

5. WHICH OF THE FOLLOWING BONES CONTRIBUTE(S) TO THE FORMATION OF THE NASAL SEPTUM?
   1. Maxillary
   2. Sphenoid
   3. Palatine
   4. Vomer   A B C D E

------------------ ANSWERS, NOTES AND EXPLANATIONS --------------------

1. **A.** <u>1, 2, and 3 are correct</u>. The olfactory nerve is distributed to the olfactory region of the nasal mucosa, located in the roof of the nasal cavity and on the superior nasal concha and upper third of the nasal septum. Collective-

ly, these nerve fibers are called the olfactory nerve; they pierce the cribriform plate of the ethmoid bone to end in the olfactory bulb.

2. **A.** <u>1, 2, and 3 are correct</u>. The frontonasal duct from the frontal sinus opens at the most anterior aspect of the semilunar hiatus in the middle nasal meatus. The anterior ethmoidal air cells open into the most anterior aspect of the hiatus, and the maxillary sinus opens posterior to it. The middle ethmoidal air cells open at the summit of the bulla ethmoidalis. The posterior ethmoidal air cells open into the superior nasal meatus and the sphenoid sinus into the sphenoethmoidal recess.

3. **E.** <u>All are correct</u>. The nasal cavity receives a rich blood supply from various sources. The sphenopalatine, a branch of the maxillary artery, and the anterior ethmoidal, a branch of the ophthalmic, are the most important arteries supplying the nasal cavity. In addition, the nasal cavity is supplied by the septal branch of the superior labial artery from the facial artery, the nasal branch of the greater palatine artery via the incisive foramen, the posterior superior lateral nasal branches of the maxillary artery, and the lateral inferior nasal branches of the greater palatine artery. Nasal bleeding most often occurs at the junction of the septal branches of the sphenopalatine and superior labial arteries. Nasal bleeding may also occur from the venous plexus located at the anteroinferior region of the nasal septum.

4. **A.** <u>1, 2, and 3 are correct</u>. Lymphatic vessels of the posterior nasal areas drain posteriorly into the superior deep cervical, the jugulodigastric, and the retropharyngeal nodes. Lymph from the anterior nasal area drains into the submandibular nodes.

5. **E.** <u>All are correct</u>. The nasal septum is formed by the perpendicular plate of the ethmoid bone, the vomer, and the septal cartilage. In addition, processes of the palatine, maxillary, frontal, sphenoid, and nasal bones make minor contributions. A deflection of the nasal septum ("deviated septum") is common and may cause blockage of the nasal passage on the affected side.

6. **A.** <u>1, 2, and 3 are correct</u>. The lateral bony wall of the nasal cavity is formed by the following bones: (1) superior and middle conchae of the ethmoid bone; (2) inferior concha as a separate bone; (3) nasal bone; (4) frontal process and the nasal surface of the maxilla; (5) lacrimal bone; (6) perpendicular plate of the palatine bone; and (7) medial pterygoid plate of the sphenoid bone.

FIVE-CHOICE ASSOCIATION QUESTIONS

DIRECTIONS: Each group of questions below consists of a numbered list of descriptive words or phrases accompanied by a diagram with certain parts indicated by letters, or by a list of lettered headings. For each numbered word or phrase, SELECT THE LETTERED PART OR HEADING that matches it correctly. Then insert the letter in the space to the right of the appropriate number. Sometimes more than one numbered word or phrase may be correctly matched to the same lettered part or heading.

ASSSOCIATION QUESTIONS

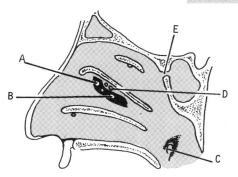

1. ___ Opening of a paranasal sinus in close relationship to the hypophysis

2. ___ Opening of the largest paranasal sinus

3. ___ Opening of a paranasal sinus located deep to the superciliary arch

4. ___ Opening of the maxillary air sinus

------------------ ANSWERS, NOTES AND EXPLANATIONS --------------------

1. E. The sphenoidal sinus is situated in the body of the sphenoid bone, therefore it is in close relationship to the hypophysis (pituitary gland). This gland is located in the hypophyseal fossa of the sella turcica. The opening of the sphenoidal sinus, located in the superior aspect of its anterior wall, leads into the sphenoethmoidal recess of the nasal cavity. Because of this location of the orifice, accumulated fluids in inflammatory conditions are not easily drained in the upright position.

2. B. The largest paranasal sinus, the maxillary sinus, is present at birth, whereas other paranasal sinuses are sometimes rudimentary or nonexistent at this time. The maxillary sinus is situated in the maxilla and opens into the posterior aspect of the hiatus semilunaris in the middle nasal meatus. Its orifice is located at the anterosuperior aspect of the medial wall of the sinus; thus, it is not easy to drain this sinus when the head is upright. The floor of the maxillary sinus is separated from the maxillary teeth by only a thin sheath of alveolar bone and the mucous membrane; therefore, infection of this sinus (sinusitis) may cause apparent toothache. Similarly, infection of a maxillary tooth may spread into the maxillary sinus, or vice versa.

3. A. The frontal sinus is located within the frontal bone, deep to the superciliary arch on each side of the medial line. A septum separates the right and left sinuses, but it may often be deviated to one side. The frontal sinus drains by the frontonasal duct which opens into the anterior part of the hiatus semilunaris in the middle nasal meatus.

4. B. The maxillary sinus opens into the posterior part of the hiatus semilunaris in the middle nasal meatus. Anterior to it is the opening of the anterior ethmoidal sinus. The middle ethmoidal sinus opens at the summit of the bulla ethmoidalis (D); the posterior ethmoidal sinus opens into the superior nasal meatus. C is the orifice of the auditory tube.

ORAL REGION

F I V E - C H O I C E   C O M P L E T I O N   Q U E S T I O N S

DIRECTIONS: Each of the following questions or incomplete statements is followed by five suggested answers or completions. SELECT THE ONE BEST ANSWER in each case and then underline the appropriate letter at the lower right of each question.

SELECT THE ONE BEST ANSWER

1. CONCERNING THE ORAL CAVITY, WHICH STATEMENT IS CORRECT?
   A. The oral cavity comprises the vestibule, oral cavity proper, and oropharynx
   B. The oral cavity is bounded anteriorly and laterally by the alveolar arches, teeth, gingivae, and palatine tonsils.
   C. The total number of deciduous teeth is 16 in each jaw
   D. The oral cavity communicates with the pharynx through the faucial isthmus
   E. The floor of the mouth is supported by the oral diaphragm consisting of the geniohyoid and mylohyoid muscles on each side
   A B C D E

2. WHICH OF THE FOLLOWING STATEMENTS ABOUT THE HARD PALATE IS INCORRECT?
   A. Formed by the maxilla and palatine bones and a lining mucous membrane
   B. Incisive papilla indicates the site of the incisive foramen
   C. Has rugae that aid in gripping food
   D. The greater palatine vessels and the nasopalatine nerve pass through the greater palatine foramen
   E. Has mucous glands innervated by the facial nerve
   A B C D E

3. THE MOST COMMON CONGENITAL MALFORMATION OF THE FACE, OR PALATE, OR BOTH, IS:
   A. Cleft lip with cleft palate   D. Median cleft lip
   B. Bilateral cleft lip   E. Unilateral cleft palate
   C. Unilateral cleft lip
   A B C D E

-------------------- ANSWERS, NOTES AND EXPLANATIONS ---------------------

1. D. The oral cavity communicates with the oropharynx through the faucial (oropharyngeal) isthmus, demarcated by the palatoglossal arch. The oral cavity comprises the vestibule and the oral cavity proper. It is bounded anteriorly and laterally by the alveolar arches, teeth, and gingivae. The palatine tonsil is located in the tonsillar fossa, formed by the palatoglossal and palatopharyngeal arches, which form the lateral wall of the oropharynx. The floor of the mouth is supported by the oral diaphragm consisting of the mylohyoid muscle. There are 10 deciduous and 16 permanent teeth in each jaw. The permanent teeth in one half of each jaw consist of two incisors, one cuspid, two bicuspids (premolars) and three molars; in the deciduous dentition, there are no bicuspid or third molar (wisdom) teeth.

2. D. The greater palatine foramen is located in the posterior part of the hard palate and transmits the greater palatine artery, vein, and nerve. The nasopalatine nerve reaches the incisive foramen, indicated by the incisive papilla, located anterior to the palatine raphe. Posterior to the incisive papilla, several transverse palatine folds (rugae) extend laterally and aid in gripping food against the tongue during mastication. The palate separates the oral cavity and the nasal cavity and consists of the hard palate (anterior two-thirds) and the soft palate (posterior one-third). The bony hard palate consists of the palatine processes of the maxillae and the horizontal plates of the palatine bones. The palate is lined by mucous membrane containing a number of mucous palatine glands. These glands are innervated by parasympathetic fibers derived from the facial (seventh cranial) nerve which are carried by the greater petrosal nerve.

3. **C.** Cleft lip is an important abnormality because it is common and affects the face. It occurs about once in 900 births, but there is a wide racial variation. Unilateral cleft lip results from failure of the maxillary prominence (process) on the affected side to merge with the intermaxillary segment (merged medial nasal prominences) during the seventh week of development. Most cases have a multifactorial origin.

MULTI-COMPLETION QUESTIONS

DIRECTIONS: In each of the following questions or incomplete statements, ONE OR MORE of the completions given is correct. At the lower right of each question, underline A if 1, 2, and 3 are correct; B if 1 and 3 are correct; C if 2 and 4 are correct; D if only 4 is correct; and E if all are correct.

1. THE FACIAL NERVE SUPPLIES:
   1. The mylohyoid muscle
   2. Muscle masses in the embryonic hyoid arch
   3. Taste fibers to the root of the tongue
   4. All muscles of facial expression

   A B C D E

2. WHICH OF THE FOLLOWING STATEMENTS ABOUT THE SOFT PALATE IS (ARE) CORRECT?
   1. Continuous laterally with the palatoglossal and palatopharyngeal folds
   2. Contributes to a partition between the nasopharynx and the oropharynx
   3. Depressed by contractions of the palatoglossus, palatopharyngeus, and tensor veli palatini muscles
   4. Contains mucous glands and taste buds

   A B C D E

3. THE TENSOR VELI PALATINI MUSCLE:
   1. Inserts by a tendon into the palatine aponeurosis
   2. Is innervated by motor branches of the pharyngeal plexus of nerves
   3. Tends to open the auditory tube during swallowing
   4. Elevates the uvula

   A B C D E

4. WHICH OF THE FOLLOWING NERVES SUPPLY(IES) GENERAL SENSORY INNERVATION TO THE SOFT AND HARD PALATES?
   1. Glossopharyngeal
   2. Mandibular
   3. Facial
   4. Maxillary

   A B C D E

5. WHICH OF THE FOLLOWING MUSCLES IS (ARE) SUPPLIED BY THE HYPOGLOSSAL NERVE?
   1. Longitudinal lingual
   2. Styloglossus
   3. Genioglossus
   4. Palatoglossus

   A B C D E

6. CONCERNING THE FORAMEN CECUM OF THE TONGUE:
   1. The limbs of the sulcus terminalis run laterally and forwards from it
   2. Marks the site of origin of the cranial end of the thyroid diverticulum in the embryo
   3. Indicates the site of origin of the thyroglossal duct in the embryo
   4. Gives passage to vessels and nerves supplying the tongue

   A B C D E

A	B	C	D	E
1,2,3	1,3	2,4	only 4	all correct

7. THE TONGUE IS SUPPLIED BY WHICH OF THE FOLLOWING ARTERIES?
   1. Deep lingual
   2. Sublingual
   3. Dorsal lingual
   4. Facial

   A B C D E

------------------ ANSWERS, NOTES AND EXPLANATIONS ---------------------

1. **C.** <u>2 and 4 are correct</u>. The mylohyoid muscle is supplied by the trigeminal (fifth cranial) nerve. The pharyngeal (posterior third) part of the tongue and the vallate (circumvallate) papillae are supplied by the lingual branch of the glossopharyngeal nerve for both general sensation and taste.

2. **E.** <u>All are correct</u>. The soft palate (velum palatinum) occupies the posterior third of the palate and is continuous laterally with the palatoglossal and palatopharyngeal folds. Its surface is lined by a mucous membrane containing palatine (mucous) glands and taste buds, innervated by fibers from the greater petrosal nerve. The soft palate is a movable fibromuscular fold, consisting of the palatine aponeurosis and the intermingling palatine muscles. The levator palatine elevates the soft palate; the palatoglossus, palatopharyngeus, and tensor veli palatini depress it; the musculus uvulae forms a projection in the median plane, and raises the uvula. The tensor veli palatini muscle is innervated by a branch of the mandibular nerve, whereas all other palatine muscles are supplied by motor branches of the pharyngeal plexus, contributed by the vagus nerve (containing fibers from the cranial root of the accessory nerve). During swallowing, sucking, and oral speech, the soft palate contributes to closing of the communication between the oral and nasal cavities, and during mastication the soft palate partitions the oral cavity from the oropharynx by narrowing the faucial (oropharyngeal) isthmus.

3. **B.** <u>1 and 3 are correct</u>. The tensor veli palatini muscle arises from the base of the medial pterygoid plate, the spine of the sphenoid bone, and the lateral side of the auditory tube. It descends vertically and its tendon winds around the pterygoid hamulus to insert into the palatine aponeurosis. The tensor veli palatini muscle tightens the soft palate, and this muscle, together with the salpingopharyngeus muscle, tends to open the auditory tube during swallowing. It is supplied by a branch of the mandibular nerve. The uvula is elevated by the musculus uvulae.

4. **D.** <u>Only 4 is correct</u>. General sensory innervation to the soft palate is supplied by the lesser palatine nerve, a branch of the maxillary nerve. The hard palate receives its general sensory innervation mainly from the greater palatine nerve, and partly from the nasopalatine nerve in its premaxillary region. The nasopalatine and greater palatine nerves are branches of the maxillary nerve.

5. **A.** <u>1, 2, and 3 are correct</u>. All muscles of the tongue are innervated by the hypoglossal (twelfth cranial) nerve. The extrinsic muscles include the genioglossus, styloglossus and hyoglossus; the intrinsic muscles include the superior and inferior longitudinal, transverse, and vertical. Although the palatoglossus muscle inserts into the side of the tongue, it is innervated by the pharyngeal plexus of nerves, and it is considered as one of the palatine muscles.

6. **A.** <u>1, 2, and 3 are correct</u>. The foramen cecum is a median pit, sometimes absent, on the dorsum of the posterior part of the tongue. It was the site of origin of the thyroid diverticulum that gives rise to the thyroid gland. Later the developing thyroid gland is temporarily connected to the tongue by the

thyroglossal duct. When this duct degenerates, its opening in the tongue persists as the foramen cecum. When present, the levator glandulae thyroideae represents the persistent caudal end of the thyroglossal duct.

7. **B.** <u>1 and 3 are correct</u>. The tongue receives its blood supply from branches of the lingual artery, a branch of the external carotid. The lingual artery runs forward deep to the hypoglossal nerve and the hyoglossus muscle, where it gives rise to the dorsal lingual artery to the dorsum of the tongue and the palatine tonsils. Anterior to the hyoglossus muscle, the lingual artery gives rise to the sublingual artery to the sublingual gland and the floor of the mouth. The deep lingual, the terminal artery of the lingual artery, supplies blood to the anterior part of the tongue.

FIVE-CHOICE ASSOCIATION QUESTIONS

DIRECTIONS: Each group of questions below consists of a numbered list of descriptive words or phrases accompanied by a diagram with certain parts indicated by letters, or by a list of lettered headings. For each numbered word or phrase, SELECT THE LETTERED PART OR HEADING that matches it correctly. Then insert the letter in the space to the right of the appropriate number. Sometimes more than one numbered word or phrase may be correctly matched to the same lettered part or heading.

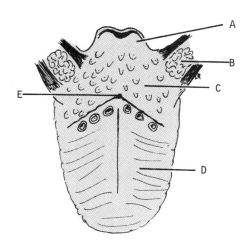

1. ___ Receives both general sensory and taste innervation from the glossopharyngeal nerve

2. ___ Receives both general sensory and taste innervation from the vagus nerve

3. ___ Receives sensory innervation from the trigeminal and glossopharyngeal nerves

4. ___ Receives general sensory innervation from the lingual nerve

5. ___ Receives taste innervation from the facial nerve

------------------- ANSWERS, NOTES AND EXPLANATIONS --------------------

1. **C.** The glossopharyngeal nerve supplies both general sensory and taste innervation to the posterior third of the tongue and the vallate papillae, located anterior to the sulcus terminalis on the dorsum of the tongue. E indicates the foramen cecum. If you chose D, you are not wrong, but C is the better answer, because C is the major area supplied by the glossopharyngeal nerve.

2. **A.** An area on the dorsum of the base of the tongue and the epiglottis receive their general sensory and taste innervation from the internal branch of the superior laryngeal nerve, a branch of the vagus nerve.

3. **B.** The palatine tonsil receives its sensory innervation from the lesser palatine nerve, a branch of the trigeminal, and from tonsillar branches of the

glossopharyngeal nerve.

4. E. General sensation of the anterior two-thirds of the dorsum of the tongue is supplied by the lingual, a branch of the mandibular nerve.

5. E. Taste buds on the dorsum of the anterior two-thirds of the tongue are innervated by the chorda tympani, a branch of the facial nerve conveyed by the lingual nerve.

PHARYNX AND LARYNX

FIVE-CHOICE COMPLETION QUESTIONS

DIRECTIONS: Each of the following questions or incomplete statements is followed by five suggested answers or completions. SELECT THE ONE BEST ANSWER in each case and then underline the appropriate letter at the lower right of each question.

1. WHICH OF THE FOLLOWING STATEMENTS ABOUT THE TONSILS IS FALSE?
   A. The pharyngeal tonsils are embedded in the wall of the nasopharynx.
   B. The tubal tonsil is located in the nasopharynx.
   C. The palatine tonsils are situated in the oral cavity.
   D. The base of the tongue contains the lingual tonsil.
   E. The adenoids are enlarged pharyngeal tonsils.   A B C D E

2. WHICH STATEMENT IS INCORRECT?
   A. All constrictor pharyngeus muscles insert into the median pharyngeal raphe.
   B. The superior constrictor pharyngeus muscle originates from the sphenoid bone, the pterygomandibular raphe, and the mandible.
   C. The middle constrictor muscle takes origin from the stylohyoid ligament and the hyoid bone.
   D. The inferior constrictor muscle arises from the thyroid, cricoid and arytenoid cartilages.
   E. All constrictor pharyngeus muscles are innervated by the pharyngeal plexus.   A B C D E

-------------------- ANSWERS, NOTES AND EXPLANATIONS --------------------

1. C. The palatine tonsils are lodged in the tonsillar fossa located in the oropharynx. The oropharynx may be separated from the nasopharynx by closing the nasopharyngeal isthmus, bounded by the soft palate, the palatopharyngeal arches, and the posterior wall of the pharynx. The pharyngeal tonsils are embedded in the posterior wall and roof of the nasopharynx; when enlarged, they are called adenoids. The tubal tonsils are embedded in the mucous membrane around the pharyngeal orifice of the auditory tube in the nasopharynx. The lingual tonsils are embedded in the posterior third (base) of the tongue. The lingual, palatine, tubal, and pharyngeal tonsils encircle the nasal and oral parts of the pharynx.

2. D. The inferior constrictor pharyngeus muscle originates from the thyroid cartilage (forming the thyropharyngeal part) and from the cricoid cartilage (forming the cricopharyngeal part), but not from the arytenoid cartilage. The middle constrictor pharyngeus muscle originates from the angle between the greater and lesser horns of the hyoid bone, and from the stylohyoid ligament.

The superior constrictor pharyngeus muscle arises from the side of the tongue, the mylohyoid line of the mandible, the pterygomandibular raphe, the pterygoid hamulus, and sometimes from the lowest point of the posterior margin of the medial pterygoid plate of the sphenoid bone. From these fixed origins, all constrictor muscles expand posteriorly and overlap one another from the inferior constrictor upward, and insert into the median (tendinous) pharyngeal raphe. The superior constrictor also attaches to the pharyngeal tubercle of the occipital bone. The muscular wall of the pharynx is invested by the buccopharyngeal fascia and lined by the pharyngobasilar fascia. All constrictor muscles are innervated by motor fibers from the pharyngeal plexus of nerves, derived from the vagus.

MULTI-COMPLETION QUESTIONS

DIRECTIONS: In each of the following questions or incomplete statements, ONE OR MORE of the completions given is correct. At the lower right of each question, underline A if 1, 2, and 3 are correct; B if 1 and 3 are correct; C if 2 and 4 are correct; D if only 4 is correct; and E if all are correct.

1. WHICH OF THE FOLLOWING MUSCLES IS (ARE) INNERVATED BY THE GLOSSOPHARYNGEAL NERVE?
   1. Superior constrictor pharyngeus
   2. Palatopharyngeus
   3. Palatoglossus
   4. Stylopharyngeus   A B C D E

2. WHICH OF THE FOLLOWING STATEMENTS ABOUT THE LARYNGOPHARYNX IS (ARE) CORRECT?
   1. Continuous with the esophagus
   2. Contains the piriform recess
   3. Related posteriorly to the fourth to sixth cervical vertebrae
   4. Receives sensory innervation from the glossopharyngeal nerve   A B C D E

3. WHICH OF THE FOLLOWING STATEMENTS ABOUT THE LARYNX IS (ARE) CORRECT?
   1. The mucous membrane is supplied by branches of the internal laryngeal nerve
   2. The cricothyroid muscle is innervated by the inferior laryngeal nerve
   3. The vocal cords are abducted by the posterior cricoarytenoid muscle
   4. The thyroarytenoid muscle is innervated by the internal laryngeal nerve   A B C D E

4. WHICH OF THE STATEMENTS CONCERNING SWALLOWING IS (ARE) CORRECT?
   1. The thyroid cartilage is elevated
   2. The hyoid bone is moved anterosuperiorly
   3. The constrictor pharyngeus muscles contract peristaltically
   4. The epiglottis is pressed backwards and downwards   A B C D E

5. WHICH OF THE FOLLOWING STATEMENTS ABOUT THE LARYNX IS (ARE) CORRECT?
   1. The rima vestibuli is the opening between the false vocal cords
   2. The aryepiglottic fold contains the corniculate and cuneiform cartilages
   3. The thyroid cartilage articulates with the cricoid cartilage by a synovial joint
   4. The true vocal cords are located at the level of the cricothyroid membrane   A B C D E

A	B	C	D	E
1,2,3	1,3	2,4	only 4	all correct

6. WHICH OF THE INTRINSIC MUSCLES OF THE LARYNX ADDUCT(S) THE VOCAL FOLDS?
   1. Vocalis
   2. Arytenoid
   3. Cricothyroid
   4. Posterior cricoarytenoid

   A B C D E

------------------ ANSWERS, NOTES AND EXPLANATIONS ----------------------

1. **D.** <u>Only 4 is correct</u>. The only somatic motor branch of the glossopharyngeal nerve supplies the stylopharyngeus muscle. The superior constrictor pharyngeus, palatopharyngeus, and palatoglossus muscles are innervated by motor fibers from the pharyngeal plexus of nerves. These motor fibers are carried by the vagus nerve, but most of them are derived from the cranial root of the accessory nerve.

2. **E.** <u>All are correct</u>. The mucous membrane of the laryngopharynx, including the piriform recess, receives sensory innervation from the internal laryngeal branch of the vagus nerve. It also receives fibers from the pharyngeal plexus of nerves contributed by the glossopharyngeal nerve. The laryngopharynx extends from the upper border of the epiglottis to the lower border of the cricoid cartilage, where it becomes continuous with the esophagus and is related to the fourth to sixth cervical vertebrae.

3. **B.** <u>1 and 3 are correct</u>. The internal laryngeal nerve of the superior laryngeal from the vagus nerve pierces the thyrohyoid membrane and supplies the mucous membrane of the larynx. The intrinsic muscles of the larynx, except the cricothyroid, are innervated by the inferior laryngeal nerve, a continuation of the recurrent laryngeal. The cricothyroid muscle is supplied by the external laryngeal branch of the superior laryngeal nerve of the vagus. Of the intrinsic laryngeal muscles, only the posterior cricoarytenoid is an abductor of the vocal cords (folds).

4. **E.** <u>All are correct</u>. During swallowing, the hyoid bone is moved anterosuperiorly by contraction of the suprahyoid muscles, as the lower jaw is fixed by closing the mouth. Due to this forward movement of the hyoid bone, the thyroid cartilage and the larynx are moved forwards and upwards and the epiglottis is pressed backwards and downwards by the base of the tongue. The laryngeal inlet is then approximated by the aryepiglottic folds so that a bolus of food or fluids do not enter the larynx. The bolus slides down the pharynx towards the esophagus, assisted by peristaltic movements of the constrictor pharyngeus muscles.

5. **A.** <u>1, 2, and 3 are correct</u>. The rima vestibuli is the space between the two vestibular (ventricular) folds, often referred to as the false vocal cords. Inferior and medial to the ventricular folds, the vocal folds (true vocal cords) extend from the angle of the thyroid cartilage anteriorly to the vocal processes of the arytenoid cartilages posteriorly. Thus, the vocal folds are located superior to the cricothyroid membrane. The rima glottidis is the interval between the true vocal cords and the narrowest part of the laryngeal cavity. The thyroid cartilage articulates with the cricoid cartilage by a hinge-type synovial joint. The cricoarytenoid articulation is also synovial, permitting gliding movements. The corniculate and cuneiform cartilages are contained in the aryepiglottic fold on each side. This fold of mucous membrane stretches between the side of the epiglottis and the apex of the arytenoid cartilage. It forms the free edge of the lateral border of the inlet of the larynx which is completed anteriorly by the upper edge of the epiglottis and posteriorly by a fold of mucous membrane stretching between the arytenoid cartilages.

6. **A. 1, 2, and 3 are correct.** The posterior cricoarytenoid muscle is the only abductor of the vocal folds; all other intrinsic laryngeal muscles adduct them. The cricothyroid muscle is the only intrinsic muscle innervated by the external branch of the superior laryngeal nerve; all other muscles are innervated by the inferior laryngeal nerve. The vocalis, thyroarytenoid and cricothyroid muscles shorten and adduct the vocal folds. The lateral cricoarytenoid and the oblique and transverse arytenoid muscles approximate the rima glottidis and the vocal folds.

4. THE THORAX

                    O B J E C T I V E S

<u>Thoracic Wall</u>

Be Able To:

  * Describe how the thoracic cage is formed and moved during respiration.

  * Sketch diagrams showing the cutaneous nerves and blood vessels of the thoracic wall.

  * Draw a diagram showing the contents of a typical intercostal space.

  * Recognize the clavicle, sternum, scapula, vertebrae, ribs, and soft tissue structures in radiographs of the chest.

<u>Diaphragm</u>

Be Able To:

  * Discuss the development, structure, position, and actions of the diaphragm.

  * Briefly describe the innervation, blood supply, and venous drainage of the diaphragm.

<u>Thoracic Cavity and Mediastinum</u>

Be Able To:

  * Indicate in simple diagrams the subdivisions of the thoracic cavity and list their contents.

  * Describe the subdivisions of the mediastinum and their contents.

  * Discuss the lymphatic drainage of the thorax and describe the course of the thoracic duct.

  * Make a diagram showing the distribution of the vagus nerves in the thorax.

  * Briefly describe the thoracic sympathetic trunk, ganglia, and splanchnic nerves.

  * Discuss the relations of the esophagus in the thorax.

  * Outline the venous return to the heart.

Pleura and Lungs

Be Able To:

* Describe the pleural cavity giving details of the parietal pleura and its reflections, including their surface markings.

* Draw simple sketches to show the surfaces, the borders, the lobes and the fissures of the lungs, and their relationship to the thoracic wall.

* Make diagrams showing the bronchial tree and the bronchopulmonary segments, indicating their clinical importance.

* Describe the blood supply and the lymphatic drainage of the pleura and lungs.

* Identify the trachea, hili, and lungs in chest radiographs.

Pericardium and Heart

Be Able To:

* Describe the surface markings, position, and chambers of the heart.

* Discuss the pericardial reflections and their attachments to the diaphragm.

* Describe the blood supply of the heart.

* Compare and contrast the walls and internal features of the atria and the ventricles.

* Briefly describe the conducting system and extrinsic innervation of the heart.

* Point out the main features of the heart and great vessels in chest radiographs.

THORACIC WALL

FIVE-CHOICE COMPLETION QUESTIONS

DIRECTION: Each of the following questions or incomplete statements is followed by five suggested answers or completions. SELECT THE ONE BEST ANSWER in each case and then underline the appropriate letter at the lower right of each question.

1. THE STERNAL ANGLE EXISTS AT THE ARTICULATION BETWEEN THE:
    A. Sternum and clavicle
    B. Manubrium and body of the sternum
    C. Manubrium and xiphoid process
    D. Body of the sternum and xiphoid process
    E. None of the above
                                                                A B C D E

2. WHICH OF THE FOLLOWING VEINS DRAINS THE FIRST INTERCOSTAL SPACE?
    A. Azygos
    B. Hemiazygos
    C. Superior intercostal
    D. Brachiocephalic
    E. None of the above
                                                                A B C D E

SELECT THE ONE BEST ANSWER

3. DURING INSPIRATION WHICH OF THE FOLLOWING DIAMETERS OF THE THORAX IS (ARE) INCREASED?
   A. Anteroposterior
   B. Transverse
   C. Vertical
   D. All of the above
   E. None of the above

   A B C D E

4. THE SUBCOSTAL NERVE:
   A. Is the ventral primary ramus of the twelfth thoracic nerve
   B. Lies behind the kidney
   C. Can be found within the rectus sheath
   D. Does not innervate intercostal muscles
   E. All of the above

   A B C D E

5. THE NEUROVASCULAR BUNDLE WITHIN EACH INTERCOSTAL SPACE LIES:
   A. Between the external and middle layers of intercostal muscles
   B. Between the middle and internal layers of intercostal muscles
   C. Deep to the internal layer of intercostal muscles
   D. Superficial to the external intercostal muscles
   E. None of the above

   A B C D E

6. THE INTERCOSTAL SPACES RECEIVE BLOOD FROM THE:
   A. Costocervical trunk
   B. Internal thoracic artery
   C. Descending thoracic aorta
   D. Musculophrenic artery
   E. All of the above

   A B C D E

7. HOW MANY PAIRS OF RIBS ARE USUALLY ATTACHED DIRECTLY TO THE STERNUM BY THEIR COSTAL CARTILAGES?
   A. Five
   B. Seven
   C. Nine
   D. Ten
   E. Twelve

   A B C D E

8. THE STRUCTURE LABELLED _____ IS AN INTERCOSTAL NERVE.

   A B C D E

-------------------- ANSWERS, NOTES AND EXPLANATIONS --------------------

1. **B.** The sternum consists of three parts: from above downwards a manubrium, a large body, and a small xiphoid process. The manubrium articulates with both the clavicle and the first rib. The sternal angle (of Louis) is an important anatomical landmark, formed by the articulation between the lower border of the manubrium and the upper border of the body of the sternum. The second costal cartilage articulates with the sternum at this angle which corresponds posteriorly to the disc between the 4th and 5th thoracic vertebrae. The sternal angle is used as a point of reference for counting the ribs, e.g., for determining the surface markings of the heart. The small cartilaginous xiphoid process articulates with the lower border of the body of the sternum; it

usually becomes ossified after the fifteenth year.

2. **D.** The intercostal spaces are drained by the intercostal veins. The first intercostal vein on each side drains into the corresponding brachiocephalic vein. Venous drainage of the remaining intercostal spaces is via the azygos, hemiazygos, and left superior intercostal veins.

3. **D.** During inspiration all three diameters of the thorax are increased. The anteroposterior diameter is increased because of the elevation of the sternal ends of the upper ribs ("pump-handle" movement), in particular the second to the sixth. Elevation of the middle part of the lower ribs ("bucket-handle" movement) causes the transverse diameter of the thorax to increase. Contraction of the diaphragm causes it to descend, thus increasing the vertical diameter. The "bucket-handle" movement occurs towards the end of deep inspiration, but not during quiet inspiration; and takes place only in the ribs of the costal margin (7-10).

4. **E.** The ventral primary ramus of the twelfth thoracic spinal nerve is called the subcostal nerve. It enters the abdomen anterior to the quadratus lumborum muscle, posterior to the kidney. The subcostal nerve pierces the posterior wall of the rectus sheath and exits superficially between the umbilicus and the symphysis pubis. The subcostal nerve supplies the transversus, the internal and external oblique, and the rectus abdominis muscles, as well as the skin and subcutaneous tissues of the gluteal region and on the lateral aspect of the thigh.

5. **B.** The neurovascular bundle, consisting of the intercostal vein, artery and nerve, is located between the middle and poorly developed internal (innermost) layer of muscles. These structures are lodged in the costal groove on the inferior aspect of the rib with the vein above, the artery in the middle, and the nerve below. Thus the intercostal nerve is in a vulnerable position during surgical incision or drainage (aspiration) of the pleural cavity. For this reason, during aspiration the needle is inserted into the lower part of the intercostal space.

6. **E.** The intercostal spaces are supplied by: anterior and posterior intercostal arteries which are derived from the costocervical trunk of the subclavian artery; the descending thoracic part of the aorta; the internal thoracic artery; and the musculophrenic artery, a terminal branch of the internal thoracic. The anterior intercostal arteries are branches of the internal thoracic and musculophrenic arteries. The lower nine posterior intercostal arteries arise directly from the descending thoracic aorta. However, the upper two posterior intercostal arteries are derived from the highest intercostal artery, a branch of the costocervical trunk. The anterior and posterior intercostal arteries anastomose with each other and supply the intercostal spaces.

7. **B.** There are twelve pairs of ribs, but supernumerary ribs, particularly in the cervical region, are not uncommon. Usually the first seven pairs (true ribs) are attached anteriorly to the sternum by their corresponding costal cartilages. The lower five pairs are known as "false" ribs. The costal cartilages of the eighth to the tenth ribs are attached to the costal cartilages of the ribs above. The anterior ends of the remaining two ribs are free and for this reason are called "floating" ribs.

8. **C.** The intercostal vein (A), artery (B), and nerve (C), lying in the plane between the middle and internal (innermost) layers of intercostal muscles, are lodged from above downwards in the costal (subcostal) groove. Each of the twelve thoracic spinal nerves gives rise to dorsal and ventral primary rami

and a meningeal branch. The ventral primary rami of the first eleven thoracic spinal nerves are the intercostal nerves; the twelfth is called the subcostal nerve. The first intercostal nerve is relatively small, but a large branch from it makes an important contribution to the brachial plexus of nerves. Intercostal nerves supply the skin, muscles, and serous membranes of both the thoracic and abdominal wall; the distribution of the nerves is segmental with considerable overlapping. Each intercostal nerve gives rise to a lateral cutaneous nerve and a collateral branch (E). The collateral branch passes forward in the lower part of the intercostal space and ends as a lower anterior cutaneous nerve. The muscles and skin of the anterior abdominal wall are innervated by the lower five intercostal nerves and the subcostal nerves.

MULTI-COMPLETION QUESTIONS

DIRECTIONS: In each of the following questions or incomplete statements, ONE OR MORE of the completions given is correct. At the lower right of each question, underline A if 1, 2 and 3 are correct; B if 1 and 3 are correct; C if 2 and 4 are correct; D if only 4 is correct; and E if all are correct.

1. THE VENTRAL PRIMARY RAMI OF THORACIC SPINAL NERVES:
    1. Innervate segmentally the intercostal muscles
    2. Supply skin of the chest wall
    3. Have an overlapping sensory distribution
    4. Innervate the anterior abdominal wall   A B C D E

2. THE "PUMP-HANDLE" MOVEMENT OF THE RIBS INVOLVES:
    1. Elevation of the sternum
    2. Movement at the costovertebral joint
    3. Depression of the sternum
    4. An increase in the anteroposterior diameter   A B C D E

3. THE TRANSVERSUS THORACIS MUSCLE:
    1. Originates from the posterior surface of the xiphoid process
    2. Inserts into the inner surfaces of the upper costal cartilages
    3. Depresses the costal cartilages
    4. Is innervated only by intercostal nerves   A B C D E

4. THE SUBCOSTAL MUSCLE:
    1. Originates from the inner surface of the ribs
    2. Is poorly developed in the upper thorax
    3. Is innervated by the intercostal nerves
    4. Inserts into the ribs below it   A B C D E

5. THE FIBERS OF THE EXTERNAL INTERCOSTAL MUSCLE:
    1. Are directed inferiorly and anteriorly between the ribs
    2. Originate from the inferior border of the rib above
    3. Insert on the superior border of the rib below
    4. Cause elevation of the rib   A B C D E

6. THE SUPERIOR THORACIC APERTURE (INLET) IS:
    1. Inclined downwards
    2. Occupied by the apices of the lungs
    3. Bounded by the first thoracic vertebra posteriorly
    4. Bounded anteriorly by the lower border of the manubrium   A B C D E

7. COMPARED TO OTHER RIBS, THE FIRST RIB IS THE:
    1. Longest   3. Most slender
    2. Flattest   4. Most curvaceous   A B C D E

A	B	C	D	E
1,2,3	1,3	2,4	only 4	all correct

8. A TYPICAL THORACIC VERTEBRA HAS:
   1. Articular facets on the sides of the body
   2. Articular facets on the transverse processes
   3. A heart-shaped body
   4. A long, slender, downward-sloping spine   A B C D E

9. WHICH OF THE FOLLOWING STRUCTURES CONTRIBUTE(S) TO FORMATION OF THE THORACIC CAGE?
   1. Sternum   3. Ribs
   2. Intervertebral discs   4. Clavicle   A B C D E

---------------------- ANSWERS, NOTES AND EXPLANATIONS ----------------------

1. E. All are correct. The ventral primary rami of the first eleven thoracic spinal nerves (intercostal nerves) are lodged in the subcostal grooves, inferior to the intercostal vessels, and pass anteriorly between the middle and internal (innermost) layers of intercostal muscles. The ventral primary ramus of the twelfth thoracic nerve is called the subcostal nerve. The intercostal nerves not only supply the intercostal muscles, but also the skin and subcutaneous tissues of the thoracic wall through their lateral and anterior cutaneous branches. Together with the subcostal nerves, the lower five intercostal nerves also innervate the anterior abdominal wall. The distribution of the thoracic spinal nerves is both segmental and overlapping.

2. E. All are correct. During this movement, occurring from the second to the sixth ribs at the costovertebral joints, the sternal ends of the upper ribs are either elevated or depressed, resulting in a "pump-handle" movement. The anteroposterior diameter of the thorax is increased when the sternum moves because of the downward inclination of the ribs. The "pump-handle" movement is involved in active inspiration.

3. E. All are correct. The internal layer of intercostal muscles is poorly developed and discontinuous. It consists of three muscles: anteriorly, the transversus thoracic (sternocostalis); laterally, the innermost intercostalis; and posteriorly, the subcostals. The transversus thoracic muscle originates from the posterior surface of the xiphoid process and the body of the sternum, and inserts into the inner surface of the second to sixth costal cartilages. Contraction of this muscle, innervated by the intercostal nerves, depresses the costal cartilages.

4. A. 1, 2 and 3 are correct. The subcostal muscle, together with the innermost intercostal and transverse thoracic muscles, constitutes the internal (innermost) layer of the thoracic wall. It is well developed only in the lower part of the thorax. The subcostal muscle originates from the inner surfaces of the ribs, and inserts into two or three ribs below. It is innervated by the intercostal or thoracoabdominal nerves. Contraction of this muscle helps the external intercostal muscles to elevate the ribs.

5. E. All are correct. The muscles of the thoracic wall are arranged in three layers. The outermost layer, the external intercostal, occupies the space between the ribs from the vertebral column to the costochondral junction. It is completed anteriorly by the anterior (external) intercostal membrane between the costal cartilages. The fibers run downwards and forwards and on contraction elevate the ribs, thereby increasing the volume of the thoracic cavity. The external intercostal muscle is supplied segmentally by branches from the anterior primary rami of the intercostal nerves.

6. **A.** <u>1, 2, and 3 are correct.</u> The superior thoracic aperture (inlet) permits communication between the cervical region and the thoracic cavity. Because the first rib is inclined obliquely downwards, the thoracic aperture is sloped downwards and forwards, thus exposing the apices of the lungs and pleura. The boundaries of the thoracic aperture are the upper border of the first thoracic vertebra, the upper border of the manubrium anteriorly, and the first pair of ribs and their corresponding costal cartilages laterally.

7. **C.** <u>2 and 4 are correct.</u> The first rib is not only the shortest, flattest and most curved, but is also the strongest rib. It contributes to the formation of the thoracic aperture (inlet) and has a prominent tubercle for the insertion of the scalenus anterior muscle on its inner border. The subclavian vein is located anterior to the attachment of this muscle. Behind the scalene tubercle, both the subclavian artery and the lowest trunk of the brachial plexus of nerves groove the first rib (subclavian groove). The brachial plexus may be blocked with local anesthetic at this site.

8. **E.** <u>All are correct.</u> The second to eighth thoracic vertebrae are considered typical because they show all these anatomical characteristics. The articular facets on each side of the body are for the heads of ribs; those on the transverse processes articulate with the tubercles of corresponding ribs. In addition, the fifth to eighth thoracic vertebrae may also have an impression on their left sides caused by the descending aorta.

9. **A.** <u>1, 2, and 3 are correct.</u> The skeleton of the thorax is formed by the sternum anteriorly, the ribs and costal cartilages laterally, and the thoracic vertebrae and their intervertebral discs posteriorly. The clavicle, not a contributor to the formation of the thoracic cage, is part of the shoulder girdle and serves as a strut for the upper limb.

F I V E - C H O I C E   A S S O C I A T I O N   Q U E S T I O N S

DIRECTIONS: Each group of questions below consists of a numbered list of descriptive words or phrases accompanied by a diagram with certain parts indicated by letters, or by a list of lettered headings. For each numbered word or phrase, SELECT THE LETTERED PART OR HEADING that matches it correctly. Then insert the letter in the space to the right of the appropriate number. Sometimes more than one numbered word or phrase may be correctly matched to the same lettered part or heading.

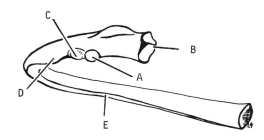

1. ____ The costal groove

2. ____ Articulates with the transverse process of a vertebra

3. ____ The angle of the rib

4. ____ The site of attachment for the lateral costotransverse ligament

------------------ ANSWERS, NOTES AND EXPLANATIONS ------------------

1. **E.** In a typical rib, the costal groove is along the lower part of its internal surface, just above the inferior border. It contains, from above down-

wards, the intercostal vein, artery, and nerve. The nerve often lies below the lower border of the rib. This fact should be remembered because the nerve may be damaged when the pleural cavity is tapped (pleural puncture) immediately below the rib.

2. A. The articular facet of the tubercle of the rib lies at the junction between the neck and the shaft (body). It articulates with the facet on the transverse process of the vertebra with which it corresponds numerically (i.e., its own vertebra), forming a synovial costotransverse joint.

3. D. The angle of the rib is marked by an oblique ridge on its external convex surface. It represents the change in direction of the shaft of the rib from backwards and laterally to forwards, medially and downwards. The iliocostalis muscle is attached to the oblique ridge at the "turn of the shaft". The axis of the "bucket-handle" movement passes through the angle of the rib.

4. C. The lateral costotransverse ligament (ligament of the tubercle) is attached to the nonarticular surface of the tubercle of the rib. This ligament, together with the medial costotransverse ligament (of the neck) and the superior and inferior costotransverse ligaments, strengthens the joint capsule between the tubercle of the rib and the transverse process of the vertebra.

DIAPHRAGM

FIVE-CHOICE COMPLETION QUESTIONS

DIRECTION: Each of the following questions or incomplete statements is followed by five suggested answers or completions. SELECT THE ONE BEST ANSWER in each case and then underline the appropriate letter at the lower right of each question.

1. WHICH OF THE FOLLOWING STRUCTURES IS (ARE) TRANSMITTED THROUGH THE CENTRAL TENDON OF THE DIAPHRAGM?
   A. Aorta
   B. Azygos vein
   C. Thoracic duct
   D. Inferior vena cava
   E. All of the above

   A B C D E

2. WHICH STRUCTURE IS NOT INVOLVED IN DEVELOPMENT OF THE DIAPHRAGM?
   A. Esophageal mesentery
   B. Septum transversum
   C. Pleuropericardial membrane
   D. Lateral body wall
   E. Pleuroperitoneal membrane

   A B C D E

3. THE AORTIC OPENING LIES OPPOSITE WHICH OF THE FOLLOWING THORACIC VERTEBRAE?
   A. 8
   B. 9
   C. 10
   D. 11
   E. 12

   A B C D E

4. WHICH OF THE FOLLOWING STATEMENTS ABOUT THE DIAPHRAGM IS (ARE) CORRECT?
   A. Involved in respiration
   B. Intrapleural pressure is decreased following contraction
   C. Is bilaterally innervated
   D. Contraction promotes venous return
   E. All of the above

   A B C D E

SELECT THE ONE BEST ANSWER

5. THE DIAPHRAGM IS SUPPLIED BY:
   A. Branches from the aorta
   B. Lower intercostal nerves
   C. Branches of the internal thoracic artery
   D. Phrenic nerves (C3, 4 and 5)
   E. All of the above

   A B C D E

6. THE _____ PART OF THE DIAPHRAGM IS NOT MUSCULAR:
   A. Central   D. Lumbar
   B. Sternal   E. None of the above
   C. Costal

   A B C D E

-------------------- ANSWERS, NOTES AND EXPLANATIONS --------------------

1. D. The three major openings in the diaphragm are the foramen for the inferior vena cava, the aortic opening, and the esophageal hiatus. The inferior vena cava, together with branches of the right phrenic nerve and lymphatic vessels, is transmitted through the vena caval foramen in the right half of the central tendon, at the level of the 8th thoracic vertebra. The aorta, azygos vein, and thoracic duct pass through the aortic opening (hiatus) behind the median arcuate ligament, at the level of the 12th thoracic vertebra.

2. C. The pleuropericardial membranes are not involved in development of the diaphragm; they give rise to the fibrous pericardium. All other structures listed form part of the diaphragm. The septum transversum contributes most to the formation of the diaphragm, giving rise to the central part called the central tendon.

3. E. The aortic opening (hiatus) lies opposite the 12th thoracic vertebra behind the median arcuate ligament of the diaphragm, somewhat to the left of the midline. In addition to the aorta, this opening transmits the thoracic duct, the azygos vein, and the greater splanchnic nerves. Some authors state that the aorta does not pierce the diaphragm, but passes behind the median arcuate ligament.

4. E. The diaphragm is the most important muscle of respiration. Contraction of the diaphragm causes the central tendon to descend, reducing the intrapleural (intrathoracic) pressure and increasing the intra-abdominal pressure. Because of the changes in intrapleural and intra-abdominal pressures, the venous return of blood to the heart is facilitated. The diaphragm contracts as a unit; however, as each half is independently innervated, paralysis of one side does not significantly affect the other.

5. E. The diaphragm is chiefly supplied by the phrenic nerves (C3, 4 and 5), containing motor and sensory fibers. During the fifth week of development, the phrenic nerves grow into the septum transversum of the developing diaphragm when it is opposite the cervical segments. When the diaphragm descends, it takes the phrenic nerves with it. The intercostal nerves supply sensory fibers to peripheral parts of the diaphragm which are stripped off the lateral body wall during excavation by the developing pleural cavities. The diaphragm is supplied by pericardiacophrenic and musculophrenic arteries from the internal thoracic artery, and intercostal and phrenic branches from the aorta.

6. A. The diaphragm is a musculofibrous partition between the thoracic and abdominal cavities. The sternal, costal, and lumbar parts consist of muscular fibers inserted into the central tendon, consisting of dense collagenous con-

nective tissue. The central tendon is fused with the pericardium above. The central tendon is derived from the septum transversum in the embryo, the earliest recognizable part of the developing diaphragm.

MULTI-COMPLETION QUESTIONS

DIRECTIONS: In each of the following questions or incomplete statements, ONE OR MORE of the completions given is correct. At the lower right of each question, underline A if 1, 2 and 3 are correct; B if 1 and 3 are correct; C if 2 and 4 are correct; D if only 4 is correct; and E if all are correct.

1. THE MUSCLE FIBERS OF THE DIAPHRAGM ORIGINATE FROM THE:
   1. Xiphoid process of the sternum
   2. Lower costal cartilages
   3. Lower ribs
   4. Lumbocostal arches and crura   A B C D E

2. THE DIAPHRAGM IS INNERVATED BY WHICH OF THE FOLLOWING NERVES?
   1. Phrenic   3. Thoracoabdominal
   2. Vagus   4. Upper intercostals   A B C D E

3. THE DIAPHRAGM IS SUPPLIED BY WHICH OF THE FOLLOWING ARTERIES?
   1. Phrenic   3. Musculophrenic
   2. Pericardiacophrenic   4. Intercostal   A B C D E

4. WHICH OF THE FOLLOWING STRUCTURES IS (ARE) TRANSMITTED THROUGH THE ESOPHAGEAL HIATUS IN THE DIAPHRAGM?
   1. Azygos vein   3. Thoracic duct
   2. Vagus nerve   4. Esophagus   A B C D E

5. WHICH STATEMENTS ABOUT THE DIAPHRAGM ARE TRUE?
   1. Each half has a separate nerve supply
   2. Develops from four embryonic structures
   3. Has a depressed median portion
   4. Posterolateral defects usually occur on the right side   A B C D E

---------------------- ANSWERS, NOTES AND EXPLANATIONS ----------------------

1. E. All are correct. The dome-shaped diaphragm forms a partition between the thoracic and abdominal cavities. It consists of a central tendinous part which represents the insertion of muscle fibers originating from the dorsal aspect of the xiphoid process of the sternum anteriorly (sternal part), the lower six costal cartilages (costal part), the lower four ribs, and posteriorly from the lateral arcuate ligaments (lumbocostal arches) and the right and left crura (lumbar part).

2. B. 1 and 3 are correct. The only motor nerve innervating the diaphragm is the phrenic. It is derived primarily from the fourth cervical nerve with contributions from the third and fifth. The phrenic nerve, as well as the thoracoabdominal (lower intercostal) nerves, carries both sensory and vasomotor fibers.

3. E. All are correct. The diaphragm is supplied by the pericardiacophrenic and musculophrenic arteries, branches of the internal thoracic artery, and the intercostal and the superior and inferior phrenic arteries, branches of the aorta.

4. **C.** <u>2 and 4 are correct</u>. The esophageal opening (hiatus) lies within the right crus of the diaphragm, at the level of the 10th thoracic vertebra. It transmits the esophagus and the right and left vagus nerves. The azygos vein and thoracic duct pass with the aorta through the aortic opening (hiatus).

5. **A.** <u>1, 2, and 3 are correct</u>. The diaphragm is the most important muscle of respiration. Each half is supplied by a phrenic nerve (C3, 4, and 5), its only motor nerve. Sensory fibers to the diaphragm are also contained in the phrenic nerves, except for peripheral parts of the diaphragm supplied by the lower intercostal (thoracoabdominal) nerves. Posterolateral defects are usually on the left side and result from failure of the pleuroperitoneal membrane to fuse with the septum transversum and the dorsal mesoesophagus, closing the pericardioperitoneal canal. Herniation of abdominal viscera through this defect constitutes a congenital diaphragmatic hernia.

FIVE-CHOICE ASSOCIATION QUESTIONS

DIRECTIONS: Each group of questions below consists of a numbered list of descriptive words or phrases accompanied by a diagram with certain parts indicated by letters, or by a list of lettered headings. For each numbered word or phrase, SELECT THE LETTERED PART OR HEADING that matches it correctly. Then insert the letter in the space to the right of the appropriate number. Sometimes more than one numbered word or phrase may be correctly matched to the same lettered part or heading.

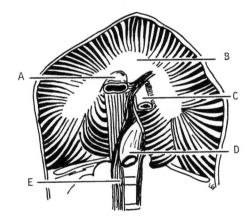

1. ___ The tendinous part of the diaphragm

2. ___ Passes behind the median arcuate ligament

3. ___ Crural attachment of the diaphragm

4. ___ Pierces the diaphragm at the level of the 12th thoracic vertebra

5. ___ Structure passing through the central tendon of the diaphragm

-------------------- ANSWERS, NOTES AND EXPLANATIONS --------------------

1. **B.** The diaphragm is a musculotendinous partition between the thoracic and abdominal cavities. The muscle fibers of the diaphragm (sternal, costal, and lumbar parts) insert radially into the central tendon (B) which is fused with the pericardium above.

2. **D.** The lumbar part of the diaphragm arises from two crura which are attached to the upper lumbar vertebrae and from two fibrous thickenings, the medial and

lateral arcuate ligaments, crossing the psoas major and quadratus lumborum muscles, respectively. The crura are bridged by a fibrous arch, the median arcuate ligament. The aortic opening (hiatus) lies in the gap between the crura, behind the median arcuate ligament.

3. **E.** The lumbar part of the diaphragm arises by two muscular crura which have tendinous attachments to the lumbar vertebrae. The right crus (E) arises from the upper three or four lumbar vertebrae; the left from the upper two or three. The median arcuate ligament unites the two crura in front of the aortic opening (hiatus). The esophagus (C) passes through the esophageal hiatus which lies within the right crus of the diaphragm.

4. **D.** The descending aorta (D) passes through the aortic opening (hiatus) which lies behind the median arcuate ligament at the level of the 12th thoracic vertebra. The aortic opening also transmits the thoracic duct, the azygos vein, and the greater splanchnic nerves.

5. **A.** The opening for the inferior vena cava lies within the right half of the central tendon of the diaphragm at the level of the 8th thoracic vertebra. In addition to the inferior vena cava, it transmits the right phrenic nerve and lymphatic vessels from the liver.

THORACIC CAVITY AND MEDIASTINUM

FIVE-CHOICE COMPLETION QUESTIONS

DIRECTION: Each of the following questions or incomplete statements is followed by five suggested answers or completions. SELECT THE ONE BEST ANSWER in each case and then underline the appropriate letter at the lower right of each question.

1. WHICH OF THE FOLLOWING SUBDIVISIONS OF THE THORACIC CAVITY IS THE LARGEST?
    A. Superior mediastinum
    B. Anterior mediastinum
    C. Inferior mediastinum
    D. Middle mediastinum
    E. Posterior mediastinum

    A B C D E

2. THE POSTERIOR MEDIASTINUM CONTAINS THE:
    A. Azygos and hemiazygos veins
    B. Sympathetic trunk
    C. Vagus nerves
    D. Splanchnic nerves
    E. All of the above

    A B C D E

3. THE _____ ARTERY IS THE FIRST LARGE BRANCH OF THE AORTA.
    A. Left common carotid
    B. Left subclavian
    C. Brachiocephalic
    D. Right common carotid
    E. Right subclavian

    A B C D E

4. THE LEFT PRIMARY BRONCHUS IS:
    A. Longer than the right
    B. Passes below the aortic arch
    C. Located in front of the esophagus
    D. Lies below the left pulmonary artery
    E. All of the above

    A B C D E

5. THE THORACIC DUCT RECEIVES LYMPH FROM THE:
    A. Lower limbs
    B. Abdomen
    C. Pelvis
    D. Left side of the thorax
    E. All of the above

    A B C D E

SELECT THE ONE BEST ANSWER

6. WHICH OF THE FOLLOWING ARTERIES DOES NOT SUPPLY THE ESOPHAGUS?
   A. Inferior thyroid
   B. Bronchial
   C. Right gastric
   D. Direct branches from the aorta
   E. Left gastric

   A B C D E

7. THE RIGHT VAGUS NERVE SIGNIFICANTLY CONTRIBUTES TO WHICH OF THE FOLLOWING PLEXUSES?
   A. Esophageal
   B. Cardiac
   C. Pulmonary
   D. All of the above
   E. None of the above

   A B C D E

8. THE GREATER SPLANCHNIC NERVE IS USUALLY DERIVED FROM WHICH OF THE FOLLOWING THORACIC GANGLIA?
   A. 1st - 5th
   B. 5th - 9th
   C. 9th - 10th
   D. Below the 10th
   E. All of the above

   A B C D E

9. THE HEART RECEIVES POSTGANGLIONIC SYMPATHETIC FIBERS VIA WHICH OF THE FOLLOWING NERVES?
   A. Cervical
   B. Cervicothoracic
   C. Thoracic cardiac
   D. All of the above
   E. None of the above

   A B C D E

-------------------- ANSWERS, NOTES AND EXPLANATIONS --------------------

1. C. The space between the mediastinal pleura is called the mediastinum. It extends from the sternum anteriorly to the vertebral column posteriorly. The superior mediastinum lies above the plane extending from the sternal angle to the intervertebral disc between the fourth and fifth thoracic vertebrae. The inferior mediastinum, the space below this plane, is divided into an anterior mediastinum, lying in front of the pericardium; a middle mediastinum occupied by the heart and pericardium; and a posterior mediastinum, behind the pericardium. Thus, the largest subdivision of the thoracic cavity is the inferior mediastinum.

2. E. The posterior mediastinum extends from the level of the sternal angle to the diaphragm, and lies posterior to the fibrous pericardium and the diaphragm in its lower region. It contains the azygos and hemiazygos veins, the sympathetic trunk, the vagus and splanchnic nerves, as well as the descending aorta, posterior intercostal vessels, thoracic duct, and esophagus.

3. C. The aorta is subdivided into an ascending part, an arch, and a descending part. The first large branch from the aorta is the brachiocephalic artery (trunk). It arises from the arch behind the manubrium, ascends to the right and divides into the right common carotid and right subclavian arteries, behind the right sternoclavicular joint. After the brachiocephalic artery, the left common carotid and left subclavian arteries are the next arterial trunks to arise from the aortic arch.

4. E. The right and left primary bronchi are approximately 2.5 and 5.0 cm, respectively. Compared to the right primary bronchus, the left is not only longer, but it is also narrower and less vertical. Aspirated foreign bodies therefore tend to pass more often into the right bronchus. The left primary bronchus lies in front of the esophagus and causes a constriction at this point. It is located first posterior, then inferior, to the pulmonary artery. In addition, the left primary bronchus is related to the descending aorta lying posterior to it.

5. **E.** The thoracic duct drains the lower limbs, pelvis, abdomen, left upper limb, and the left side of the head and thorax. It usually empties into the junction of the left internal jugular and subclavian veins. Sometimes it empties into the left internal jugular vein. Although most of the lymphatic drainage into the venous system occurs via the thoracic duct, there are no serious consequences following ligation of the thoracic duct because of the numerous anastomotic lymphatic channels in the thorax and neck. The right side of the upper half of the body is drained by the short right lymphatic duct. It is formed by the union of the right jugular and subclavian lymphatic trunks, and empties into the right brachiocephalic trunk.

6. **C.** The upper segment of the esophagus is supplied by: a branch of the inferior thyroid; direct esophageal branches from the aorta; and bronchial arteries. The left gastric and inferior phrenic arteries supply the lower segment. The right gastric artery, a branch of the common hepatic, lies along the lesser curvature of the stomach; however, it does not supply the esophagus.

7. **D.** The two vagus nerves descend behind the corresponding hilum of the lung and give branches to form the esophageal, cardiac, and pulmonary plexuses. From the plexuses formed by the vagus nerves, delicate fibers arise and innervate the heart, lungs and esophagus. The right vagus nerve enters the abdomen on the dorsal surface of the esophagus, the left ventrally.

8. **B.** The three splanchnic nerves consist primarily of preganglionic sympathetic fibers and afferent branches from the viscera. The greater splanchnic nerve is usually derived from the 5th to 9th thoracic ganglia; the lesser splanchnic nerve usually arises from the 9th and 10th; and the least splanchnic nerve from the lowest thoracic ganglion. The greater splanchnic nerves pierce the diaphragm and end in the celiac ganglion. The greater splanchnic nerves send twigs to the esophagus and aorta.

9. **D.** The white ramus communicans from the spinal nerve contains preganglionic fibers from the corresponding spinal nerve. The grey ramus contains postganglionic fibers and those from the first to fifth thoracic ganglia reach the heart via the cervical, cervicothoracic, and thoracic cardiac nerves.

MULTI-COMPLETION QUESTIONS

DIRECTIONS: In each of the following questions or incomplete statements, ONE OR MORE of the completions given is correct. At the lower right of each question, underline A if 1, 2 and 3 are correct; B if 1 and 3 are correct; C if 2 and 4 are correct; D if only 4 is correct; and E if all are correct.

1. WHICH OF THE FOLLOWING STRUCTURES IS (ARE) NOT PRESENT IN THE ANTERIOR MEDIASTINUM?
   1. Thymus gland
   2. Thoracic duct
   3. Lymph nodes
   4. Phrenic nerve

   A B C D E

2. THE ESOPHAGUS IS DRAINED BY WHICH OF THE FOLLOWING VEINS?
   1. Azygos
   2. Gastric
   3. Hemiazygos
   4. Inferior thyroid

   A B C D E

3. THE THORACIC SYMPATHETIC TRUNK:
   1. Is located in the posterior mediastinum
   2. Descends anterior to the heads of the ribs
   3. Pierces the crus of the diaphragm to enter the abdomen
   4. Is associated with approximately six ganglia

   A B C D E

A	B	C	D	E
1,2,3,	1,3	2,4	only 4	all correct

4. THE LEFT VAGUS NERVE:
   1. Is anterior to the aortic arch
   2. Lies in front of the hilum of the lung
   3. Gives rise to a recurrent laryngeal branch
   4. Lies on the dorsal surface of the esophagus   A B C D E

5. THE THORACIC DUCT:
   1. Begins in the abdomen
   2. Lies between the aorta and azygos vein
   3. Passes through the aortic hiatus of the diaphragm
   4. Drains into the external jugular vein   A B C D E

6. WHICH OF THE FOLLOWING ARTERIES SUPPLY(IES) THE THORACIC PART OF THE ESOPHAGUS?
   1. Inferior thyroid   3. Branches from the aorta
   2. Left gastric   4. Bronchial   A B C D E

7. THE RIGHT PRIMARY BRONCHUS IS NORMALLY:
   1. Smaller in diameter than the left
   2. More vertical than the left
   3. Superior to the azygos vein
   4. Shorter than the left   A B C D E

8. THE TRACHEA IS SUPPLIED BY WHICH OF THE FOLLOWING ARTERIES?
   1. Inferior thyroid   3. Bronchial
   2. Superior thyroid   4. Internal thoracic   A B C D E

9. THE TRACHEA:
   1. Is approximately 20-25 cm long
   2. Divides at the level of the 8th thoracic vertebra
   3. Has incomplete bony rings
   4. Is an extremely mobile structure   A B C D E

10. THE SUPERIOR VENA CAVA:
    1. Is thin-walled
    2. Contains no valves
    3. Receives blood from the azygos vein
    4. Is formed by the brachiocephalic veins   A B C D E

11. WHICH OF THE FOLLOWING STRUCTURES LIE(S) ANTERIOR TO THE ESOPHAGUS IN THE THORAX?
    1. Thoracic duct   3. Right vagus nerve
    2. Pericardium   4. Left bronchus   A B C D E

12. WHICH OF THE FOLLOWING STRUCTURES ARE PRESENT IN THE SUPERIOR MEDIASTINUM?
    1. Aortic arch   3. Esophagus
    2. Trachea   4. Sympathetic trunk   A B C D E

13. THE THORACIC AORTA:
    1. Descends on the right side of the vertebral column
    2. Gives rise to all posterior intercostal arteries
    3. Passes through the diaphragm at the level of the 10th thoracic vertebra
    4. Lies behind the esophagus   A B C D E

A	B	C	D	E
1,2,3	1,3	2,4	only 4	all correct

14. THE ESOPHAGUS:
    1. Begins at the level of the sixth cervical vertebra
    2. Is 10-12 cm long
    3. Is constricted by the left bronchus
    4. Lies in front of the trachea in the neck

    A B C D E

------------------- ANSWERS, NOTES AND EXPLANATIONS -------------------

1. **C. 2 and 4 are correct**. The anterior mediastinum represents a relatively small area which lies behind the sternum and anterior to the pericardium. It contains remnants of the thymus gland, lymph nodes, fat, and connective tissue. The phrenic nerve enters the thorax and descends between the pericardium and the mediastinal parietal pleura to innervate the diaphragm. The thoracic duct ascends through the posterior mediastinum between the aorta and azygos vein.

2. **E. All are correct**. The esophagus is drained from above downwards by the inferior thyroid, azygos, hemiazygos, and gastric veins. These veins form superficial and submucosal plexuses in the esophagus, thereby establishing a communication between the portal (gastric) and systemic (azygos, hemiazygos and inferior thyroid) venous systems. In portal hypertension, this alternate route of venous drainage often becomes dilated and functional. In addition, rupture of these dilated varicose submucosal veins may cause severe esophageal hemorrhage, resulting in vomiting of blood or hematemesis. The venous drainage of the esophagus is therefore of considerable clinical importance.

3. **A. 1, 2, and 3 are correct**. The thoracic sympathetic trunk is the most laterally placed structure in the posterior mediastinum. Each trunk lies anterior to the heads of the ribs and is associated with approximately 10-13 ganglia. These ganglia receive white rami communicantes from their corresponding spinal nerves. The thoracic sympathetic trunk becomes the lumbar sympathetic trunk after piercing the crus of the diaphragm on that side, or passing behind the medial arcuate ligament.

4. **B. 1 and 3 are correct**. The left vagus nerve descends in the thorax anterior to the aortic arch, then it runs between the aorta and the left pulmonary trunk. At the lower border of the aortic arch, it gives off the recurrent laryngeal nerve which curves around the arch of the aorta and ascends into the neck between the trachea and the esophagus. Both vagus nerves pass behind the corresponding hilum of the lung and send branches to the heart (cardiac plexus), lungs (pulmonary plexus), and esophagus (esophageal plexus). The right and left vagus nerves enter the abdomen on the dorsal and ventral surfaces of the esophagus, respectively.

5. **A. 1, 2, and 3 are correct**. The thoracic duct begins in the abdomen at a sac-like dilatation, the cisterna chyli, into which the intestinal, lumbar and descending intercostal lymphatic trunks drain. It passes through the aortic hiatus of the diaphragm and ascends through the posterior mediastinum, between the aorta and the azygos vein. The thoracic duct then crosses obliquely to the left behind the esophagus, curves above the subclavian artery in the neck, and terminates by emptying into the junction between the left internal jugular vein and the subclavian vein. Occasionally it empties into the left internal jugular vein.

6. **E. All are correct**. As the esophagus descends through the posterior mediastinum, it is supplied by the following arteries: the inferior thyroid from the thyrocervical trunk; bronchial; an ascending branch from the left gastric; and

esophageal branches from the aorta. There are extensive anastomoses between these vessels.

7. **C. 2 and 4 are correct.** Normally, the right primary bronchus is considerably wider, shorter and more vertical, compared to the left. This is clinically important because foreign bodies in the trachea pass more commonly into the right bronchus and lung. The azygos vein arches over the right bronchus and the pulmonary artery lies below it.

8. **E. All are correct.** The trachea is supplied by the following arteries: inferior thyroid, superior thyroid, bronchial, and internal thoracic. Its venous drainage is chiefly via the inferior thyroid vein. Lymph from the trachea enters the cervical, tracheal, and tracheobronchial group of lymph nodes.

9. **D. Only 4 is correct.** The trachea begins at the lower border of the cricoid cartilage at the level of the 6th cervical vertebra. It terminates at the level of the sternal angle (T4/5), where it bifurcates into the right and left primary bronchi. The inside of the trachea at the bifurcation is marked by a ridge, the carina, which is an important landmark during bronchoscopy. The trachea is an extremely mobile structure. The rigidity of its wall is due to the presence of a series of approximately 16-20 incomplete hyaline cartilaginous rings which keep it patent. The trachea is about 9 to 15 cm in length.

10. **E. All are correct.** The superior vena cava is a major venous channel lying on the right side of the ascending aorta and, like the inferior vena cava, empties into the right atrium. The superior vena cava is formed where the right and left brachiocephalic veins join behind the right second costal cartilage. On each side, the internal jugular and subclavian veins join posterior to the sternoclavicular joint, forming the brachiocephalic vein. Like all veins, the superior vena cava is thin-walled; however, unlike most veins it is valveless. The azygos vein empties into the superior vena cava. In obstruction of the superior vena cava above the entry of the azygos vein, an alternate route of venous return from the upper half of the trunk to the heart may become functional via the internal thoracic vein, the intercostal veins, and the azygos systems.

11. **C. 2 and 4 are correct.** The pericardium and left bronchus are related to the ventral aspect of the esophagus. The thoracic duct and the right vagus nerve lie behind and on the right side of the esophagus, respectively. Because the left atrium is located posteriorly and attached to the posterior pericardium, enlargement of the left atrium may cause a deviation of the esophagus which can be demonstrated in a lateral radiograph of the thorax following ingestion of barium.

12. **E. All are correct.** The space between the two pleural cavities is called the mediastinum. The heart within the fibrous pericardium occupies the central (middle) mediastinum; the regions above, in front of, and behind the pericardium are called the superior, the anterior, and the posterior mediastina, respectively. The anterior, middle, and posterior mediastinum together are known as the inferior mediastinum. The plane of subdivision between the superior and inferior mediastinum lies at a level between the sternal angle and the intervertebral disc between the fourth and fifth thoracic vertebrae. In addition to the aortic arch, trachea, esophagus, and sympathetic trunk, the superior mediastinum contains all other structures passing between the neck and thorax. These include the brachiocephalic veins, superior vena cava, three major branches from the aortic arch (brachiocephalic, left common carotid, and left subclavian arteries), phrenic and vagus nerves, thoracic duct, and the thymus gland (or the remains of it).

13. **D.** <u>Only 4 is correct</u>. The thoracic aorta descends through the posterior mediastinum on the left side of the bodies of the thoracic vertebrae, and gradually inclines towards the midline. The lower part of the descending thoracic aorta lies in front of the vertebral column and behind the esophagus. It gives rise to the lower nine pairs of posterior intercostal arteries; the upper two are derived from the costocervical trunk of the subclavian artery. The aorta passes behind the median arcuate ligament of the diaphragm at the level of the 12th thoracic vertebra.

14. **B.** <u>1 and 3 are correct</u>. The esophagus extends from the pharynx, at the level of the sixth cervical vertebra, to the cardia of the stomach at the level of the tenth thoracic vertebra. It is approximately 20-25 cm long. The lumen of the esophagus is constricted at its junction with the pharynx, where it is crossed by the left bronchus, and at its entry into the cardia of the stomach. The esophagus lies on the cervical and thoracic vertebrae, posterior to the trachea. The recurrent laryngeal branches of the vagus nerves ascend to the larynx in the groove between the trachea and esophagus.

F I V E - C H O I C E   A S S O C I A T I O N   Q U E S T I O N S

DIRECTIONS: Each group of questions below consists of a numbered list of descriptive words or phrases accompanied by a diagram with certain parts indicated by letters, or by a list of lettered headings. For each numbered word or phrase, SELECT THE LETTERED PART OR HEADING that matches it correctly. Then insert the letter in the space to the right of the appropriate number. Sometimes more than one numbered word or phrase may be correctly matched to the same lettered part or heading.

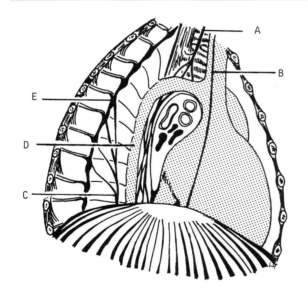

1. _____ Phrenic nerve
2. _____ Greater splanchnic nerve
3. _____ Vagus nerve
4. _____ Sympathetic trunk
5. _____ Azygos vein

---------------------- ANSWERS, NOTES AND EXPLANATIONS ----------------------

1. **B.** The phrenic nerve provides the entire motor innervation to the diaphragm. It is derived mainly from the fourth cervical nerve with contributions from the third and fifth nerves. Injury of the phrenic nerve on one side produces paralysis and elevation of the corresponding half of the diaphragm because each half is innervated by a phrenic nerve. The phrenic nerves enter the thorax in front of the scalenus anterior muscle, deep to the fascia of this

muscle. Each nerve is accompanied by a pericardiacophrenic artery, a branch of the internal thoracic. On the right side, the nerve lies anterior to the hilum of the lung, between the pericardium and the mediastinal pleura. On the left, the phrenic nerve lies between the left subclavian and the left common carotid arteries, lateral to the arch of the aorta. It then descends anterior to the hilum of the lung between the pericardium and the mediastinal pleura. In addition to the diaphragm, the phrenic nerves give off branches to the pericardium, the mediastinal pleura, and the central portion of the diaphragmatic pleura.

2. **C.** Preganglionic nerve fibers from the fifth to the twelfth thoracic nerves form the greater (T5-9), the lesser (T9-10), and the least (T11) splanchnic nerves. The splanchnic nerves descend through the crura of the diaphragm, medial to the sympathetic trunk, and end in the celiac ganglia and plexuses. From here, postganglionic fibers pass to the abdominal viscera. The aorta, esophagus, and pleura receive branches from the greater splanchnic nerves.

3. **A.** The vagus (10th cranial) nerve passes vertically downwards through the neck and enters the thorax where it contributes to the formation of the pulmonary, cardiac, and esophageal plexuses. Each vagus nerve gives rise to: (1) a recurrent laryngeal branch, supplying the esophagus, trachea and larynx; (2) cervical cardiac branches, joining branches from the cervical sympathetic ganglia; (3) cervicothoracic cardiac branches; and (4) branches to the bronchi and esophagus from the pulmonary and esophageal plexuses. The course of the vagus nerve differs on the two sides. The right vagus nerve gives off a recurrent laryngeal branch as it crosses the subclavian artery. It then passes through the superior mediastinum, close to the great veins, and descends posterior to the hilum of the right lung. The left vagus nerve enters the thorax and lies lateral to the common carotid artery. It then crosses the arch of the aorta and gives off the left recurrent laryngeal branch. Like the right vagus nerve, the left also descends posterior to the hilum of the lung. Both vagus nerves leave the thorax by passing through the esophageal opening in the diaphragm.

4. **E.** The sympathetic trunk is the most lateral structure in the posterior mediastinum. It descends through the thorax anterior to the heads of the ribs, the posterior intercostal arteries and veins, and accompanying nerves. A variable number (10-13) of ganglia is associated with the thoracic part of the sympathetic trunk on each side. The sympathetic trunk and ganglia are joined with the ventral rami of thoracic nerves by white and grey rami communicantes. The inferior cervical ganglion frequently joins the first thoracic ganglion to form the stellate ganglion. The thoracic sympathetic trunk passes behind the medial arcuate ligament of the diaphragm to become the lumbar sympathetic trunk in the abdomen.

5. **D.** The azygos vein, together with the hemiazygos and accessory hemiazygos veins, drains most of the thoracic wall and back. It is formed at the junction of the right subcostal and right ascending lumbar veins. The azygos vein ascends through the posterior and superior parts of the mediastinum, arches over the hilum of the lung and empties into the superior vena cava. The azygos vein is inconstant in origin. From its development, it may be expected to arise from the posterior aspect of the inferior vena cava, at or below the level of the renal veins. Such a vein, called a lumbar azygos, is often present and ascends in front of the upper lumbar vertebrae.

In obstruction of the venae cavae, the azygous and hemiazygous veins and the vertebral venous plexuses form an important way by which the venous circulation is carried on.

PLEURA AND LUNGS

FIVE-CHOICE COMPLETION QUESTIONS

DIRECTION: Each of the following questions or incomplete statements is followed by five suggested answers or completions. SELECT THE ONE BEST ANSWER in each case and then underline the appropriate letter at the lower right of each question.

1. THE LYMPHATIC DRAINAGE OF THE LUNGS IS TO WHICH OF THE FOLLOWING GROUPS OF NODES?
   A. Paratracheal
   B. Tracheobronchial
   C. Bronchopulmonary
   D. Pulmonary
   E. All of the above

   A B C D E

2. THE LUNGS ARE INNERVATED BY ALL OF THE FOLLOWING EXCEPT THE:
   A. Vagus nerve
   B. Phrenic nerve
   C. Thoracic sympathetic
   D. Anterior pulmonary plexus
   E. Posterior pulmonary plexus

   A B C D E

3. BLOOD IN THE FETUS IS OXYGENATED IN THE:
   A. Maternal lungs
   B. Liver via the ductus venosus
   C. Fetal lungs
   D. Aorta via the ductus arteriosus
   E. Placenta

   A B C D E

4. WHICH OF THE FOLLOWING STATEMENTS REGARDING THE LEFT LUNG IS FALSE?
   A. Has only two lobes
   B. Is subdivided into eight bronchopulmonary segments
   C. The cardiac notch is present in its upper lobe
   D. Has a groove on its mediastinal surface for the azygos vein
   E. Is supplied by the bronchial arteries

   A B C D E

5. THE PULMONARY LIGAMENT IS A COMPONENT OF THE _____ PLEURA.
   A. Visceral
   B. Costal
   C. Diaphragmatic
   D. Mediastinal
   E. Cervical

   A B C D E

6.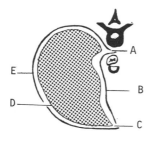

   IN THIS HORIZONTAL SECTION OF THE UPPER PART OF THE RIGHT LUNG, THE COSTOMEDIASTINAL RECESS IS INDICATED BY _____.

   A B C D E

---------------------- ANSWERS, NOTES AND EXPLANATIONS ----------------------

1. E. The lymphatics of the lungs are arranged as superficial and deep plexuses, beneath the visceral pleura and in the submucosa of the bronchi, respectively. From both plexuses, lymph flows to the pulmonary (within the lung) and the bronchopulmonary (at the hilum) groups of nodes. From here, efferent channels proceed to the tracheobronchial nodes at the bifurcation of the trachea and subsequently to the nodes at the sides of the trachea (tracheal or paratracheal).

2. **B.** The phrenic nerve supplies the diaphragm, not the lungs. The anterior and posterior pulmonary plexuses are derived from fibers of both vagus and sympathetic trunks. Efferent fibers from the pulmonary plexus supply the bronchial musculature; afferent fibers from the alveoli and the mucous membranes of the bronchioles pass to the plexuses.

3. **E.** The fetal blood is oxygenated in the placenta. Oxygen passes across the placental membrane (barrier) by simple diffusion. Although movements like those of respiration may occur in utero, the lungs do not begin to function until birth. Before birth the lungs are filled with fluid. Although some blood in the pulmonary trunk passes to the lungs, most of it passes through the ductus arteriosus into the aorta.

4. **D.** The azygos vein makes an impression on the mediastinal surface of the right lung as it arches over the hilum to enter the superior vena cava. The left lung is divided by an oblique fissure into upper and lower lobes and into eight bronchopulmonary segments. The left upper lobe has a lingular segment supplied by a branch from the primary upper bronchus. The deep impression (cardiac notch) caused by the heart and pericardium is located on its upper lobe. Both lungs are supplied by small bronchial arteries. The two bronchial arteries to the left lung are derived directly from the aorta, whereas the right lung receives its arterial supply from a single branch of the third posterior intercostal artery. The pulmonary arteries and veins form a capillary (pulmonary) plexus in the alveolar walls which anastomose with the bronchial vessels.

5. **D.** The pulmonary ligament is a double layer of mediastinal pleura extending below the root of the lung. Above, it is continuous with the pleural reflection around the root of the lung and the remaining upper portions of the mediastinal pleura. The other components of the parietal pleura reflect onto the diaphragm, ribs, and costal cartilages, and into the thoracic inlet; these are called the diaphragmatic, costal, and cervical components, respectively. The visceral pleura covers the lobes and fissures of the lungs.

6. **C.** The costomediastinal recess is the part of the pleural cavity lying between the costal and mediastinal pleurae anteriorly. The retroesophageal recess (A) is the part of the pleural cavity lying behind the esophagus; it is formed by reflection of pleura on the right and left sides. The mediastinal pleura (B) is the parietal pleura forming the lateral wall of the mediastinum. The costal pleura (E) is the parietal pleura which adheres to the ribs and costal cartilages. The visceral pleura (D) covers the lobes and the fissures of the lung. The potential space between the visceral and parietal pleurae is the pleural cavity.

MULTI-COMPLETION QUESTIONS

DIRECTIONS: In each of the following questions or incomplete statements, ONE OR MORE of the completions given is correct. At the lower right of each question, underline A if 1, 2 and 3 are correct; B if 1 and 3 are correct; C if 2 and 4 are correct; D if only 4 is correct; and E if all are correct.

1. THE LEFT LUNG:
   1. Is divided by an oblique fissure
   2. Consists of eight bronchopulmonary segments
   3. Shows a large impression caused by the pericardium
   4. Is below the level of the first sternocostal joint

   A B C D E

A	B	C	D	E
1,2,3	1,3	2,4	only 4	all correct

2. THE ROOT OF THE LUNG IS A SHORT, THICK PEDICLE MADE UP OF THE:
   1. Bronchi
   2. Pulmonary artery and veins
   3. Lymph and bronchial vessels
   4. Pulmonary plexus of nerves

   A B C D E

3. THE INFERIOR LOBE OF THE LEFT LUNG IS SUBDIVIDED INTO THE FOLLOWING BRONCHOPULMONARY SEGMENTS:
   1. Superior
   2. Anterior basal
   3. Lateral basal
   4. Posterior basal

   A B C D E

4. WHICH OF THE FOLLOWING STATEMENTS ABOUT THE FISSURES AND LOBES OF THE LUNGS IS (ARE) CORRECT?
   1. The right oblique fissure originates at the head of the fourth or fifth rib
   2. The horizontal fissure ends anteriorly at the level of the fourth costal cartilage
   3. The right middle lobe is triangular in shape
   4. The oblique fissure follows the line of the fifth rib

   A B C D E

5. THE PARIETAL PLEURA RECEIVES ITS BLOOD SUPPLY FROM WHICH OF THE FOLLOWING ARTERIES?
   1. Internal thoracic
   2. Superior phrenic
   3. Posterior intercostal
   4. Bronchial

   A B C D E

6. WHICH STATEMENT(S) ABOUT A BRONCHOPULMONARY SEGMENT IS (ARE) TRUE?
   1. The term bronchopulmonary segment is applied to the largest segment within a lobe
   2. Surgical removal of a bronchopulmonary segment in disease is feasible
   3. Bronchopulmonary segments are separated from each other by connective tissue septa
   4. Subdivision of the lobes of the lungs into bronchopulmonary segments is useful anatomically, but is of little clinical importance

   A B C D E

7. THE RIGHT LUNG:
   1. Is divided into three lobes
   2. Is shorter than the left
   3. Has both oblique and horizontal fissures
   4. Has eight bronchopulmonary segments

   A B C D E

8. WHICH STATEMENT(S) ABOUT THE PLEURA IS (ARE) CORRECT?
   1. Is a mucous membrane
   2. Encloses a potential space
   3. Parietal pleura is insensitive to pain
   4. Lines the mediastinum

   A B C D E

9. THE AREA LABELLED 'A' IN THE DIAGRAM IS:
   1. The apical bronchopulmonary segment
   2. A common site of pulmonary tuberculosis
   3. Auscultated over the supraclavicular region
   4. Supplied by the left apical segmental bronchus

   A B C D E

A	B	C	D	E
1,2,3	1,3	2,4	only 4	all correct

10.

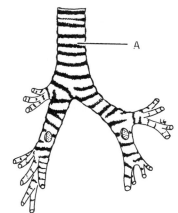

THE STRUCTURE LABELLED 'A' IN THE DIAGRAM:
1. Begins at the level of the 5th cervical vertebra
2. Enters the superior mediastinum
3. Bifurcates at the lower border of the 5th thoracic vertebra
4. Lies anterior to the esophagus

A B C D E

---------------------- ANSWERS, NOTES AND EXPLANATIONS ----------------------

1. **A.** <u>1,2, and 3 are correct</u>. The apices of both lungs project 2-5 cm above the level of the first sternocostal joint. The left lung is divided by an oblique fissure into two lobes, the superior and inferior. The left lung is subdivided into eight bronchopulmonary segments each of which receives a segmental (tertiary) bronchus. The mediastinal surface of the left lung has a deep impression (the cardiac notch) formed by the heart and pericardium.

2. **E.** <u>All are correct</u>. The root of the lung, formed by the structures entering and leaving the lung at the hilum, connects the medial surface of the lung to the heart and trachea. The main structures in the root of the lung are the bronchi and pulmonary vessels. Other structures are nerves, bronchial vessels, lymphatic vessels and nodes. All these structures are embedded in connective tissue and are enveloped in pleura.

3. **E.** <u>All are correct</u>. The subdivision of the lungs into bronchopulmonary segments is related to the arrangement of the bronchial tree. Each bronchopulmonary segment is supplied by a tertiary bronchus. The left lung is subdivided into eight bronchopulmonary segments, named as follows: apicoposterior and anterior (upper lobe bronchus); superior and inferior (lingular bronchus); apical, anteromedial basal, lateral basal, and posterior basal (lower lobe) bronchus). As bronchopulmonary segments may be resected, they are clinically important in cases where the lesion (tumor or infection) is confined or localized in a segment.

4. **A.** <u>1, 2, and 3 are correct</u>. The right lung is divided into upper, middle, and lower lobes by oblique and horizontal fissures, whereas the left lung is divided into upper and lower lobes by the oblique fissure. On the right, the oblique fissure originates at the head of the fourth or fifth rib and follows the line of the <u>sixth</u> rib anteriorly. The left oblique fissure usually originates a little <u>higher</u> than on the right. The horizontal fissure originates at the midaxillary line at the level of the sixth rib, and passes anteriorly to the level of the fourth costal cartilage. The upper and lower lobes of each lung are large, whereas the middle lobe, present only in the right lung, is small and triangular in shape.

5. **A.** <u>1, 2, and 3 are correct</u>. The parietal pleura is supplied by branches of the internal thoracic, superior phrenic, and posterior intercostal arteries,

and in addition, receives contributions from the superior intercostal arteries. The visceral pleura is supplied by the bronchial arteries.

6. **A.** <u>1, 2, and 3 are correct</u>. Each bronchopulmonary segment receives a segmental bronchus, an artery, and a vein. The arrangement of these segments is of clinical importance because pulmonary disorders may be localized in, or confined to, a bronchopulmonary segment. The involved segment (tumor or infection) may be located by radiography or bronchoscopy and treated, e.g., by surgical removal.

7. **A.** <u>1, 2, and 3 are correct</u>. The right lung is divided by oblique and horizontal fissures into upper, middle, and lower lobes. The right lung is shorter because the large right lobe of the liver pushes the diaphragm upwards. However, the right lung is heavier and has a greater volume, compared to the left lung. The right lung is subdivided into 10 bronchopulmonary segments as follows: apical, posterior, and anterior (upper lobe); medial and lateral (middle lobe); superior, medial basal, anterior basal, lateral basal, and posterior basal (lower lobe).

8. **C.** <u>2 and 4 are correct</u>. The pleura is a serous membrane. The portion lining the thoracic wall and mediastinum is called the parietal pleura. The pleura reflecting from the mediastinum onto the surface of the lung is called the visceral pleura. The two parts of the pleura enclose a potential space, the pleura cavity. The parietal pleura is very sensitive to pain, especially the costal pleura; however the visceral pleura is insensitive to pain.

9. **A.** <u>1, 2, and 3 are correct</u>. The region indicated in the diagram is the apical bronchopulmonary segment of the right lung. It is supplied by the right apical segmental bronchus of the right bronchus. This segment is a common site of pulmonary tuberculosis. Examination of the apex of the lung is conducted by placing the chest piece of the stethoscope over the supraclavicular region for evidence of adventitious (abnormal) sounds. In the left lung, there is a composite apicoposterior bronchopulmonary segment but not an apical segment.

10. **C.** <u>2 and 4 are correct</u>. The trachea is the continuation of the larynx and begins at the level of the 6th cervical vertebra. It terminates at the lower border of the fourth thoracic vertebra by dividing into the right and left primary bronchi. The esophagus forms a posterior relation to the trachea. The left recurrent laryngeal nerve lies in the groove between the trachea and the esophagus. The right recurrent laryngeal nerve also has a similar relationship, but it is shorter.

FIVE-CHOICE ASSOCIATION QUESTIONS

DIRECTIONS: Each group of questions below consists of a numbered list of descriptive words or phrases accompanied by a diagram with certain parts indicated by letters, or by a list of lettered headings. For each numbered word or phrase, SELECT THE LETTERED PART OR HEADING that matches it correctly. Then insert the letter in the space to the right of the appropriate number. Sometimes more than one numbered word or phrase may be correctly matched to the same lettered part or heading.

ASSOCIATION QUESTIONS

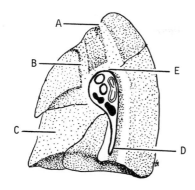

1. ____ An impression for an artery supplying the upper limb.
2. ____ Related to a chamber that forms the right border of the heart.
3. ____ A groove accommodating a venous channel draining the thoracic cage.
4. ____ An impression for a vessel that drains blood from the upper half of the body.

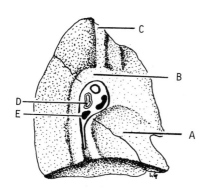

5. ____ An impression for a large vessel arising from the left ventricle.
6. ____ Is related to the chamber of the heart with the thickest wall.
7. ____ Is related to a blood vessel supplying the upper extremity.
8. ____ A structure forming the basis for the division of lung tissue into segments.

---------------------- ANSWERS, NOTES AND EXPLANATIONS --------------------

1. **A.** The anterior border immediately below the apex of the lung has an impression for the right subclavian artery which supplies blood to the upper limb.

2. **C.** The large depression in front of and below the root of the right lung is produced by the right atrium of the heart. This chamber receives the venous return from the entire body via the superior and inferior vena cavae. The right atrium forms the right border of the heart.

3. **E.** The prominent groove located superior and posterior to the root of the right lung is caused by the azygos vein. The vein directly drains the second to eleventh intercostal spaces of the right side; and indirectly, via the accessory (superior) and inferior hemiazygos veins, the fourth to eleventh intercostal spaces on the left. Thus, the venous return of the thorax is primarily via the azygos vein except for the following: the first intercostal space on each side drains into the corresponding brachiocephalic veins; the second and third intercostal spaces on the left drain into the left brachiocephalic vein via the left superior intercostal vein.

4. **B.** The short wide vertical impression extending from the anterior margin of the lung to the depression for the right atrium (C) is caused by the superior vena cava. The superior vena cava is formed by the union of the right and left brachiocephalic veins behind the right first sternocostal junction and descends to enter the upper and posterior aspect of the right atrium at the level of the third sternocostal junction. The azygos vein joins the superior

vena cava at the level of the second rib.

5. B. The medial surface of the left lung, anterior, superior and posterior to the root, has a well marked impression for the arch of the aorta, the continuation of the ascending aorta. This arch extends from the second right costal cartilage to the second left costal cartilage, where it continues as the descending aorta. The arch of the aorta occupies the lower half of the superior mediastinum.

6. A. The large concave area located in front of the root of the lung and the pulmonary ligament accommodates the apex of the left ventricle. The left ventricle has the thickest wall of all the heart chambers as it has to pump blood through the systemic circulation and must overcome the peripheral resistance offered by the blood vessels.

7. C. The short, vertical depression superior to the impression created by the aortic arch is occupied by the left subclavian artery, supplying blood to the upper extremity. A vertical depression anterior to the groove for the subclavian artery is for the brachiocephalic artery.

8. D. The wall of the bronchus contains cartilage and is the most posterior structure in the hilum of the lung. This is the primary left bronchus (D) which divides into two secondary (lobar) bronchi for the two lobes of the left lung. The secondary bronchi divide into tertiary (segmental) bronchi. Each segmental bronchus and the portion of the lung tissue supplied by it is called a bronchopulmonary segment. The bronchopulmonary segment is an anatomical, pathological, and surgical unit. A branch of the pulmonary vein is indicated by E.

PERICARDIUM AND HEART

FIVE-CHOICE COMPLETION QUESTIONS

DIRECTION: Each of the following questions or incomplete statements is followed by five suggested answers or completions. SELECT THE ONE BEST ANSWER in each case and then underline the appropriate letter at the lower right of each question.

1. ON THE ANTERIOR CHEST WALL, THE MITRAL VALVE IS MOST AUDIBLE OVER THE:
   A. Left second intercostal space
   B. Right junction of the sternal body and the costal margin
   C. Left fifth intercostal space
   D. Left junction of the sternal body and the costal margin
   E. Right second intercostal space   A B C D E

2. THE EPICARDIUM IS FORMED BY THE:
   A. Fibrous pericardium
   B. Parietal layer of serous pericardium
   C. Visceral layer of serous pericardium
   D. Muscle fibers
   E. None of the above   A B C D E

3. THE VENAE CORDIS MINIMAE DRAIN INTO THE:
   A. Right atrium   D. Left ventricle
   B. Left atrium   E. All of the above
   C. Right ventricle   A B C D E

SELECT THE ONE BEST ANSWER

4.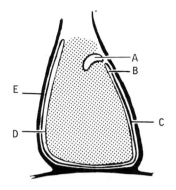

THE OBLIQUE SINUS IS INDICATED BY _____.

A B C D E

5. THE CORONARY SINUS IS DERIVED FROM THE:
   A. Sinus venosus
   B. Right anterior cardinal and right common cardinal veins
   C. Right horn of the sinus venosus
   D. Left horn of the sinus venosus
   E. None of the above

   A B C D E

6. THE RIGHT ATRIUM RECEIVES ALL THE FOLLOWING VESSELS EXCEPT THE:
   A. Superior vena cava   D. Pulmonary veins
   B. Inferior vena cava   E. Coronary sinus
   C. Small coronary veins

   A B C D E

7. WHICH OF THE FOLLOWING STATEMENTS REGARDING THE RIGHT VENTRICLE IS FALSE?
   A. Part of its wall is smooth
   B. Communicates with the right atrium via the mitral valve
   C. Is triangular in shape
   D. Contains the moderator band
   E. Has papillary muscles attached to its internal wall

   A B C D E

8.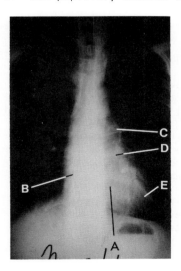

THE PULMONARY TRUNK IS LABELLED _____.

A B C D E

SELECT THE ONE BEST ANSWER

9. WHICH OF THE FOLLOWING STATEMENTS IS FALSE?
   A. The right coronary artery is smaller than the left
   B. The left coronary artery arises from the posterior aortic sinus
   C. The right coronary artery is located in the atrioventricular groove
   D. The circumflex branch arises from the right coronary artery
   E. The marginal branch arises from the right coronary artery

   A B C D E

10. WHICH OF THE FOLLOWING STATEMENTS ABOUT THE CONDUCTING SYSTEM OF THE HEART IS FALSE?
    A. The sinuatrial node is located in the wall of the right atrium
    B. The atrioventricular node lies in the interatrial septum
    C. The atrioventricular bundle is a collection of specialized nerve fibers
    D. Fibers from the right crus of the atrioventricular bundle pass through the moderator band
    E. Cardiac impulses are initiated in the sinuatrial node

    A B C D E

11.

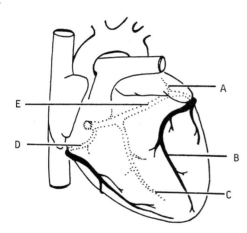

THE MIDDLE CARDIAC VEIN IS LABELLED ____.

A B C D E

---------------------- ANSWERS, NOTES AND EXPLANATIONS ----------------------

1. C. The sound of the mitral valve is most audible over the left fifth intercostal space on the anterior chest wall. This heart sound, like all other valve sounds, does not correspond to the anatomical location of the valve. It represents the area where the cavity in which the valve lies is nearest to the body surface and as far away as possible from the other valves. The pulmonary and aortic valve sounds are located over the left and right second intercostal spaces, respectively. The tricuspid valve sound is most audible over the lower right part of the body of the sternum.

2. C. The pericardium consists of an outer conical fibrous pericardium which encloses an inner double-layered serous sac, the serous pericardium. The parietal layer of the serous pericardium is reflected from the roots of the great vessels of the heart onto its surface as the visceral layer or epicardium. The pericardial sac is therefore a closed serous cavity surrounding the heart. It contains a small amount of fluid which lubricates the two serous

layers, thereby preventing friction during movements of the heart.

3. **E.** The venae cordis minimae (Thebesian veins) are tiny venous channels which begin in the walls of the heart and drain directly into all four chambers. Compared to the coronary sinus, the venous return to the right atrium via these small direct channels is considerable.

4. **B.** The oblique sinus of the pericardium lies between the entrances of the right and left pulmonary veins into the left atrium on the posterior aspect of the heart. The sinus is limited superiorly by a reflection of the serous layer into the inner surface of the fibrous pericardium (E) and is called the parietal pericardium (C). The serous layer of the pericardium reflected onto the heart is the visceral pericardium (D) or epicardium. The transverse sinus of the pericardium (A) is located between the aorta and the pulmonary trunk anteriorly, and the left atrium and superior vena cava posteriorly. This sinus is formed by the serous pericardium covering the above structures. The outer fibrous pericardium, consisting of a network of collagen and elastic fibers, blends with the central tendon of the diaphragm.

5. **D.** The left horn of the sinus venosus forms the coronary sinus; the right horn becomes incorporated into the wall of the right atrium. Hence if you chose A, your answer is correct, but it is not the best answer because it is not so specific as D. The right anterior cardinal and right common cardinal veins become the superior vena cava.

6. **D.** The interior of the right atrium is both smooth and trabeculated because of its dual embryological origin from the sinus venosus and the primitive atrium, respectively. The smooth part contains openings for the superior vena cava above; the inferior vena cava, below; and the coronary sinus between the opening of the inferior vena cava and the right atrioventricular orifice. The small coronary veins drain directly into the right atrium. The pulmonary veins, carrying oxygenated blood, enter the left atrium.

7. **B.** The right atrium communicates with the right ventricle through the right atrioventricular orifice, guarded by the tricuspid valve. Like the left ventricle, most of its internal wall is trabeculated and has an extensive, irregular network of muscular elevations (trabeculae carneae), but the infundibulum leading into the pulmonary trunk is smooth-walled. The right branch of the atrioventricular bundle (of His) descends through the interventricular septum into the musculature of the right ventricle and reaches the anterior papillary muscle via the moderator band. This band is a muscular elevation extending from the interventricular septum to the anterior wall. It has been suggested that the moderator band helps to prevent overdistension of the right ventricle.

8. **D.** In this posterior-anterior radiograph of the chest, the pulmonary trunk is labelled D. It lies immediately below the "aortic knuckle" (C) of the aortic arch, which usually becomes more prominent with increasing age. The aortic arch and the pulmonary trunk together with the left ventricle (E) form the left border of the heart silhouette. The right ventricle (A) and the right atrium (B) are also indicated. The right atrium forms the entire right border of the heart, whereas the sternocostal surface of the heart is represented almost entirely by the right ventricle (A).

9. **D.** The circumflex branch, as well as the anterior interventricular, arises from the left coronary artery. The heart is supplied by the coronary arteries. The right coronary artery is smaller than the left; originates from the posterior aortic sinus; and is located in the right part of the atrioventricular groove.

10.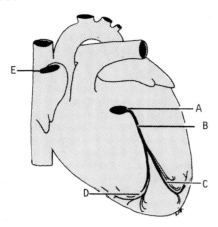

C. The atrioventricular bundle (B) consists of specialized conducting cardiac muscle (and not nerve) fibers, passing from the atrioventricular node (A) through the interventricular septum where it divides into left and right crura (C & D). The right and left branches of the atrioventricular bundle descend in the interventricular septum, deep to the endocardium, to supply the ventricular walls. The right crus (D) reaches the anterior papillary muscle of the right ventricle via the moderator band. The left crura (C) supplies the myocardium and the anterior and posterior papillary muscles of the left ventricle. The sinuatrial (sinoatrial) node (E) lies in the wall of the right atrium at the upper end of the crista terminalis, near to the opening of the superior vena cava. The atrioventricular node (A) is located in the interatrial septum, above the opening of the coronary sinus.

11. C. The coronary sinus (E) is the largest vein draining the heart. It lies in the coronary (atrioventricular) sulcus and opens directly into the right atrium. The coronary sinus receives the following tributaries: the great (B), middle (C), and small (D) cardiac veins. The great and middle cardiac veins are located in the anterior and posterior interventricular grooves, respectively. The small cardiac vein lies with the right marginal artery along the lower right border of the heart. These vessels account for approximately two-thirds of the venous drainage of the heart; the remaining third is via small venous channels, the venae cordis minimae (Thebesian veins), which empty directly into all heart chambers. In addition, the small oblique vein (A), a remnant of the left common cardinal vein, drains the posterior wall of the left atrium.

MULTI-COMPLETION QUESTIONS

DIRECTIONS: In each of the following questions or incomplete statements, ONE OR MORE of the completions given is correct. At the lower right of each question, underline A if 1, 2 and 3 are correct; B if 1 and 3 are correct; C if 2 and 4 are correct; D if only 4 is correct; and E if all are correct.

1. THE PERICARDIAL SAC:
   1. Extends from the level of the 2nd to 6th costal cartilages
   2. Lies immediately in front of the esophagus
   3. Fuses with the diaphragm
   4. Lies posterior to the phrenic nerves   A B C D E

2. THE BASE OF THE HEART IS FORMED BY THE:
   1. Left atrium   3. Right atrium
   2. Left ventricle   4. Right ventricle   A B C D E

3. WHICH STRUCTURE(S) CONTRIBUTE(S) TO THE WALL OF THE HEART?
   1. Endocardium   3. Myocardium
   2. Visceral pericardium   4. Pleuropericardial membrane   A B C D E

4. THE DIAPHRAGMATIC SURFACE OF THE HEART IS FORMED BY THE:
   1. Right atrium   3. Left atrium
   2. Right ventricle   4. Left ventricle   A B C D E

A	B	C	D	E
1,2,3	1,3	2,4	only 4	all correct

5. WHICH OF THE FOLLOWING STATEMENTS ABOUT THE AURICLES OF THE HEART IS (ARE) TRUE?
   1. Have a rough trabeculated appearance
   2. Form part of the atria of the heart
   3. Represent the remains of the primitive atrium
   4. Are derived from the sinus venosus   A B C D E

6. CONCERNING THE FOSSA OVALIS IN THE RIGHT ATRIUM:
   1. Its floor is derived from the septum primum
   2. It is bounded by the limbus (annulus) fossae ovalis
   3. Its floor may be recognized in the right atrium
   4. The limbus fossae ovalis is derived from the septum secundum   A B C D E

7. WHICH STATEMENT(S) ABOUT THE HEART IS (ARE) CORRECT?
   1. The interior of the right atrium is divided into two parts by the sulcus terminalis
   2. The floor of the fossa ovalis represents the persistent portion of the septum primum of the embryonic heart
   3. Structural defects of the ventricular septum are rare
   4. The infundibulum of the right ventricle is derived from the bulbus cordis of the embryonic heart   A B C D E

8. WHICH STATEMENT(S) ABOUT THE HEART IS (ARE) CORRECT?
   1. The atria form its base
   2. The long axis of the heart is commonly transverse in late pregnancy
   3. Blood flow from the atria to the ventricles is horizontal and forward
   4. Diaphragmatic movements have little to do with heart positioning   A B C D E

9. CONCERNING THE LEFT VENTRICLE:
   1. Its wall is thicker than the right
   2. The mitral valve has only two cusps
   3. The aortic vestibule is smooth
   4. Its internal surface is trabeculated   A B C D E

10. WHICH STATEMENT(S) ABOUT THE ATRIA OF THE HEART IS (ARE) CORRECT?
    1. Left atrial wall is roughened by musculi pectinati
    2. Interatrial wall of the right atrium is smooth
    3. Intervenous tubercle lies between the pulmonary veins
    4. Are in communication throughout prenatal life   A B C D E

11. WHICH STRUCTURE(S) IS (ARE) LOCATED IN THE VENTRICULAR WALLS OF THE HEART?
    1. Trabeculae carneae   3. Chordae tendineae
    2. Musculi pectinati   4. Crista terminalis   A B C D E

12. THE LEFT ATRIUM:
    1. Receives the pulmonary veins
    2. Communicates with the left ventricle via the tricuspid valve
    3. Has smooth and trabeculated portions on its internal wall
    4. Is demarcated from the right atrium by the coronary sulcus   A B C D E

A	B	C	D	E
1,2,3	1,3	2,4	only 4	all correct

13. CONCERNING THE CORONARY ARTERIES:
    1. Lie superficial to the myocardium
    2. Flow of blood through them occurs during diastole
    3. Their branches are end arteries
    4. Are the first branches of the aorta   A B C D E

14. WHICH OF THE FOLLOWING VEINS DRAIN DIRECTLY INTO THE CORONARY SINUS?
    1. Great cardiac   3. Small cardiac
    2. Middle cardiac   4. Anterior cardiac   A B C D E

15. THE FOSSA OVALIS:
    1. Has a membranous floor
    2. Is present only after birth
    3. Is located in the interatrial septum
    4. Permits flow of blood from the right to the left atrium   A B C D E

16. THE LEFT CORONARY ARTERY:
    1. Arises from the anterior aortic sinus
    2. Lies between the left auricle and the pulmonary trunk
    3. Is smaller than the right coronary artery
    4. Is located in the atrioventricular groove   A B C D E

17. WHICH OF THE FOLLOWING VESSELS DRAIN(S) INTO THE RIGHT ATRIUM?
    1. Superior vena cava   3. Coronary veins
    2. Inferior vena cava   4. Coronary sinus   A B C D E

18. THE CORONARY SINUS:
    1. Is a thin-walled venous channel
    2. Opens directly into the right atrium
    3. Lies in the coronary sulcus
    4. Is 3-5 cm long   A B C D E

19. THE MAJOR NAMED BRANCH(ES) OF THE LEFT CORONARY ARTERY IS (ARE) THE:
    1. Anterior interventricular   3. Circumflex
    2. Marginal   4. Posterior interventricular   A B C D E

20. THE INTRINSIC CONDUCTING SYSTEM OF THE HEART CONSISTS OF THE:
    1. Sinuatrial node   3. Atrioventricular bundle
    2. Atrioventricular node   4. Purkinje fibers   A B C D E

21. THE EXTRINSIC INNERVATION OF THE HEART IS VIA WHICH OF THE FOLLOWING?
    1. Vagus nerve   3. Thoracic and cervical ganglia
    2. Superficial cardiac plexus   4. Deep cardiac plexus   A B C D E

22. POSTGANGLIONIC SYMPATHETIC FIBERS REACH THE HEART VIA THE FOLLOWING NERVES:
    1. Cervical   3. Thoracic cardiac
    2. Inferior cervical cardiac   4. Vagus   A B C D E

23. THE RIGHT CORONARY ARTERY:
    1. Arises from the anterior aortic sinus
    2. Passes between the right auricle and the pulmonary trunk
    3. Is located in the coronary sulcus
    4. Anastomoses with the left coronary artery   A B C D E

---------------------- ANSWERS, NOTES AND EXPLANATIONS ----------------------

1. **A.** <u>1,2, and 3 are correct</u>. The pericardium is a fibroserous sac containing the heart and the roots of the great vessels. It lies posterior to the sternum and the 2nd to 6th costal cartilages. The esophagus and descending thoracic aorta lie immediately behind the pericardium; its base fuses with the central tendon of the diaphragm. The phrenic nerve is in contact with the lateral wall of the pericardium as it descends between the pericardial sac and the mediastinal pleura on each side, anterior to the roots of the lungs.

2. **B.** <u>1 and 3 are correct</u>. The cone-shaped heart has an apex and a base. The base is formed chiefly by the left atrium with the right atrium forming its right border. The great veins enter the base of the heart, which is directed posteriorly, upwards, and to the right. The apex of the heart usually lies at the left fifth intercostal space, approximately 8 cm from the midsternal line. Its pulsation (apex beat) is clinically useful for determining the position of the left border of the heart and for the auscultation of the mitral valve.

3. **A.** <u>1, 2, and 3 are correct</u>. The wall of the heart consists of three layers: the inner layer or endocardium; the middle layer or myocardium; and the outer layer or epicardium (also called the visceral pericardium). The pleuropericardial membrane in the embryo gives rise to the fibrous pericardium. Between the pericardium and the epicardium is a potential space, the pericardial cavity.

4. **C.** <u>2 and 4 are correct</u>. The heart may be described as having three surfaces: diaphragmatic, sternocostal, and pulmonary. The diaphragmatic surface is formed by the ventricles and lies chiefly against the central tendon of the diaphragm. The sternocostal surface is formed primarily by the right ventricle with smaller contributions from the left ventricle and right atrium. The pulmonary or left surface is formed by the left ventricle. The cardiac impression on the medial surface of the left lung is caused by pressure from the left ventricle. The right atrium makes a similar impression on the mediastinal surface of the right lung.

5. **A.** <u>1, 2, and 3 are correct</u>. The auricles are the parts of the atria derived from the primitive atrium; they contain bundles of muscle fibers called musculi pectinati. The smooth-walled portions of the right and left atria are derived from the sinus venosus and the primitive pulmonary vein, respectively. The terms atria and auricles should not be used interchangeably because they are not synonymous. The auricles are the small conical appendages to the atria.

6. **E.** <u>All are correct</u>. The foramen ovale, an opening in the septum secundum, usually closes functionally at birth. Later, anatomical closure occurs and the valve of the foramen ovale, derived from the septum primum, becomes the floor of the fossa ovalis. The lower edge of the septum secundum forms a rounded fold, the limbus fossae ovalis (annulus ovalis). The fossa ovalis is most evident in the right atrium and may also be recognized in the left atrium because the thin floor of the fossa (derived from the septum primum) appears as a translucent region of the interatrial septum.

7. **C.** <u>2 and 4 are correct</u>. The interior of the right atrium is divided into two parts by the crista terminalis, and externally by the sulcus terminalis. When the foramen ovale closes at birth, its valve derived from the septum primum forms the floor of the fossa ovalis. The smooth-walled portion of the right ventricle, known as the infundibulum is derived from the bulbus cordis of the embryonic heart. In the left ventricle, the bulbus cordis is represented by the aortic vestibule. Ventricular septal defects are relatively common,

ranking first in frequency on all lists of cardiac defects. Defects in the membranous upper part of the septum are most common because of its complex embryological origin.

8. **A. 1, 2, and 3 are correct.** The base of the heart is formed by the atria and its long axis is normally oriented obliquely in the thoracic cage. The long axis may lie horizontal in infancy, obesity and late pregnancy. The diaphragm and its movements are the most important factors determining the position of the heart. The mitral and tricuspid valves lie in a vertical plane and blood flowing through the atrioventricular canals is in an anterior horizontal direction.

9. **E. All are correct.** The left ventricle is longer, narrower, and has a thicker wall, than the right. Its internal wall shows an extensive meshwork of muscular ridges, the trabeculae carneae, except for the fibrous aortic vestibule, which is smooth. The two cusps of the mitral (bicuspid atrioventricular) valve are attached by chordae tendineae to the papillary muscles, two prominent conical projections from the wall of the cavity. The aorta arises from the left ventricle; the aortic orifice is surrounded by a fibrous ring to which the three semilunar cusps of the aortic valve are attached.

10. **C. 2 and 4 are correct.** The walls of the atria are relatively smooth with the exception of the musculi pectinati of the right atrium and the auricles. A muscular projection of the atrial wall, the intervenous tubercle, lies between the openings of the superior and inferior venae cavae into the right atrium. Throughout prenatal life, the atria are in communication with each other through an opening in the interatrial septum, the foramen ovale.

11. **B. 1 and 3 are correct.** Most of the internal surfaces of the ventricles are irregular due to muscular projections, the trabeculae carneae. These muscular trabeculae take various shapes: the pillar-shaped ones, called papillary muscles, are attached by tendinous projections (chordae tendineae) to the cusps of the atrioventricular valves. The chordae tendineae prevent eversion of the valves during contraction of the ventricle. The musculi pectinati are small muscular projections of the wall of the auricles, whereas the crista terminalis is a muscular ridge extending between the orifices of the superior and inferior venae cavae of the right atrium.

12. **B. 1 and 3 are correct.** The left atrium lies posterior to the right atrium and together they form the base of the heart. The left atrium is demarcated from the left ventricle externally by the coronary sulcus. The internal wall of the left atrium is smooth; the auricle, however, is trabeculated (muscular). The four pulmonary veins enter its smooth posterior wall. Blood from the left atrium passes into the left ventricle via the mitral (bicuspid) valve. This valve was likened to a bishop's miter by Vesalius (16th century); thus it is commonly called the mitral valve.

13. **E. All are correct.** The heart is supplied by right and left coronary arteries, arising from the anterior and posterior aortic sinuses, respectively. The coronary arteries are the first vessels to arise from the aorta with their principal branches located in the interventricular and atrioventricular grooves. The flow of blood through these vessels occurs during diastole because the branches penetrating the myocardium are compressed during systole. The branches of the coronary arteries are considered to be "end arteries" without any overlap. For this reason, occlusion of these vessels leads to ischemia of heart muscle and subsequent infarction.

14. **A. 1, 2, and 3 are correct.** The following veins drain directly into the coronary sinus: the great, middle, and small cardiac veins; and the oblique

vein of the left atrium. The anterior cardiac veins are several small vessels draining the anterior wall of the right ventricle and enter the right atrium directly. The venae cordis minimae drain into all chambers of the heart.

15. **A.** <u>1, 2, and 3 are correct</u>. The interatrial septum is set obliquely, facing forward and to the right. An oval depression, the fossa ovalis, is present in this septum. Its floor is membranous and represents the remnants of the valve for the opening (foramen ovale) which permitted the flow of blood from the right into the left atrium before birth. Complete anatomical closure of the foramen ovale fails to occur in up to 25 per cent of persons, thus a small opening (probe patency) remains.

16. **C.** <u>2 and 4 are correct</u>. The left coronary artery is larger than the right one. It arises from the left posterior aortic sinus, just above the aortic semilunar valve, and descends between the left auricle and the pulmonary trunk to reach the atrioventricular groove. The two principal branches of the left coronary artery are the anterior interventricular and the circumflex.

17. **E.** <u>All are correct</u>. The main part of the right atrium of the heart develops from the sinus venosus and is smooth-walled. The atrial appendage or auricle is trabeculated (muscular) and is derived from the primitive atrium. A groove, the sulcus terminalis, on the surface of the right atrium corresponds to a ridge, the crista terminalis, on its internal wall. The sulcus terminalis and the crista terminalis demarcate the two embryologically different parts of the atrium described above. The superior and inferior venae cavae, the small coronary veins, and the large coronary sinus receiving blood from the cardiac veins enter the right atrium. The oblique interatrial septum represents the medial wall of this chamber and contains an oval depression, the fossa ovalis. A thick fold, the limbus or annulus, is present at the upper margin of the fossa ovalis.

18. **E.** <u>All are correct</u>. The coronary sinus is a relatively short (3-5 cm), wide, thin-walled venous channel, lying in the coronary sulcus between the left atrium and ventricle. The coronary sinus opens directly into the right atrium between the inferior vena cava and the atrioventricular (tricuspid) openings. The valve of the coronary sinus guards its entry into the atrium.

19. **B.** <u>1 and 3 are correct</u>. The left coronary artery gives rise to an anterior interventricular and a circumflex branch as it passes between the left auricle and the pulmonary trunk. The anterior interventricular branch, a direct continuation of the left coronary artery, enters the anterior interventricular groove and descends towards the apex of the heart. It anastomoses with the posterior interventricular branch of the right coronary artery. The circumflex branch passes to the left in the coronary sulcus between the left atrium and ventricle, ending by anastomosing with the transverse branch of the right coronary artery. The marginal branch of the right coronary artery courses along the lower border and anterior aspect of the heart to reach the apex. Thus, the major branches of the coronary arteries form two important anastomoses, one in the posterior coronary (atrioventricular) sulcus, the other in the posterior interventricular groove.

20. **E.** <u>All are correct</u>. The intrinsic conducting system of the heart is responsible for the coordination of the orderly sequence of contraction and relaxation of the heart musculature. It consists of highly specialized conducting cardiac muscle fibers arranged in a sinuatrial (sinus) node, an atrioventricular node, the atrioventricular bundle (of His), and Purkinje fibers.

21. **E.** <u>All are correct</u>. Autonomic nerve fibers from the sympathetic trunks and the vagus nerves (vagi) form plexuses, from which fibers proceed to the heart,

modulating its intrinsic rhythm. The thoracic and cervical ganglia provide sympathetic fibers which, when stimulated, increase the heart rate; stimulation of the vagal nerves (parasympathetic) produces the opposite effect. The superficial and deep cardiac plexuses are located on the ligamentum arteriosum and at the bifurcation of the trachea, respectively.

22. **A.** <u>1, 2, and 3 are correct</u>. The white rami communicantes entering the ganglia contain preganglionic fibers. The grey rami communicantes contain postganglionic fibers to the heart from the first cervical to the fifth thoracic spinal cord segments, and are carried via the cervical, inferior cervical thoracic (cervicothoracic), and thoracic cardiac nerves. The vagus nerve provides parasympathetic innervation to the heart.

23. **E.** <u>All are correct</u>. The right coronary artery is smaller than the left and arises from the anterior aortic sinus, just above the aortic semilunar valve. The right coronary artery descends between the right auricle and the pulmonary trunk, then it continues in the coronary sulcus to reach the posterior surface of the heart, where it anastomoses with the left coronary artery. Its principal branches are the right marginal and the posterior interventricular.

F I V E - C H O I C E   A S S O C I A T I O N   Q U E S T I O N S

DIRECTIONS: Each group of questions below consists of a numbered list of descriptive words or phrases accompanied by a diagram with certain parts indicated by letters, or by a list of lettered headings. For each numbered word or phrase, SELECT THE LETTERED PART OR HEADING that matches it correctly. Then insert the letter in the space to the right of the appropriate number. Sometimes more than one numbered word or phrase may be correctly matched to the same lettered part or heading.

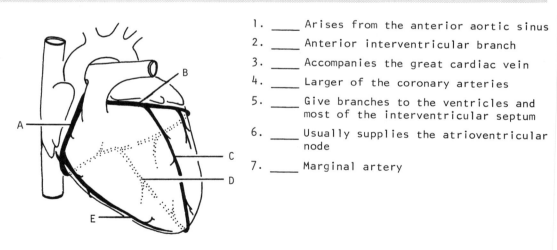

1. ____ Arises from the anterior aortic sinus
2. ____ Anterior interventricular branch
3. ____ Accompanies the great cardiac vein
4. ____ Larger of the coronary arteries
5. ____ Give branches to the ventricles and most of the interventricular septum
6. ____ Usually supplies the atrioventricular node
7. ____ Marginal artery

---------------------- ANSWERS, NOTES AND EXPLANATIONS --------------------

1. **A.** The right coronary artery (A) arises from the right anterior aortic sinus and passes forwards between the right auricle and pulmonary trunk. The left coronary artery (B) arises from the left posterior aortic sinus and passes behind the pulmonary trunk; then it courses between the left auricle and the pulmonary trunk to reach the anterior coronary (atrioventricular) sulcus.

2. **C.** The anterior interventricular branch is derived from the left coronary artery and is located in the anterior interventricular groove. It curves

around the apex of the heart and anastomoses with the posterior interventricular branch (D) of the right coronary artery.

3. **C.** Both the anterior interventricular branch of the left coronary artery and the great cardiac vein lie in the anterior interventricular groove. The middle and small cardiac veins accompany the posterior (descending) interventricular (D) and the transverse branches (not labelled) of the right coronary artery, respectively.

4. **B.** The heart is supplied by branches of the left (B) and right (A) coronary arteries, the first vessels to arise from the aorta. The left coronary artery is larger than the right. Its main branches are the anterior interventricular (C) and the circumflex (unlabelled) which anastomose with posterior interventricular and the transverse branches of the right coronary artery.

5. **C.** The anterior interventricular artery supplies both ventricles and most of the interventricular septum. It is an important branch of the left coronary artery and descends in the anterior interventricular groove, together with the great cardiac vein, lymphatic vessels, and cardiac nerves.

6. **D.** The atrioventricular node, like the sinuatrial (sinus) node, is a meshwork of specialized muscle fibers that conduct impulses. It lies beneath the endocardium in the septum of the right atrium. The atrioventricular node usually receives its blood supply from the posterior interventricular artery. The sinuatrial node, the "pacemaker" of the heart, is supplied by the sinus node artery, usually a branch of the right coronary artery.

7. **E.** The marginal branch of the right coronary artery is the larger of its two principal branches. It passes along the lower border of the right ventricle to the apex of the heart, accompanied by the small cardiac vein.

5. THE ABDOMEN

O B J E C T I V E S

Abdominal Walls

Be Able To:

* Describe the origin, course, insertion, and innervation of the muscles of the anterolateral abdominal wall.

* Illustrate the formation of the rectus sheath above the level of the costal margin; below the midpoint between the umbilicus and the symphysis pubis; and in the intermediate area between the above mentioned levels.

* List the contents of the rectus sheath.

* Describe the inguinal canal and its contents, including a discussion of its development.

* Differentiate between direct and indirect inguinal hernias.

* Describe the anatomical basis of femoral hernias.

* Draw an outline of the anterior abdominal wall indicating the position of the following surgical incisions: (a) supraumbilical midline incision; (b) infra-umbilical midline incision; (c) paramedian incisions; (d) McBurney's incisions; (e) Kocher's incisions; (f) incision for inguinal hernias.

* Describe the different regions of the abdominal cavity.

* List the principal arterial supply, venous return, and lymphatic drainage of the anterior abdominal wall.

* Give a brief account of the muscles of the posterior abdominal wall: psoas major; quadratus lumborum; transversus abdominis, iliacus and crura of the diaphragm.

Peritoneum and Peritoneal Cavity

Be Able To:

* Draw the greater and lesser sacs of the peritoneum in a sagittal section of the abdomen and pelvis.

* Trace the continuity of the peritoneum of the greater and lesser sacs at the level of the epiploic foramen (foramen of Winslow).

* Locate and identify the following peritoneal folds: lesser omentum; greater

omentum; transverse mesocolon; sigmoid (pelvic) mesocolon; falciform ligament; left and right triangular ligaments; upper and lower layers of the coronary ligaments.

  * Discuss the clinical importance of the hepatorenal recess (Morison's pouch); the paracolic gutters; the rectouterine pouch (of Douglas).

Hollow Abdominal Viscera

Be Able To:

  * Comment on the importance of the gastroesophageal junction in preventing reflux of the stomach contents.

  * Discuss the anatomy of the stomach giving its parts, relations, blood and nerve supply, and lymphatic drainage.

  * Describe the parts of the duodenum, giving their relations and blood supply.

  * List the typical characteristics of the jejunum and ileum.

  * Compare the external features that distinguish the large from the small intestine.

  * Write short notes on: cecum; vermiform appendix; ascending, transverse, and descending colon; celiac axis; superior mesenteric artery; inferior mesenteric artery; stomach bed; Meckel's diverticulum; paraduodenal and ileocolic fossae.

  * Identify the parts of the gastrointestinal tract in a radiograph after the use of a contrast medium.

Solid Abdominal Viscera

Be Able To:

  * Describe the surfaces, lobes, relations, and peritoneal reflections of the liver.

  * Discuss the course and relations of the extrahepatic biliary passages.

  * Give an account of the relations and blood supply of the pancreas.

  * Draw a diagram of the spleen indicating its relations and blood supply.

  * Illustrate, using a sagittal section, the coverings of the kidney and suprarenal (adrenal) gland.

  * Describe the relations, arterial supply, and venous drainage of the right and left kidneys.

  * Draw and label a coronal section of the kidney showing its internal structure.

  * Give an account of the right and left suprarenal glands (adrenals), discussing their shape, relations, and vascular supply.

  * Write short notes on: ligamentum teres; ligamentum venosum; ureter; ectopic

kidney; horseshoe kidney; renal angle; and renal colic.

Blood Vessels, Nerves and Lymphatics

Be Able To:

* Describe the abdominal aorta and its paired and unpaired branches.

* Discuss the formation of the inferior vena cava and its tributaries.

* Describe the formation of the portal vein and list its tributaries.

* List the sites of portal and systemic venous anastomosis and discuss their clinical significance.

* Briefly describe the formation and location of the lumbar plexus of nerves.

* Write short notes on: celiac plexus of nerves; superior and inferior mesenteric plexuses of nerves; cisterna chyli; and lumbar lymph nodes.

* Discuss briefly the autonomic innervation of the abdominal viscera.

ABDOMINAL WALLS

FIVE-CHOICE COMPLETION QUESTIONS

DIRECTIONS: Each of the following questions or incomplete statements is followed by five suggested answers or completions. SELECT THE ONE BEST ANSWER in each case and then underline the appropriate letter at the lower right of each question.

1. WHICH OF THE FOLLOWING STRUCTURES PASS THROUGH THE INGUINAL CANAL IN A MALE FETUS?
   A. Cremaster muscle
   B. Testicular artery
   C. Ductus deferens
   D. Processus vaginalis
   E. All of the above

   A B C D E

2. THE PART OF THE ABDOMINAL CAVITY DELINEATED BY THE XIPHOID PROCESS AND THE ADJACENT COSTAL MARGINS, THE TWO VERTICAL PLANES, AND THE SUBCOSTAL PLANE IS KNOWN AS _____ REGION:
   A. Hypochondriac
   B. Hypogastric
   C. Iliac
   D. Epigastric
   E. Suprapubic

   A B C D E

3. WHICH OF THE FOLLOWING STATEMENTS ABOUT THE RIGHT VERTICAL PLANE IS INCORRECT?
   A. Drawn upwards from the right pubic tubercle
   B. Meets the ninth right costal cartilage
   C. Meets the transpyloric plane
   D. Passes through the fundus of the gall bladder
   E. Used for dividing the abdominal cavity into zones

   A B C D E

SELECT THE ONE BEST ANSWER

4. WHICH OF THE FOLLOWING STATEMENTS ABOUT THE TRANSPYLORIC PLANE IS FALSE?
   A. Passes through the pylorus of the stomach
   B. Is at the level of the second lumbar vertebra
   C. Is not used for dividing the abdominal cavity into zones
   D. Corresponds to the level of origin of the superior mesenteric artery
   E. Passes through the fundus of the gall bladder   A B C D E

5. WHICH OF THE FOLLOWING STATEMENTS ABOUT THE INTERTUBERCULAR PLANE (IS) ARE TRUE?
   A. Corresponds to the highest point of the iliac crest
   B. Is one of the planes used in dividing the abdomen into zones
   C. Is at the level of the fifth lumbar vertebra
   D. Is used as a guide when performing a lumbar puncture
   E. All of the above   A B C D E

6. WHICH OF THE FOLLOWING STATEMENTS ABOUT AN INGUINAL HERNIA IS FALSE?
   A. May be present at birth
   B. Are of two types
   C. Is associated with the processus vaginalis
   D. May be associated with weak abdominal musculature
   E. More common in females   A B C D E

7. WHICH OF THE FOLLOWING STATEMENTS REGARDING THE SUBCOSTAL PLANE IS FALSE?
   A. Indicates the level of origin of the inferior mesenteric artery
   B. Passes through the lowest part of the inferior costal margin
   C. Is at the level of the third lumbar vertebra
   D. Passes horizontally through the trunk
   E. None of the above   A B C D E

---------------------ANSWERS, NOTES AND EXPLANATIONS---------------------

1. E. In addition to the four structures listed, the following structures also traverse the inguinal canal: the artery and vein of the ductus deferens; the nerves to the epididymis; the testicular plexus of nerves; the genital branch of the genitofemoral nerve, and the ilioinguinal nerve. Most of the processus vaginalis usually degenerates during the perinatal period, losing its connection with the peritoneum. The tunica vaginalis testis is the normal remnant of the processus vaginalis.

2. D. The epigastric region (epigastrium) is the centrally located, uppermost region of the abdominal cavity and is flanked on each side by the right and left hypochondriac regions. The hypogastric region (hypogastrium), also called the suprapubic region, is the lowest centrally located area, and is flanked by the right and left iliac regions. Pain or tenderness is often felt in the epigastrium in diseases of the stomach and duodenum. Pain felt in the hypogastrium is often related to diseases of the urinary bladder. Discomfort is felt in this region when the bladder is distended. Inflammation of the vermiform appendix causes pain to be felt in the right iliac region, whereas pain is felt in the left iliac fossa when there are disorders of the descending colon.

3. A. The right vertical plane is drawn upwards from the right midinguinal point; not the pubic tubercle. This vertical line meets the ninth costal cartilage

where it bisects the transpyloric plane. The fundus of the gall bladder is usually located behind the right ninth costal cartilage at the transpyloric plane. The right and left vertical lines along with the subcostal and intertubercular planes are used for dividing the abdominal cavity into nine zones.

4. B. The transpyloric plane is at the level of the first lumbar vertebra. This plane passes through the pylorus of the stomach, the fundus of the gall bladder, the origin of the superior mesenteric artery, and the commencement of the portal vein. The transpyloric plane is not used for dividing the abdominal cavity into zones.

5. E. All the statements are true. The intertubercular plane is drawn around the trunk and corresponds to a line drawn between the highest points of the iliac crest. The intertubercular plane, together with the subcostal plane, is used to divide the cavity of the abdomen into three artificial zones: costal, umbilical, and hypogastric. The intertubercular plane lies at the level of the fifth lumbar vertebra and is used as a guide in performing a lumbar puncture. Lumbar puncture or spinal tap is the term used to describe the technique of entering the vertebral canal for inducing spinal anesthesia or for removal of cerebrospinal fluid for clinical examination.

6. E. Inguinal hernias are more common in males because of the greater size of the inguinal canal. Two kinds of inguinal hernia are described: an oblique or indirect type where the hernia passes through the deep ring, the inguinal canal, and superficial ring to descend into the scrotum; a direct hernia occurs through the posterior wall of the inguinal canal, but does not descend into the scrotum. Indirect hernias are associated with persistence of the processus vaginalis, whereas direct hernias are associated with weak abdominal musculature.

7. E. None of the statements is false. The subcostal plane is drawn through the trunk at the lowest part of the inferior costal margin of the tenth costal cartilage. This plane passes through the third lumbar vertebra posteriorly. The inferior mesenteric artery arises from the aorta at the level of the third lumbar vertebra.

MULTI-COMPLETION QUESTIONS

DIRECTIONS: In each of the following questions or incomplete statements, ONE OR MORE of the completions given is correct. At the lower right of each question, underline A if 1, 2 and 3 are correct; B if 1 and 3 are correct; C if 2 and 4 are correct; D if only 4 is correct; and E if all are correct.

1. THE ILIACUS MUSCLE:
   1. Is related to the medial aspect of the psoas major muscle
   2. Inserts into the lesser trochanter
   3. Takes origin from the hip bone below the arcuate line
   4. Is one of the muscles of the floor of the femoral triangle   A B C D E

2. THE RECTUS SHEATH CONTAINS THE:
   1. Musculophrenic artery   3. Intercostal nerves T3-T6
   2. Inferior epigastric artery 4. Rectus abdominis muscle   A B C D E

3. THE PSOAS MAJOR MUSCLE:
   1. Lies posterior to the kidney
   2. Is innervated by the femoral nerve
   3. Is a powerful flexor of the thigh
   4. Inserts into the greater trochanter of the femur   A B C D E

A	B	C	D	E
1,2,3	1,3	2,4	only 4	all correct

4. THE ARTERIAL SUPPLY OF THE ANTERIOR ABDOMINAL WALL IS FROM THE:
   1. Superior epigastric
   2. Lateral thoracic
   3. Inferior epigastric
   4. Superficial external pudendal   A B C D E

5. THE QUADRATUS LUMBORUM MUSCLE:
   1. Has the subcostal, ilioinguinal, and iliohypogastric nerves on its anterior aspect
   2. Is a lateral flexor of the trunk
   3. Is attached inferiorly to the iliac crest
   4. Forms a posterior relation of the kidney   A B C D E

6. THE PSOAS MAJOR MUSCLE:
   1. Takes origin from the twelfth rib
   2. Contains the lumbar plexus within its substance
   3. Enters the thigh within the inguinal ligament
   4. Is one of the muscles of the posterior abdominal wall   A B C D E

7. THE UMBILICAL REGION IS OF CLINICAL SIGNIFICANCE IN THAT IT SEPARATES THE TERRITORIES:
   1. Of arterial supply
   2. Of venous drainage
   3. Of lymphatic drainage
   4. Innervated by the 9th and 11th thoracic segments   A B C D E

8. THE ANATOMICAL FEATURES OF THE VENOUS DRAINAGE OF THE ANTERIOR ABDOMINAL WALL INCLUDE:
   1. Above the level of the umbilicus via the lateral thoracic vein
   2. Below the level of the umbilicus via the great saphenous vein
   3. Extensive connections between the above two venous territories
   4. An alternate (collateral) route of venous return to the heart in vena caval obstruction   A B C D E

9. THE RECTUS ABDOMINIS MUSCLE:
   1. Has a segmental innervation
   2. Originates from the pubic crest
   3. Inserts into the 5th, 6th, and 7th costal cartilages
   4. Is covered anteriorly in its entire length by the aponeurosis of the three abdominal muscles   A B C D E

10. THE INGUINAL CANAL:
    1. Is about 10 cm (about four inches) long
    2. Begins at the deep inguinal ring
    3. Is located posterior to the transversus abdominis muscle
    4. Terminates at the superficial inguinal ring   A B C D E

11. THE ANTERIOR WALL OF THE INGUINAL CANAL:
    1. Is strongest in its lateral half
    2. Is composed entirely of the external oblique aponeurosis in its lateral half
    3. Is composed of the aponeurosis of the external oblique and the internal oblique muscles in its lateral half
    4. Is strongest in its medial half   A B C D E

A	B	C	D	E
1,2,3	1,3	2,4	only 4	all correct

12. THE INGUINAL CANAL PRESENTS:
    1. A superficial inguinal ring in the aponeurosis of the external oblique muscle
    2. A floor formed by the grooved inner surface of the inguinal ligament
    3. A roof formed by arching fibers of the internal oblique muscle
    4. A deep inguinal ring in the fascia transversalis   A B C D E

13. CONCERNING HERNIAS IN THE GROIN:
    1. Inguinal hernia is more common in the female
    2. Inguinal hernia is more common in the male
    3. Femoral hernia is more common in the male
    4. Femoral hernia is more common in the female   A B C D E

14. THE RIGHT PARAMEDIAN INCISION:
    1. Is made through the anterior and posterior walls of the rectus sheath
    2. Provides an adequate exposure of most abdominal structures
    3. Is the incision of choice for explorative laparotomy
    4. Is often used for operations on the stomach   A B C D E

15. THE INGUINAL CANAL IS:
    1. Associated developmentally with descent of the testis
    2. Associated with the gubernaculum in the embryo
    3. The site of indirect hernias
    4. Present in females   A B C D E

16. A DIRECT INGUINAL HERNIA:
    1. Rarely descends into the scrotum
    2. Passes through the posterior wall of the inguinal canal
    3. Results from weakness of the posterior wall of the inguinal canal
    4. May be prevented from descending by applying pressure over the deep inguinal ring   A B C D E

17. THE INTERNAL OBLIQUE MUSCLE OF THE ANTERIOR ABDOMINAL WALL:
    1. Takes origin from the lateral half of the inguinal ligament
    2. Originates from the intermediate part of the anterior two-thirds of the iliac crest
    3. Inserts into the linea alba and the lower ribs
    4. Forms the conjoint tendon supporting the lateral half of the inguinal canal   A B C D E

18. THE EXTERNAL OBLIQUE MUSCLE OF THE ABDOMEN:
    1. Originates from the external surface of the lower eight ribs
    2. Forms the anterior wall of the inguinal canal
    3. Forms the inguinal ligament
    4. Inserts exclusively into the linea alba   A B C D E

-------------------- ANSWERS, NOTES AND EXPLANATIONS --------------------

1. C, 2 and 4 are correct. The iliacus muscle originates primarily from the upper part of the iliac fossa of the hip bone. It is inserted into the lateral side of the tendon of the psoas major forming a composite muscle, known as the iliopsoas, which is inserted into the lesser trochanter. The floor of the femoral triangle, from lateral to medial, is formed by the iliacus, psoas major, pectineus, and adductor longus muscles. A large portion of the hip bone

172

below the arcuate line provides origin for the obturator internus muscle.

2. **C. 2 and 4 are correct.** The rectus sheath extends from the costal margin to the symphysis pubis and encloses the rectus abdominis and sometimes the pyramidalis muscles. It also contains the superior and inferior epigastric arteries. The intercostal nerves T3-T6 innervate the third to sixth intercostal spaces, but not the abdominal wall. The intercostal (thoracoabdominal) nerves 7-11 and the subcostal nerve innervate the rectus abdominis muscle and the abdominal skin. The musculophrenic artery, a branch of the internal thoracic, runs along the costal margin.

3. **B. 1 and 3 are correct.** The psoas major muscle is related to the posteromedial surface of the kidney. It originates from the lumbar vertebrae and intervertebral discs, and inserts into the lesser trochanter of the femur. It is a powerful flexor of the thigh and flexes the trunk, when the thigh is fixed. Contraction of the psoas major also bends the vertebral column to one side (lateral flexion). This muscle is innervated directly by branches of the lumbar plexus of nerves. The femoral nerve, a major branch of the lumbar plexus, provides motor innervation to the quadriceps femoris, pectineus, and sartorius muscles.

4. **B. 1 and 3 are correct.** The superior epigastric artery, a branch of the internal thoracic, is found in the upper part of the rectus sheath, posterior to the rectus abdominis muscle. It communicates with the inferior epigastric artery which arises from the external iliac. The lateral thoracic artery, a branch of the second part of the axillary, runs along the lateral border of the pectoralis minor muscle to supply the mammary gland. The superficial external pudendal artery takes origin from the femoral artery and supplies skin of the external genitalia.

5. **E. All are correct.** The quadratus lumborum muscle, the psoas major and the iliacus constitute the muscles of the posterior abdominal wall. The quadratus lumborum muscle takes origin from the posterior third of the inner lip of the iliac crest and inserts into the lower border of the twelfth rib. It is enclosed between the anterior and middle layers of the lumbar fascia. Unilateral contraction of this muscle results in lateral flexion, whereas bilateral contraction probably stabilizes the trunk. The quadratus lumborum forms an important posterior relation of the kidney along with the psoas major medially and the transversus abdominis laterally. The subcostal, iliohypogastric, and ilioinguinal nerves intervene between the muscle and the kidney.

6. **C. 2 and 4 are correct.** The psoas major muscle originates from the intervertebral discs and the bodies of the lumbar vertebrae. The lumbar plexus of nerves, formed principally by the anterior primary rami of the first to fourth lumbar spinal nerves, is located within the substance of the psoas major muscle. The psoas major and iliacus muscles enter the thigh behind the inguinal ligament, lateral to the femoral sheath. The psoas major, quadratus lumborum, and iliacus muscles constitute the three major muscles of the posterior abdominal wall.

7. **E. All are correct.** The arterial supply of the anterior abdominal wall above the umbilicus is by the superior epigastric and musculophrenic arteries, branches of the internal thoracic. The inferior epigastric, a branch of the external iliac artery, supplies the region below the level of the umbilicus. Similarly, the venous drainage above the umbilicus reaches the superior vena cava via the lateral thoracic, superior epigastric, and musculophrenic veins. On the other hand, the venous return below the umbilicus is to the inferior vena cava via the superficial and inferior epigastric veins. The lymphatic drainage of the anterior abdominal wall flows in two directions: the region

above the umbilicus drains into the axillary lymph nodes, whereas the region below the umbilicus drains into the superficial inguinal lymph nodes. The cutaneous innervation of the skin at the level of the umbilicus is by the tenth thoracoabdominal (intercostal) nerves and is of significance in determining the site of a lesion of the thoracic spinal cord.

8. E. All are correct. The cutaneous venous drainage above the umbilicus is principally by the lateral thoracic vein which drains into the axillary vein. On the other hand, the venous return below the umbilicus is via the superficial epigastric and superficial circumflex iliac veins into the great saphenous vein. Extensive communications exist between the lateral thoracic and superficial epigastric veins via the thoracoepigastric vein. Should the superior or inferior vena cava be obstructed, these collateral channels become dilated and provide an alternate route of venous return. It should also be realized that in obstruction of the portal vein, e.g., in cirrhosis of the liver, the anastomosis between the paraumbilical veins and the inferior epigastric veins around the umbilicus becomes dilated to form the caput medusae, providing a collateral pathway of venous return.

9. A. 1, 2, and 3 are correct. The rectus abdominis is a wide muscle that takes origin from the pubic crest and inserts into the fifth, sixth, and seventh costal cartilages. The rectus abdominis has a segmental innervation from the sixth to the eleventh intercostal and subcostal nerves. The anterior wall of the rectus sheath is constituted as follows: above the costal margin it is formed by the aponeurosis of the external oblique muscle; below the costal margin and the midpoint between the umbilicus and the symphysis pubis, it is formed by the fusion of the aponeurosis of the external oblique with the anterior layer of the aponeurosis of the internal oblique; and below the arcuate line, by the fused layers of the aponeurosis external, internal, and transversus abdominis muscles.

10. C. 2 and 4 are correct. The inguinal canal is an oblique passage through the lower part of the anterior abdominal wall; it is about 3.75 cm (1 1/2 inches) in length. It extends between the deep inguinal ring (opening) in the fascia transversalis and the superficial inguinal ring (opening) in the aponeurosis of the external oblique muscle of the abdomen. The aponeurosis of the transversus abdominis muscle is located posterior to the medial half of the inguinal canal, where it fuses with the aponeurosis of the internal oblique muscle, forming the conjoint tendon.

11. B. 1 and 3 are correct. The inguinal canal has anterior and posterior walls. The anterior wall throughout its extent is composed of the aponeurosis of the external oblique muscle. It is strengthened by the internal oblique muscle in its lateral half. The posterior wall is strongest in the medial half of the inguinal canal, where it is formed by the fascia transversalis and the conjoint tendon. The conjoint tendon is composed of the aponeurosis of the internal oblique and transversus abdominis muscles.

12. E. All are correct. The inguinal canal is an oblique passage in the lower part of the anterior abdominal wall. It contains the spermatic cord in the male and the round ligament of the uterus in the female. It has two rings or openings: a deep inguinal ring in the transversalis fascia at the midinguinal point, and a superficial inguinal ring in the aponeurosis of the external oblique muscle of the abdomen. The midinguinal point is located midway between the anterior superior iliac spine and the pubic tubercle. The floor of the canal is formed by the grooved inner surface of the inguinal ligament. The inguinal ligament is constituted by the lower, free border of the aponeurosis of the external oblique muscle and extends from the anterior superior iliac spine to the pubic symphysis. The roof of the inguinal canal

is formed by the fibers of the internal oblique muscle which arch from the anterior to the posterior wall.

13. **C. 2 and 4 are correct**. A hernia in the region of the groin may be either a femoral or an inguinal hernia. In a femoral hernia, a portion of the intestine passes through the femoral canal, the most medial compartment of the femoral sheath. The dimension of the femoral ring, the upper part of the femoral canal, is greater in the female than in the male due to the fact that the distance between the anterior superior iliac spine and the pubic tubercle is greater in the female. Therefore, femoral hernias are more common in the female. Inguinal hernias occur in the region of the inguinal canal and their incidence is higher in males because the inguinal canal is wider than in females.

14. **E. All are correct**. The right paramedian incision is one of the most common incisions for exposing the abdominal cavity; it may be supraumbilical, infraumbilical, or it may extend the whole length of the abdominal wall. It is the incision of choice for explorative laparotomy where the surgeon is not certain of the diagnosis and needs to examine the contents of the entire abdominal cavity. The skin incision is made about 2.5 cm (an inch) lateral to the midline. The anterior rectus sheath is incised; then the rectus abdominis muscle is retracted medially with its innervation intact, thereby exposing the posterior rectus sheath. The posterior wall of the sheath is then incised along with the fascia transversalis and peritoneum. After the necessary surgery, the abdomen is closed in layers.

15. **E. All are correct**. The inguinal canal is an oblique passage in the lower part of the anterior abdominal wall. Developmentally, it is associated with the gubernaculum and the processus vaginalis, the latter being a diverticulum of the peritoneal sac. During development in the male, the testis descends from the posterior abdominal wall through the inguinal canal into the scrotum. It is the processus vaginalis that aids in the formation of the inguinal canal. Inguinal hernias occur in this region. Indirect (oblique) inguinal hernia is associated with the persistence of the processus vaginalis. The inguinal canal is also present in the female and contains the round ligament of the uterus and the ilioinguinal nerve. In the male, the inguinal canal contains the spermatic cord and the ilioinguinal nerve.

16. **A. 1, 2, and 3 are correct**. Inguinal hernias are commonly seen in the male. There are two kinds: direct and indirect. Direct inguinal hernia results from the passage of the contents of the abdominal cavity through a weakness in the posterior wall of the inguinal canal, lateral to the conjoint tendon. As this type of hernia is not associated with the existence of a congenital defect, such as the presence of a processus vaginalis, the swelling is confined to the inguinal region and does not usually enter the scrotum. However, in rare cases where the muscular tone of the abdominal wall is extremely weak, the hernia may descend into the scrotum. Clinical examination to distinguish a direct from an indirect inguinal hernia is conducted in the following manner. First, the hernia is reduced by pushing it back into the abdominal cavity. The examiner then places his thumb over the deep inguinal ring, located at the midinguinal point, about 3 cm above the inguinal ligament. The patient is then asked to cough in order to increase the intraabdominal pressure. In indirect inguinal hernias, the thumb prevents the descent of the hernia. On the other hand, in the direct variety, the hernia becomes evident medial to the thumb.

17. **A. 1, 2, and 3 are correct**. The internal oblique muscle has a two-fold origin from the lateral half of the inguinal ligament, and from the anterior two-thirds of the intermediate area of the iliac crest. The muscle inserts

into the linea alba and the lower three ribs. The lowest part of the aponeurosis of the internal oblique fuses with the aponeurosis of the transversus abdominis muscle to form the conjoint tendon, strengthening the posterior wall of the medial half of the inguinal canal.

18. **A. 1, 2, and 3 are correct.** The external oblique muscle of the abdomen arises by fleshy fibers from the external surface of the lower eight ribs. The muscle is inserted into the anterior half of the outer lip of the iliac crest and the linea alba. Its free lower border, extending between the anterior superior iliac spine and pubic tubercle, forms the inguinal ligament. The external oblique muscle forms the anterior wall of the inguinal canal along with fibers of the internal oblique muscle which take origin from the lateral half of the inguinal ligament.

FIVE-CHOICE ASSOCIATION QUESTIONS

DIRECTIONS: Each group of questions below consists of a numbered list of descriptive words or phrases accompanied by a diagram with certain parts indicated by letters, or by a list of lettered headings. For each numbered word or phrase, SELECT THE LETTERED PART OR HEADING that matches it correctly. Then insert the letter in the space to the right of the appropriate number. Sometimes more than one numbered word or phrase may be correctly matched to the same lettered part or heading.

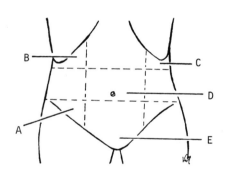

1. ___ Region of the anterior abdominal wall related to a solid organ, contraction of which releases blood with a high concentration of red blood cells

2. ___ Overlies a solid abdominal viscus, enlargement of which usually results in pain being referred to the right shoulder

3. ___ Indicates an area of the anterior abdominal wall separating territories of different arterial supply

4. ___ An area of the anterior abdominal wall containing McBurney's point

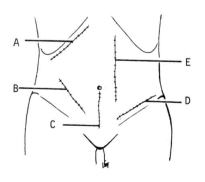

5. ___ An incision commonly used for explorative laparotomy

6. ___ The common incision for removal of an inflamed appendix

7. ___ A commonly used incision in obstetrics

8. ___ An incision used for repair of an inguinal hernia

9. ___ An incision used for removal of the gall bladder

ASSOCIATION QUESTIONS

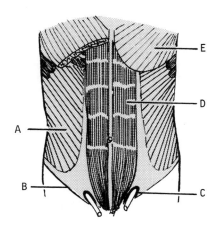

10. ___ A muscle, the aponeurosis of which forms the anterior wall of the rectus sheath
11. ___ A muscle that can be divided transversely without endangering its innervation
12. ___ An opening through which an indirect inguinal hernia enters the scrotum
13. ___ The lower border of a muscle extending between the anterior superior iliac spine and the pubic tubercle
14. ___ A muscle lying anterior to the branches of the internal thoracic and external iliac arteries

---------------------- ANSWERS, NOTES AND EXPLANATIONS ----------------------

1. **C.** The left hypochondriac region of the anterior abdominal wall is related to the spleen and the stomach. Contraction of the spleen releases blood with a high concentration of red blood cells. Enlargement of this solid organ may cause pain to be referred to the left shoulder.

2. **B.** The right hypochondriac region overlies the right lobe of the liver. In enlargement of the liver, irritation of the diaphragm usually causes pain to be referred to the right shoulder. It should be realized that the diaphragm is innervated by the phrenic nerve composed mainly of fibers from the anterior primary rami of C4 with contributions from C3 and C5. The sensory fibers of C4 are distributed to the skin over the shoulder.

3. **D.** A short horizontal segment of the skin, passing through the umbilicus and continuing laterally, demarcates important areas of arterial supply, venous and lymphatic drainage. Thus, the arterial supply above the umbilicus is via the superior and musculophrenic branches of the internal thoracic artery, whereas the inferior epigastric artery arising from the external iliac artery supplies the area below the umbilicus. The venous drainage above the umbilicus is via the thoracoepigastric and superior epigastric veins; whereas below, this area drains into the great saphenous vein via the inferior epigastric. The lymph above the umbilicus drains to the axillary lymph nodes, whereas below the umbilicus it drains into the superficial inguinal lymph nodes.

4. **A.** The lower right area of the anterior abdominal wall, called the right inguinal region, overlies the vermiform appendix. In inflammation of the appendix, known as appendicitis, the area of maximum tenderness is at McBurney's point. This point is located at the junction of the lateral third with the medial two thirds on the line joining the anterior superior iliac spine and the umbilicus. McBurney's incision for removal of the appendix (appendectomy) is made through this region. The area located between the right and left inguinal regions is called the hypogastrium (E).

5. **E.** A paramedian incision is used for an explorative laparotomy when a surgeon wishes to explore the entire abdominal cavity. This is one of the most common abdominal incisions. This is usually done in acute abdominal conditions where the diagnosis is uncertain. The paramedian incision may be either to the left or right of the midline. It may be supraumbilical, or extend the entire length of the abdominal wall.

6. B. The incision of choice in a definitive case of acute appendicitis is McBurney's incision, a muscle splitting incision. The skin incision is made at right angles to an imaginary line drawn between the umbilicus and anterior superior iliac spine and passes through a point located at the junction of the lateral and medial two thirds. The incisions through the various muscle layers are made in the direction of the muscle fibers. The right McBurney incision is made also for a temporary colostomy in which the ascending colon is opened and sutured to the skin. A permanent colostomy is made on the left side to form a permanent artificial anus in cases of carcinoma of the rectum or sigmoid colon.

7. C. The infraumbilical midline incision is commonly used in obstetrical practice, e.g., in a caesarian section. The incision passes through the linea alba and the resulting scar is usually weak because of poor vascularity in this area. Midline incisions have the advantage of permitting rapid entry into the abdominal cavity and may be used in abdominal emergencies.

8. D. Incisions for repair of inguinal hernias are made above and parallel to the inguinal ligament. The operative aims are to reduce the hernia, excise the hernial sac, and strengthen the walls of the inguinal canal in an appropriate manner.

9. A. The Kocher's incision on the right side is usually used for operations on the gall bladder. On the left side, this incision may be used to approach the spleen. The incision is made parallel to the costal margin. In gall bladder operations, a drainage tube is usually placed in the right hypochondriac region through a separate stab incision, to drain fluid which accumulates following these surgical procedures.

10. A. The external oblique muscle of the abdomen originates from the external surface of the lower eight ribs and is inserted into the outer lip of the iliac crest and the linea alba. It forms the anterior wall of the entire rectus sheath. The internal oblique and transversus abdominis muscles, also contribute, in varying degrees, to the anterior wall of the rectus sheath.

11. D. The rectus abdominis muscle is segmentally innervated by the thoracoabdominal nerves and can be sectioned transversely in the intersegmental region. It extends from the pubic crest to the fifth, sixth, and seventh costal cartilages.

12. C. The superficial inguinal ring is an opening in the aponeurosis of the external oblique muscle of the abdomen and is located lateral to the pubic tubercle. In an indirect inguinal hernia, associated with persistence of the processus vaginalis, the hernia may descend into the scrotum.

13. B. The inguinal ligament is the folded lower border of the external oblique muscle and extends between the anterior superior iliac spine and the pubic tubercle.

14. D. The rectus abdominis muscle lies anterior to the superior and inferior epigastric arteries. The superior epigastric is one of the terminal branches of the internal thoracic artery which arises from the first part of the subclavian artery. The inferior epigastric arises from the external iliac artery. The superior and inferior epigastric arteries lie between the rectus abdominis muscle and the posterior rectus sheath.

PERITONEUM AND PERITONEAL CAVITY

FIVE-CHOICE COMPLETION QUESTIONS

DIRECTIONS: Each of the following questions or incomplete statements is followed by five suggested answers or completions. SELECT THE ONE BEST ANSWER in each case and then underline the appropriate letter at the lower right of each question.

1. WHICH OF THE FOLLOWING STATEMENTS ABOUT THE MESENTERY IS FALSE?
   A. Suspends the small gut to the posterior abdominal wall
   B. Its attached border runs downwards and to the right
   C. Has an extensive free border enclosing the gut
   D. Its attached border is about 6 cm in length
   E. Has a short attached border

   A B C D E

2. WHICH OF THE FOLLOWING STATEMENTS ABOUT THE EPIPLOIC FORAMEN IS FALSE?
   A. The opening between the greater and lesser peritoneal sacs
   B. Related superiorly to the caudate lobe of the liver
   C. Bounded inferiorly by the first part of the duodenum
   D. Related posteriorly to the portal vein
   E. Related anteriorly to the gastrohepatic ligament

   A B C D E

3. WHICH OF THE FOLLOWING STATEMENTS ABOUT THE FALCIFORM LIGAMENT IS (ARE) TRUE?
   A. A sickle-shaped peritoneal fold
   B. Attached to the anterior abdominal wall
   C. Attached to the anterior and superior aspects of the liver
   D. Contains the ligamentum teres in its free border
   E. All of the above

   A B C D E

-------------------- ANSWERS, NOTES AND EXPLANATIONS --------------------

1. D. The mesentery has an attached border which is 15 cm (about six inches) in length and extends from the duodenojejunal flexure to the right iliac fossa. The mesentery is the peritoneal fold that suspends the jejunum and ileum from the posterior abdominal wall. Its attached border is also termed the root of the mesentery. Its free border enclosing the gut is about 600 cm (about twenty feet) in length. The vascular, lymphatic, and nerve supply reaches the gut by passing through the layers of the mesentery to the gut.

2. D. The epiploic foramen (foramen of Winslow) is the channel of communication between the greater and lesser sacs of the peritoneal cavity. It is related posteriorly to the inferior vena cava but not to the portal vein. A finger inserted in this foramen could palpate the following structures: superiorly, the caudate lobe of the liver; inferiorly, the first part of the duodenum (first inch); anteriorly, the free margin of the gastrohepatic ligament or the lesser omentum enclosing the portal vein, bile duct, and hepatic artery; and posteriorly, the inferior vena cava.

3. E. The falciform ligament is a sickle-shaped peritoneal fold attached in the midline to the anterior abdominal wall and the diaphragm. It is also attached to the anterior and superior aspects of the liver. Its free border, stretching from the umbilicus to the inferior border of the liver, contains the ligamentum teres. The ligamentum teres is the obliterated umbilical vein which in the fetus transported oxygenated blood from the placenta to the left branch of the portal vein.

MULTI-COMPLETION QUESTIONS

DIRECTIONS: In each of the following questions or incomplete statements, ONE OR MORE of the completions given is correct. At the lower right of each question, underline A if 1, 2 and 3 are correct; B if 1 and 3 are correct; C if 2 and 4 are correct; D if only 4 is correct; and E if all are correct.

1. THE LESSER OMENTUM:
    1. Is attached inferiorly to the lesser curvature of the stomach
    2. Is attached superiorly to the porta hepatis of the liver
    3. Has a free right margin
    4. Encloses the inferior vena cava in its free margin   A B C D E

2. THE ANTERIOR WALL OF THE OMENTAL BURSA (LESSER SAC) IS COMPOSED OF:
    1. The posterior peritoneal layer extending downwards from the greater curvature
    2. Peritoneum covering the posterior surface of the stomach
    3. Peritoneum on the posterior surface of the liver
    4. The posterior layer of the lesser omentum   A B C D E

3. THE GREATER OMENTUM:
    1. Is attached to the greater curvature of the stomach
    2. Contains the left and right gastric vessels
    3. Is the term applied to a fold of peritoneum
    4. Is attached to the lower border of the pancreas   A B C D E

-------------------- ANSWERS, NOTES AND EXPLANATIONS --------------------

1. **A.** 1, 2, and 3 are correct. The lesser omentum is a two-layered peritoneal fold, extending from the porta hepatis of the liver superiorly to the lesser curvature of the stomach inferiorly. Its lower attachment also extends onto the first 2.5 cm (inch) of the first part of the duodenum below. Its free right margin encloses the portal vein, the bile duct, and the hepatic artery. The inferior vena cava is a retroperitoneal structure. The lesser omentum forms the anterior wall of the lesser sac of the peritoneal cavity. A finger introduced into the epiploic foramen, through which the lesser sac communicates with the greater sac, separates the posteriorly situated inferior vena cava from the three structures (portal vein, hepatic artery, and bile duct) that are enclosed in the free margin of the lesser omentum.

2. **E.** All are correct. The omental bursa is essentially a closed peritoneal space which communicates with the greater sac of the peritoneal cavity through the epiploic foramen (foramen of Winslow). It is described as having an anterior and posterior wall. Its anterior wall is formed from above downwards by: the peritoneum of the posterior surface of the liver, the lesser omentum, the posterior surface of the stomach, and the greater omentum. The posterior wall of the lesser sac is contributed from below upwards by: the reflected layer of peritoneum of the greater omentum, the peritoneum in front of the transverse colon, the transverse mesocolon and peritoneum covering the posterior abdominal wall above the pancreas.

3. **B.** 1 and 3 are correct. The greater omentum is attached superiorly to the greater curvature of the stomach. It is a two-layered peritoneal fold that extends downwards and reflects back up to meet the lower border of the transverse colon. The space between the two anterior and the two posterior layers contains the lower extension of the omental bursa. However, in many instances this space is eliminated by the fusion of the anterior and posterior layers.

The left and right gastroepiploic vessels run between the two layers of the greater omentum near the greater curvature of the stomach. The left and right gastric vessels run between the layers of the lesser omentum. The transverse mesocolon is the peritoneal fold that extends from the transverse colon to the lower border of the pancreas.

FIVE-CHOICE ASSOCIATION QUESTIONS

DIRECTIONS: Each group of questions below consists of a numbered list of descriptive words or phrases accompanied by a diagram with certain parts indicated by letters, or by a list of lettered headings. For each numbered word or phrase, SELECT THE LETTERED PART OR HEADING that matches it correctly. Then insert the letter in the space to the right of the appropriate number. Sometimes more than one numbered word or phrase may be correctly matched to the same lettered part or heading.

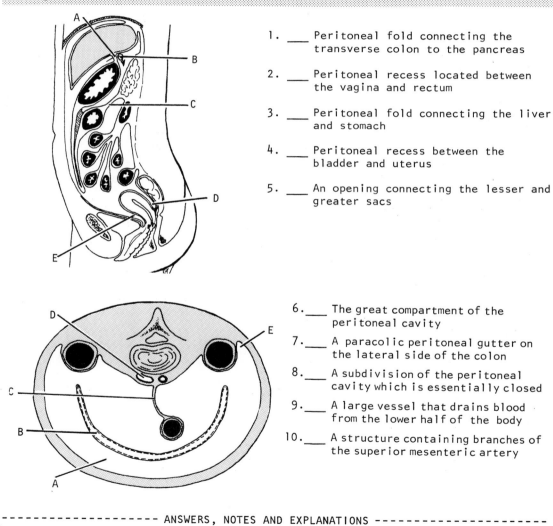

1. ___ Peritoneal fold connecting the transverse colon to the pancreas

2. ___ Peritoneal recess located between the vagina and rectum

3. ___ Peritoneal fold connecting the liver and stomach

4. ___ Peritoneal recess between the bladder and uterus

5. ___ An opening connecting the lesser and greater sacs

6. ___ The great compartment of the peritoneal cavity

7. ___ A paracolic peritoneal gutter on the lateral side of the colon

8. ___ A subdivision of the peritoneal cavity which is essentially closed

9. ___ A large vessel that drains blood from the lower half of the body

10. ___ A structure containing branches of the superior mesenteric artery

------------------ ANSWERS, NOTES AND EXPLANATIONS ---------------------

1. C. The transverse mesocolon connects the transverse colon with the lower

181

border of the pancreas. The middle colic artery, a branch of the superior mesenteric, runs in the mesocolon to supply the transverse colon.

2. D. The rectovaginal recess (pouch of Douglas) is the peritoneal recess between the upper third of the posterior vaginal wall and the middle third of the rectum. This recess is also called the rectouterine pouch, as the uterus forms a major anterior relation of this space. Accumulations of fluid in the peritoneal cavity tend to collect in this space and can be aspirated by a needle introduced through the posterior vaginal wall.

3. A. The lesser omentum is a peritoneal fold that extends from the porta hepatis of the liver to the lesser curvature of the stomach.

4. E. The uterovesical recess or pouch is the space formed by the reflection of peritoneum from the upper surface of the uterus.

5. B. The epiploic foramen (foramen of Winslow) is the channel of communication between the lesser and greater sacs of the peritoneal cavity.

6. A. This large space lined by peritoneum is the greater sac of the peritoneal cavity. The greater sac is divided by the greater omentum into supracolic and infracolic compartments.

7. E. The space located on the lateral side of the colon, lined by peritoneum reflected from the abdominal wall onto the colon, is called the paracolic gutter (sulcus). Collections of fluid in the peritoneal cavity tend to accumulate in these spaces. The right paracolic sulcus is continuous superiorly with the hepatorenal recess (pouch of Morison), which is the deepest part of the peritoneal cavity in the recumbent position except for the rectouterine pouch. The left paracolic recess is continuous inferiorly with the pelvic cavity, whereas the right recess is limited inferiorly by the attachment of the root of the mesentery to the right iliac fossa.

8. B. The lower part of the peritoneal cavity (inferior recess) of the lesser sac (omental bursa) is between the layers of the greater omentum. The omental bursa is essentially a closed sac, except for the communication with the greater sac through the epiploic foramen. The upper part (superior recess) of the omental bursa is behind the stomach, lesser omentum, and liver.

9. D. The inferior vena cava, located retroperitoneally, drains blood from the lower half of the body. It is formed by the union of the two common iliac veins and ascends up the posterior abdominal wall, piercing the central tendon of the diaphragm to enter the right atrium.

10. C. The mesentery is a peritoneal fold that suspends the jejunum and ileum. The root of the mesentery extends from the duodenojejunal flexure to the right iliac fossa and is about 15 cm in length. It contains branches not only of the superior mesenteric artery, but also venous and lymphatic vessels supplying the gut.

HOLLOW ABDOMINAL VISCERA

FIVE-CHOICE COMPLETION QUESTIONS

DIRECTIONS: Each of the following questions or incomplete statements is followed by five suggested answers or completions. SELECT THE ONE BEST ANSWER in each case and then underline the appropriate letter at the lower right of each question.

SELECT THE ONE BEST ANSWER

1. WHICH OF THE FOLLOWING STATEMENTS ABOUT THE STOMACH IS FALSE?
   A. Is divided into cardiac and pyloric portions
   B. Receives its blood supply from the celiac artery
   C. Has the greater omentum attached to its greater curvature
   D. Its fundic portion receives its blood supply from the left gastroepiploic artery
   E. Is a common site of ulcers

   A B C D E

2. WHICH OF THE FOLLOWING STATEMENTS ABOUT THE DESCENDING COLON IS FALSE?
   A. Commences at the left colic flexure
   B. Related to the lower pole of the spleen
   C. Covered by peritoneum on its front and sides
   D. Supplied by the middle colic artery
   E. Terminates at the inlet of the true pelvis

   A B C D E

3. WHICH OF THE FOLLOWING STATEMENTS ABOUT THE VENOUS DRAINAGE OF THE STOMACH IS (ARE) TRUE?
   A. The left gastric vein drains the upper half of the lesser curvature
   B. The left gastric veins communicate with the esophageal veins
   C. The left gastroepiploic vein drains into the splenic vein
   D. The right gastric vein drains into the portal vein
   E. All of the above

   A B C D E

4. THE DUODENUM IS APPROXIMATELY _____ CM LONG.
   A. 10   D. 25
   B. 15   E. 30
   C. 20

   A B C D E

5. WHICH OF THE FOLLOWING STATEMENTS ABOUT THE SECOND PART OF THE DUODENUM IS FALSE?
   A. Related anteriorly to the transverse colon
   B. Related medially to the head of the pancreas
   C. Related posteriorly to the hilum of the right kidney
   D. Related laterally to the hepatic flexure of the colon
   E. Related superiorly to the uncinate process of the pancreas

   A B C D E

6. WHICH OF THE FOLLOWING STATEMENTS ABOUT THE SUPERIOR MESENTERIC VEIN IS TRUE?
   A. Lies on the left side of the superior mesenteric artery
   B. Terminates at the level of the second lumbar vertebra
   C. Is the vein draining the midgut
   D. Joins the right renal vein
   E. Drains into the inferior vena cava

   A B C D E

7. ALL THE FOLLOWING STATEMENTS ABOUT THE VERMIFORM APPENDIX ARE TRUE EXCEPT:
   A. Opens into the apex of the cecum in fetuses
   B. Subject to considerable variation in position
   C. Retroileal appendices are the most common type
   D. Opens about two cm below the ileocecal junction in adults
   E. When the cecum is full, the free appendix usually hangs in the pelvis

   A B C D E

SELECT THE ONE BEST ANSWER

8.

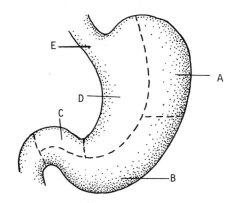

THE REGION OF THE STOMACH DRAINED BY THE PANCREATICO-LIENAL LYMPH NODES IS LABELLED _____.

A B C D E

9.

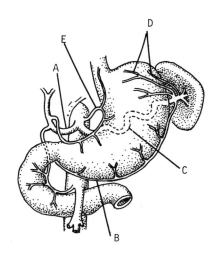

THE LEFT GASTRIC ARTERY IS INDICATED BY _____.

A B C D E

10. WHICH OF THE FOLLOWING STATEMENTS ABOUT THE SUPERIOR MESENTERIC ARTERY IS FALSE?
    A. Originates at the level of the first lumbar vertebra
    B. Gives off the middle colic artery
    C. Supplies the jejunum and ileum
    D. Gives off the left colic artery
    E. Is the artery of the midgut

A B C D E

11. WHICH STATEMENT ABOUT A MECKEL'S DIVERTICULUM IS FALSE?
    A. Always located on the antimesenteric side of the ileum
    B. May contain gastric and pancreatic tissues in its wall
    C. May become a leading point for an intussusception
    D. Occurs at a variable distance from the cecum
    E. Occurs in nearly all newborn infants

A B C D E

SELECT THE ONE BEST ANSWER

12.

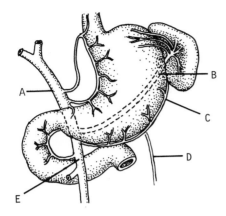

THE LEFT GASTROEPIPLOIC VEIN IS INDICATED BY ____.

A B C D E

13.

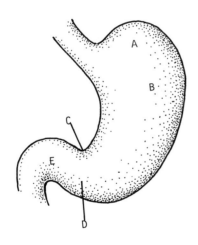

THE ANGULAR NOTCH IS LABELLED ____.

A B C D E

14. WHICH OF THE FOLLOWING STATEMENTS ABOUT THE SMALL INTESTINE IS (ARE) TRUE?
    A. Supplied by the celiac and superior mesenteric arteries
    B. Includes the duodenum, which is mostly retroperitoneal
    C. Terminates at the ileocecal junction
    D. Begins at the pylorus
    E. All of the above

A B C D E

SELECT THE ONE BEST ANSWER

15. WHICH OF THE FOLLOWING STATEMENTS ABOUT THE TRANSVERSE COLON IS FALSE?
    A. Attached to the lower border of the pancreas
    B. Mainly supplied by the artery of the midgut
    C. Gives attachment to the greater omentum
    D. The most mobile part of the colon
    E. Suspended by the lesser omentum

    A B C D E

------------------ANSWERS, NOTES AND EXPLANATIONS ---------------------

1. D. The fundus of the stomach receives its blood supply from the short gastric arteries which arise from the splenic artery. The stomach is divided into two major parts, cardiac and pyloric, by the incisura angularis. Its arterial supply is from several branches which arise from the celiac artery. The greater curvature provides an attachment for the greater omentum. The stomach is not infrequently a site of ulceration. Gastric ulcers are more prone to become malignant than duodenal ulcers. Examination of the stomach includes physical examination, gastric analysis after fractional test meal (FTM), and radiological examination following ingestion of contrast medium (barium meal).

2. D. The descending colon is supplied by the left colic branch of the inferior mesenteric artery. It begins at the left colic (splenic) flexure where it is related to the lower pole of the spleen. The descending colon is covered by peritoneum on its front and sides and terminates at the inlet of the true pelvis to become continuous with the sigmoid colon.

3. E. The left and right gastric veins drain into the portal vein. The left gastroepiploic and short gastric veins join the splenic vein. The right gastroepiploic vein drains into the superior mesenteric vein. The portal vein is formed by the union of the superior mesenteric and splenic veins. Tributaries of the left gastric vein communicate with the esophageal veins at the lower end of the esophagus. It should be recalled that esophageal veins drain into the systemic venous system via the azygos veins, whereas the left gastric vein joins the portal vein. In cases of portal venous obstruction, as in cirrhosis of the liver, the collateral channels at the lower end of the esophagus dilate to form esophageal varices. These tend to rupture, causing bleeding into the stomach, resulting in vomiting of blood (hematemesis).

4. D. The duodenum is approximately 25 cm long and is the shortest segment of the small intestine. It is divided into the following four parts: the first or superior part ascends from the gastroduodenal junction; the second part descends around the head of the pancreas and anterior to the right kidney; the third or horizontal part crosses the inferior vena cava, the aorta and vertebral column to the left; and the fourth part ascends to terminate at the duodenojejunal junction. The common bile and pancreatic ducts enter the posteromedial aspect of the second part of the duodenum.

5. E. The proximal half of the second (descending) part of the duodenum is developed from the foregut, whereas its distal half is derived from the midgut. Its vascular supply thereby is provided by the celiac and superior mesenteric arteries, which are responsible for supplying structures derived from the foregut and midgut respectively. The descending part of the duodenum is crossed anteriorly by the transverse colon: above this it is related to the right lobe of the liver; and below this to coils of small intestine. Posteriorly this part of the duodenum overlaps the medial border of the kidney and its hilum. On the lateral aspect, the hepatic flexure of the colon is related to it. The

head of the pancreas lies in the concavity formed by the first, second, and third parts of the duodenum. The uncinate process of the pancreas is related to the third part of the duodenum.

6. **C.** The superior mesenteric vein drains the ileum, the jejunum, the 4th and 3rd parts and the distal half of the 2nd part of the duodenum; thus it drains the derivatives of the embryonic midgut. Its tributaries are the ileocolic, right colic, middle colic, jejunal, ileal veins and the inferior pancreaticoduodenal vein. In its course, the superior mesenteric vein lies on the right side of the superior mesenteric artery, and terminates at the level of the first lumbar vertebra by joining the splenic vein to form the portal vein. It does not join the right renal vein nor does it drain into the inferior vena cava.

7. **C.** Retroileal appendices occur in about 0.5 per cent of persons. When the cecum is empty and contracted, a free appendix may also be retrocecal in position. Fixed appendices are attached by a short peritoneal fold or by fibrous tissue. Retrocecal appendices occur in about 64 per cent of persons.

8. **A.** The lymphatic drainage of the fundus, adjacent to the greater curvature, and the upper part of the greater curvature is to the nodes along the splenic artery and at the hilum of the spleen (pancreaticolienal nodes). The lower region of the greater curvature (B) drains into the inferior gastric and subpyloric lymph nodes. A small region including the upper part of the pylorus of the stomach (C) drains into the hepatic group of nodes. The region labelled D is the largest area of drainage, and includes most of the lesser curvature of the stomach and the cardia. This extensive area drains into superior gastric and paracardiac lymph nodes.

9. **E.** The left gastric artery is one of the three major branches of the celiac artery. The other two are the common hepatic (A) and the splenic (C). The splenic artery gives off the short gastric arteries (D) and the left gastroepiploic artery (unlabelled). The right gastroepiploic artery (B) is a branch of the gastroduodenal. It forms an arterial arcade along the greater curvature of the stomach by anastomosing with the left gastroepiploic artery.

10. **D.** The left colic artery is a branch of the inferior mesenteric, the artery of the hindgut. The superior mesenteric artery, the artery of the midgut, arises from the abdominal aorta at the level of the first lumbar vertebra. The superior mesenteric artery supplies the entire small intestine, except for the first part and the proximal half of the second part of the duodenum. The superior mesenteric artery gives off numerous branches from its convex left border, termed the jejunal and ileal branches. The branches given off from its concave, right border are from above downwards: the middle colic, the right colic, and the ileocolic.

11. **E.** About two percent of newborn infants have a Meckel's diverticulum. Very occasionally the gastric tissue located in its wall secretes acid, causing ulceration and bleeding during infancy or childhood. Meckel's diverticula are always on the antimesenteric border of the ileum because they represent persistent portions of the yolk stalk which attaches to the ventral side of the midgut loop in the embryo. If a diverticulum inverts, it may serve as a leading point for an intussusception (inversion of the diverticulum into the lumen of the ileum) and cause obstruction of the bowel.

12. **C.** The left gastroepiploic vein drains the left half of the greater curvature of the stomach below the fundus. The right gastroepiploic vein, draining the remaining portion of the greater curvature, empties into the superior mesenteric vein (E). The short gastric veins, returning blood from the fundus,

join the splenic vein (B). The splenic vein joins the superior mesenteric vein at the transpyloric plane to form the portal vein (A). In most instances, the inferior mesenteric vein (D) joins the splenic vein. The portal vein receives both the left and right gastric veins.

13. **C.** The angular notch (incisura angularis) is located in the lower right portion of the lesser curvature. A line drawn vertically downwards from the angular notch to the greater curvature divides the stomach into cardiac and pyloric portions. The cardiac part is further divided into two parts: the fundus (A) and the body (B). The fundus of the stomach is the part located above the line drawn horizontally from the cardiac orifice. The pylorus is divided by an ill-defined sulcus intermedius into a proximal pyloric antrum (D) and a distal tubular pyloric canal (E). The area between the pyloric canal and the duodenum is the pyloric sphincter. An angular notch is adjacent to the cardioesophageal junction.

14. **E.** The small intestine extends from the pylorus of the stomach to the ileocecal junction and includes the duodenum, jejunum, and ileum. The duodenum is about 25 cm long and is retroperitoneal except for the first 2.5 cm. The jejunum and ileum have a total length of about 6 m (20-22 ft.). The duodenum receives its blood supply from both the celiac and superior mesenteric arteries. The superior mesenteric artery also supplies the jejunum and ileum.

15. **E.** The transverse colon has no attachment to the lesser omentum. It gives attachment to the greater omentum and is suspended from the lower border of the pancreas by the transverse mesocolon. The blood supply to the transverse colon is mainly by the superior mesenteric artery (midgut artery) via its middle and right colic arteries. The left colic artery supplies the terminal 5 cm (2") of the transverse colon.

MULTI-COMPLETION QUESTIONS

DIRECTIONS: In each of the following questions or incomplete statements, ONE OR MORE of the completions given is correct. At the lower right of each question, underline A if 1, 2, and 3 are correct; B if 1 and 3 are correct; C if 2 and 4 are correct; D if only 4 is correct; and E if all are correct.

1. THE FEATURES OF THE STOMACH INCLUDE WHICH OF THE FOLLOWING?
    1. Its cardiac orifice is located at the level of the eighth thoracic vertebra
    2. Has a pyloric orifice located at the level of the first lumbar vertebra
    3. Has a lesser omentum containing the gastroepiploic vessels
    4. Receives its blood supply from the artery of the foregut   A B C D E

2. CONCERNING THE DUODENUM:
    1. Derived from the foregut and midgut of the embryo
    2. The beginning of its first part is called the free part
    3. Supplied by the celiac trunk and the superior mesenteric artery
    4. Is relatively fixed except for its first part   A B C D E

3. CONCERNING MECKEL'S DIVERTICULUM:
    1. Common congenital malformation of the digestive tract
    2. May produce gross bleeding from the bowel during infancy
    3. Occurs in about two per cent of persons
    4. Represents a remnant of the yolk stalk   A B C D E

A	B	C	D	E
1,2,3	1,3	2,4	only 4	all correct

4. THE POSTERIOR SURFACE OF THE STOMACH IS RELATED TO THE:
   1. Left kidney
   2. Splenic artery
   3. Left suprarenal
   4. Head of the pancreas

   A B C D E

5. THE ROOT OF THE MESENTERY:
   1. Has an oblique attachment to the posterior abdominal wall
   2. Begins on the left side of the second lumbar vertebra
   3. Contains nerves, arteries, veins, and lymphatics
   4. Terminates at the right sacroiliac joint

   A B C D E

6. THE SECOND PART OF THE DUODENUM:
   1. Begins at the neck of the gall bladder
   2. Is derived from the foregut and midgut
   3. Runs vertically downwards to the third lumbar vertebra
   4. Receives its blood supply exclusively from the superior mesenteric artery

   A B C D E

7. THE PYLORUS OF THE STOMACH:
   1. Is the junction between the stomach and duodenum
   2. Is related posteriorly to the hepatic artery
   3. Has the prepyloric vein on its anterior surface
   4. Contains the pyloric antrum

   A B C D E

8. THE ANATOMICAL FEATURES DISTINGUISHING THE JEJUNUM FROM THE ILEUM ARE:
   1. The jejunum has a thicker wall
   2. The jejunum has a greater diameter
   3. The jejunal mesentery has less fat
   4. The jejunal arterial arcades are four or five in number

   A B C D E

9. THE FIRST PART OF THE DUODENUM:
   1. Terminates at the neck of the gall bladder
   2. Commences in the transpyloric plane
   3. Is a common site for ulcers
   4. Is the most mobile part

   A B C D E

10. THE ASCENDING COLON IS:
    1. Supplied by the ileocolic artery
    2. Related laterally to the right paracolic gutter
    3. Related posteriorly to the quadratus lumborum muscle
    4. Covered by peritoneum on its anterior and posterior surfaces

    A B C D E

11. THE DUODENUM:
    1. Commences at the pylorus
    2. Is entirely retroperitoneal
    3. Is usually divided into four parts
    4. Is supplied exclusively by the celiac artery

    A B C D E

12. THE ANTERIOR RELATIONS OF THE STOMACH INCLUDE THE:
    1. Anterior abdominal wall
    2. Left lobe of the liver
    3. Diaphragm
    4. Gall bladder

    A B C D E

13. THE RIGHT FLEXURE OF THE COLON IS:
    1. Related to the right lobe of the liver
    2. Supplied by the right colic artery
    3. Anterior to the right kidney
    4. Also called the splenic flexure

    A B C D E

A	B	C	D	E
1,2,3	1,3	2,4	only 4	all correct

14. THE TRANSVERSE COLON:
    1. Is supplied mainly by the middle colic artery
    2. Terminates immediately below the spleen
    3. Begins anterior to the right kidney
    4. Is the shortest segment of the colon   A B C D E

15. CONCERNING THE GASTRIC NERVES:
    1. The right vagus forms a major part of the posterior gastric nerve
    2. The left vagus forms a major part of the anterior gastric nerve
    3. Contain both parasympathetic and sympathetic fibers
    4. Division of the vagus nerves reduces gastric acidity   A B C D E

16. THE DUODENUM IS SUPPLIED BY WHICH OF THE FOLLOWING ARTERIES?
    1. Pancreaticoduodenal   3. Gastroduodenal
    2. Superior mesenteric   4. Common hepatic   A B C D E

17. THE FIRST INCH OF THE DUODENUM:
    1. Is covered by peritoneum
    2. Is referred to as the "duodenal bulb or cap"
    3. Forms the inferior boundary of the epiploic foramen
    4. Is in contact with the neck of the gall bladder   A B C D E

------------------------ANSWERS, NOTES AND EXPLANATIONS--------------------------

1. **C.** <u>2 and 4 are correct</u>. The stomach is the first major subdivision of the abdominal alimentary tract. It has an upper cardiac orifice and a lower pyloric orifice. The cardiac orifice is the opening of the esophagus into the stomach and is located at the level of the tenth thoracic vertebra. The pyloric orifice is the communication between the stomach and the duodenum and lies in the transpyloric plane at the level of the first lumbar vertebra. The stomach receives its blood supply from the celiac artery, the artery of the foregut. The lesser omentum is attached to the lesser curvature of the stomach. The right and left gastric arteries run along the lesser curvature, whereas the gastroepiploic arteries course through the greater omentum.

2. **E.** <u>All are correct</u>. Because the duodenum is derived from the embryonic foregut and midgut, it is supplied by branches of the foregut (celiac trunk) and midgut (superior mesenteric) arteries. Most of the duodenum becomes attached to the posterior abdominal wall during fixation of the gut in the fetus. The beginning of the first part of the duodenum remains free; this mobile, free part of the duodenum is usually called the "duodenal cap" by radiologists.

3. **E.** <u>All are correct</u>. A Meckel's diverticulum is a relatively common evagination of the terminal part of the ileum. It is clinically important because bleeding may occur from an ulcer in its wall, especially during infancy and childhood. Only a small number of Meckel's diverticula become symptomatic. Bleeding is a common sign in infants and children, whereas symptoms of diverticulitis are most common in older persons.

4. **A.** <u>1, 2, and 3 are correct</u>. The posterior surface of the stomach is related to several structures which form the "stomach-bed". These structures include the: left kidney, left suprarenal, body of the pancreas, splenic artery, and spleen. They are separated from the stomach by the omental bursa. The head of the pancreas lies in the concavity of the duodenum and does not normally form a posterior relation of the stomach.

5. **E.** <u>All are correct</u>. The mesentery is a two-layered peritoneal fold enclosing the small gut, except the duodenum. It is attached to the posterior abdominal wall in an oblique manner extending from the left side of the second lumbar vertebra to the right sacroiliac joint. Its attachment is approximately 15 cm (6 in) in length and crosses the third part of the duodenum, aorta, inferior vena cava, psoas muscle, right ureter, and right gonadal vessels.

6. **A.** <u>1, 2, and 3 are correct</u>. The second part of the duodenum, also termed the descending part, begins at the neck of the gall bladder and runs downwards to the level of the third lumbar vertebra, where it becomes continuous with the third or horizontal part of the duodenum. The proximal half of the second part is derived from the foregut, whereas the distal half is from the midgut. The junction of the foregut and midgut is at the site where the common bile and pancreatic ducts open into the duodenum. This portion of the gut receives its blood supply from the vascular arcades formed by the superior and inferior pancreaticoduodenal arteries, branches of the celiac (foregut artery) and superior mesenteric (midgut artery), respectively.

7. **B.** <u>1 and 3 are correct</u>. The term pylorus is used to define the junctional area between the stomach and the duodenum and contains the pyloric orifice. The circular muscle fibers are thickened in this area to constitute the pyloric sphincter. On the anterior surface of the pylorus is the prepyloric vein which surgeons use as a guide to the gastroduodenal junction. The posterior surface of the pylorus is related to the bile duct, the gastroduodenal artery, and the portal vein. The hepatic artery runs up the free margin of the lesser omentum in company with the bile duct and the portal vein.

8. **A.** <u>1, 2, and 3 are correct</u>. The jejunum is about 2.5 m (8 ft) in length, whereas the ileum is about 4 m (12 ft). The jejunum has a thicker wall and a greater diameter than the ileum. The jejunal mesentery contains less fat than the ileal mesentery so that the arterial arcades are easier to visualize. The jejunal arterial arcades are simple or double with long vasa recti. On the other hand, the arterial arcades in the ileal region are four to five in number, with short vasa recti.

9. **E.** <u>All are correct</u>. The first part (about 5 cm) of the duodenum, also called the superior portion, extends from the pylorus in the transpyloric plane to the neck of the gall bladder. The proximal half is covered by peritoneum on its anterior and posterior surfaces; hence, it is the most mobile part of the duodenum. Ulceration occurs most commonly in the first part of the duodenum. The prepyloric vein (of Mayo), used by surgeons as a landmark to the gastroduodenal junction, aids in distinguishing a duodenal from a gastric ulcer. It is of interest that duodenal ulcers have little tendency to become malignant, whereas gastric ulcers have a greater tendency to become malignant.

10. **A.** <u>1, 2, and 3 are correct</u>. The ascending colon forms the medial boundary of the right paracolic gutter and is continuous superiorly with the hepatorenal recess (Morison's pouch). The ascending colon is supplied by the ileocolic and right colic branches of the superior mesenteric artery. The ascending colon is related posteriorly from below upwards to the iliacus muscle, iliac crest, quadratus lumborum muscle, and the right kidney. The ascending colon is usually covered by peritoneum on its anterior surface and sides; on rare occasions it may have a mesentery.

11. **B.** <u>1 and 3 are correct</u>. The duodenum is the first part of the small intestine. It is about 25 cm in length and extends from the pylorus to the duodenojejunal flexure. It is retroperitoneal, except for the first inch of the first part of the duodenum. The duodenum is a C-shaped structure which is divided into four parts for descriptive purposes. Its blood supply is derived

from both the celiac and the superior mesenteric arteries.

12. **A.** <u>1, 2, and 3 are correct</u>. The anterior surface of the stomach is in contact with the diaphragm and the visceral surface of the left lobe of the liver. The anterior abdominal wall, between the right and left costal margins, is related to the stomach. This anatomical relationship is used in performing a gastrostomy, an artificial opening between the stomach and the anterior abdominal wall, produced for the administration of nutrient fluids to patients who cannot take food by mouth, as in carcinoma of the esophagus. The gall bladder is located further to the right and is not normally an anterior relation of the stomach.

13. **A.** <u>1, 2, and 3 are correct</u>. The right (hepatic) flexure of the colon lies on the inferolateral surface of the kidney, and is supplied by the right colic artery. The hepatic flexure is covered by peritoneum on its anterior surface and is related to the inferior surface of the right lobe of the liver, lateral to the gall bladder. The splenic flexure lies in relation to the inferior aspect of the spleen and is the junction between the transverse colon and the descending colon.

14. **A.** <u>1, 2, and 3 are correct</u>. The transverse colon begins at the hepatic flexure, anterior to the right kidney, and terminates at the splenic flexure, located at the lateral border of the left kidney immediately below the spleen. It is supplied primarily by the middle colic artery; however, its right and left ends receive branches from the right and left colic arteries. The transverse colon is the longest part of the colon, about 45-50 cm (18 - 20 in) in length.

15. **E.** <u>All are correct</u>. The anterior and posterior gastric nerves run on the anterior and posterior surfaces of the stomach along the lesser curvature. The gastric nerves contain both parasympathetic and sympathetic fibers and are continuous with the anterior and posterior esophageal plexus of nerves. As a result of rotation of the stomach during fetal life, the left vagus contributes primarily to the anterior gastric nerve and the right vagus to the posterior gastric nerve. Division of the vagus nerves (vagotomy) is one of the methods for reducing gastric acidity. Quite often a vagotomy is performed in conjunction with other surgical procedures, such as gastrojejunostomy, in the treatment of gastric and duodenal ulcers.

16. **E.** <u>All are correct</u>. The duodenum is derived embryologically from both the foregut and midgut; thus it is supplied by branches from the celiac trunk and the superior mesenteric artery, as follows: the superior pancreaticoduodenal, a branch of the gastroduodenal artery; the inferior pancreaticoduodenal, the first branch of the superior mesenteric artery; and direct branches from the common hepatic artery.

17. **A.** <u>1, 2, and 3 are correct</u>. The first inch of the first part of the duodenum is covered on its anterior and posterior surfaces by peritoneum, continuous superiorly with the two layers of the lesser omentum. This part of the duodenum, often referred to as the duodenal "bulb" or "cap", is the only part of the duodenum that is not retroperitoneal. The duodenal bulb may be visualized in radiographs taken after the ingestion of a contrast medium (e.g., barium meal). It is also the most common site of ulceration of the duodenum. The epiploic foramen lies immediately above the first inch of the duodenum. The neck of the gall bladder lies at the junction of the first and second parts of the duodenum; it is therefore not related to the first inch of the superior portion of the duodenum.

FIVE-CHOICE ASSOCIATION QUESTIONS

DIRECTIONS: Each group of questions below consists of a numbered list of descriptive words or phrases accompanied by a diagram with certain parts indicated by letters, or by a list of lettered headings. For each numbered word or phrase, SELECT THE LETTERED PART OR HEADING that matches it correctly. Then insert the letter in the space to the right of the appropriate number. Sometimes more than one numbered word or phrase may be correctly matched to the same lettered part or heading.

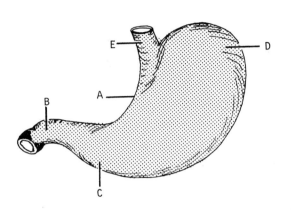

1. ___ An orifice located at the level of the tenth thoracic vertebra

2. ___ An area partly drained by the pancreaticolienal lymph nodes

3. ___ A region supplied by the right gastroepiploic artery

4. ___ A border providing an attachment for the gastrohepatic ligament

5. ___ An area of the stomach supplied by the short gastric arteries

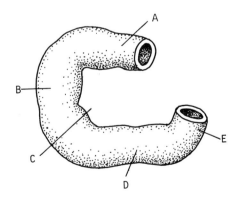

6. ___ Part of the duodenum with a dual developmental origin

7. ___ Part of the duodenum that is crossed anteriorly by an important artery

8. ___ The most mobile part of the duodenum

9. ___ Part of the duodenum where internal hernias may occur

10. ___ Part of the duodenum which is the most common site for duodenal ulcers

ASSOCIATION QUESTIONS

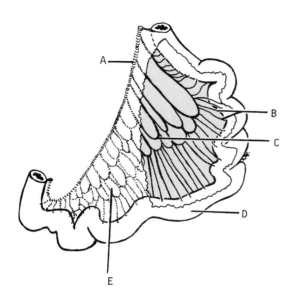

11. ____ An anastomosis between adjacent arteries forming a single arcade

12. ____ Long vasa recti

13. ____ An artery arising at the level of the first lumbar vertebra

14. ____ Numerous branches arising from the distal half of the midgut artery

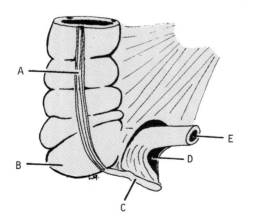

15. ____ The part of the large intestine that lies in the right iliac fossa

16. ____ Inflammation of this organ occurs frequently

17. ____ A distinguishing feature of the large intestine

18. ____ Where a Meckel's diverticulum may occur

19. ____ Is the site of an internal hernia

ASSOCIATION QUESTIONS

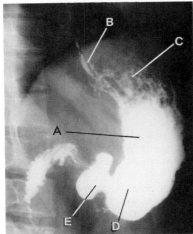

20. ___ An orifice located at the level of the tenth thoracic vertebra

21. ___ An area supplied by the short gastric arteries

22. ___ A border of the stomach along which the gastric nerves course

23. ___ Region supplied by the right gastroepiploic artery

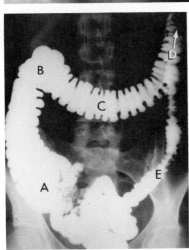

24. ___ Portion of the gut that lies anterior to the pancreas

25. ___ Region of the gut that receives its blood supply from the right colic artery

26. ___ portion of the gut related to the spleen

27. ___ Region of the gut that lies anterior to the right iliacus muscle

28. ___ Part of the gut that receives its blood supply from the lower left colic artery

------------------ ANSWERS, NOTES AND EXPLANATIONS --------------------

1. E. The cardiac orifice is the opening of the esophagus into the stomach and is at the level of the tenth thoracic vertebra. B is the pyloric orifice of the stomach.

2. D. The lymphatics of the upper left portion of the greater curvature, including the fundus and the upper part of the body, drain into the pancreaticolienal lymph nodes located along the splenic artery and the hilum of the spleen.

3. C. The lower right portion of the greater curvature of the stomach is supplied by the right gastroepiploic artery.

4. A. The lesser curvature of the stomach provides attachment for the gastrohepatic ligament or lesser omentum. Its other attachment is to the porta hepatis of the liver.

5. D. The fundus of the stomach is supplied by the short gastric arteries which arise from the splenic artery.

6. B. The second or descending part of the duodenum has a dual developmental origin: the proximal half is derived from the foregut and the distal half from the midgut.

7. D. The third (horizontal) part of the duodenum is crossed anteriorly by the superior mesenteric artery, the artery of the midgut. The superior mesenteric vein draining the midgut is also anterior to this portion of the duodenum. The third part of the duodenum crosses the midline anterior to the right psoas major, inferior vena cava, and aorta from right to left.

8. A. The first (superior) part of the duodenum is the most mobile portion as its first inch has a peritoneal covering on both its anterior and posterior surfaces. The remaining parts of the duodenum are retroperitoneal.

9. E. The fourth (ascending) part of the duodenum is the shortest and terminates at the duodenojejunal flexure on the left side of the second lumbar vertebra. The peritoneal folds in this area form several fossae, e.g., the duodenojejunal fossa and the paraduodenal fossa, which are potential sites for internal herniae.

10. A. The first part of the duodenum (duodenal bulb or cap) is the most common site of ulcers. One of the common complications of this lesion is hemorrhage and perforation of the floor of the ulcer.

11. C. The anastomosis between arteries proceeding to the small intestine are called arterial arcades. The number of arcades in the mesentery of the gut provides a valuable clue to the surgeon in distinguishing between coils of the jejunum and ileum. In the jejunal mesentery, the arcades are one or two in number, whereas there are about four to five present in the ileum.

12. B. The arteries proceeding towards the mesenteric border of the small intestine are called the straight arteries or vasa recti. In the jejunum, these are much longer than in the ileum.

13. A. The artery arising from the abdominal aorta at the level of the first lumbar vertebra is the superior mesenteric, supplying numerous jejunal and ileal branches to the small gut.

14. D. The ileal branches arise from the convex border of the distal half of the superior mesenteric (midgut) artery. In this region the arterial arcades are about four or five in number, the vasa recti are short, and the mesentery contains greater amounts of fat than the jejunal mesentery.

15. B. The cecum is the blind commencement of the large intestine. It usually lies in the right iliac fossa, but may be located anywhere between the liver and false pelvis. Abnormalities of position result from abnormal descent and fixation of the cecum during fetal development.

16. C. Inflammation of the vermiform appendix (appendicitis) occurs frequently and pain is usually referred to the right iliac region. McBurney's incision is commonly used for the removal of the appendix (appendectomy). The position of the appendix is variable: retrocecal 64%; pelvic 32%, iliac fossa 2%; and retroileal 0.5%.

17. A. The anterior taenia (taenia libera) lies on the anterior surface of the ascending colon and cecum. It serves as an important guide to the base of the

vermiform appendix. The three taeniae coli represent the longitudinal muscle fibers of the colon. The other two features distinguishing the large intestine from the small intestine are the haustra (sacculations) and appendices epiploicae (fatty appendages).

18. E. The terminal portion of the ileum opens into the cecum through the ileocecal valve. The terminal 1.5 m (five feet) may exhibit an evagination from the antimesenteric border known as Meckel's diverticulum. This is a remnant of the yolk stalk (vitelline or omphalomesenteric duct). This malformation occurs in about 2 per cent of persons. Inflammation of this diverticulum mimics symptoms of appendicitis. Ulceration and hemorrhage are significant pathological findings in a Meckel's diverticulum.

19. D. The superior and inferior ileocecal recesses are found in relation to the terminal part of the ileum. The superior ileocecal recess lies deep to the superior ileocecal fold, enclosing the anterior cecal branch of the ileocolic artery. The inferior ileocecal recess lies beneath the inferior ileocolic fold which is usually devoid of blood vessels and is sometimes called the "bloodless fold of Treves". Both these recesses are directed caudally and are potential sites for internal hernias.

20. B. The cardiac orifice is the opening of the esophagus into the stomach and is situated at the level of the tenth thoracic vertebra.

21. C. The fundus of the stomach is supplied by the short gastric arteries arising from the splenic artery.

22. A. The lesser curvature of the stomach provides attachment to the lesser omentum. The anterior and posterior gastric nerves containing parasympathetic and sympathetic fibers course along the lesser curvature on both the anterior and posterior surfaces of the stomach.

23. D. The lower right half of the greater curvature of the stomach is supplied by the right gastroepiploic artery, a branch of the gastroduodenal. E indicates a peristaltic wave at the approximate junction of the pyloric antrum and canal.

24. C. The transverse colon lies anterior to the pancreas and is attached to the posterior well of the abdomen by the transverse mesocolon.

25. B. The right colic (hepatic) flexure receives its blood supply from the right colic artery, a branch of the superior mesenteric. This part of the gut lies anterior to the right kidney and is alo related to the inferior surface of the liver.

26. D. The left colic (splenic) flexure is in contact with the inferior pole of the spleen. It is supplied by branches of the upper left colic artery.

27. A. The cecum is the first part of the large intestine and lies in the right false pelvis anterior to the right iliacus muscle. It is supplied by the ileocolic artery, a branch of the superior mesenteric.

28. E. The lower part of the descending colon is supplied by branches from the lower left colic artery which originates from the inferior mesenteric. The superior mesenteric artery supplies derivatives of the hindgut, as follows: the left one-third to half of the transverse colon; the descending colon; the sigmoid colon; the rectum; and the upper part of the anal canal.

SOLID ABDOMINAL VISCERA

FIVE-CHOICE COMPLETION QUESTIONS

DIRECTIONS: Each of the following questions or incomplete statements is followed by five suggested answers or completions. SELECT THE ONE BEST ANSWER in each case and then underline the appropriate letter at the lower right of each question.

1.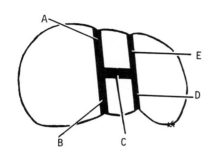

   THE FISSURE FOR THE LIGAMENTUM VENOSUM IS INDICATED BY ____.   A B C D E

2. WHICH OF THE FOLLOWING STATEMENTS ABOUT THE CAUDATE LOBE OF THE LIVER IS (ARE) TRUE?
   A. Limited on the right by the groove for the inferior vena cava
   B. Limited on the left by the fissure for the ligamentum venosum
   C. Part of the left half (lobe) of the liver
   D. Related posteriorly to the lesser sac
   E. All of the above   A B C D E

3. WHICH STATEMENT ABOUT THE BILE DUCT, AFTER IT LEAVES THE LESSER OMENTUM, IS FALSE?
   A. Related to the superior pancreaticoduodenal artery
   B. Traverses the medial part of the body of the pancreas
   C. Related medially to the gastroduodenal artery
   D. Lies behind the first part of the duodenum
   E. Related posteriorly to the portal vein   A B C D E

4. THE HILUM OF THE RIGHT KIDNEY IS COVERED BY THE:
   A. Second part of the duodenum   D. Right colic flexure
   B. Quadrate lobe of the liver   E. Gall bladder
   C. Right colic vessels   A B C D E

5. WHICH STATEMENT ABOUT THE RELATIONS OF THE LEFT SUPRARENAL GLAND IS FALSE?
   A. Posterior to the body of the pancreas
   B. Anterior to the kidney
   C. Related to the stomach
   D. Anterior to the diaphragm
   E. Posterior to the spleen   A B C D E

6. THE VISCERAL SURFACE OF THE SPLEEN IS NOT IN CONTACT WITH THE:
   A. Stomach   D. Left adrenal gland
   B. Left kidney   E. Tail of the pancreas
   C. Left colic flexure   A B C D E

SELECT THE ONE BEST ANSWER

7. THE LIGAMENTUM TERES REPRESENTS THE OBLITERATED:
   A. Right umbilical vein
   B. Left umbilical artery
   C. Right umbilical artery
   D. Left umbilical vein
   E. None of the above

   A B C D E

8. WHICH STATEMENT(S) ABOUT THE GALL BLADDER IS(ARE) TRUE?
   A. Inflammation of it is common in older obese persons
   B. Forms part of the extrahepatic biliary system
   C. Is supplied only by the cystic artery
   D. Consists of a fundus, body, and neck
   E. All of the above

   A B C D E

9.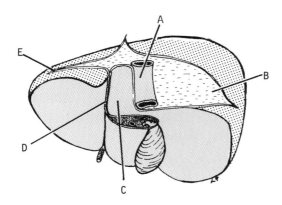

   THE BARE AREA OF THE LIVER IS INDICATED BY _____.

   A B C D E

10. THE MODE OF FORMATION OF THE RENAL FASCIA PERMITS MOVEMENT OF THE KIDNEY:
    A. Medially
    B. Laterally
    C. Superiorly
    D. Inferiorly
    E. None of the above

    A B C D E

11. WHICH STATEMENT ABOUT THE QUADRATE LOBE OF THE LIVER IS FALSE?
    A. Supplied by the left branch of the portal vein
    B. Found on the inferior surface of the liver
    C. In contact with the right colic flexure
    D. Related to the pylorus of the stomach
    E. Located medial to the gall bladder

    A B C D E

12.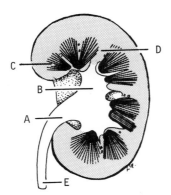

    THE RENAL PELVIS IS LABELLED _____.

    A B C D E

SELECT THE ONE BEST ANSWER

13. THE SUPRARENAL GLANDS RECEIVE THEIR ARTERIAL SUPPLY FROM:
    A. The superior suprarenal, a branch of the inferior phrenic artery
    B. The inferior suprarenal, a branch of the renal artery
    C. The middle suprarenal artery from the aorta
    D. All of the above
    E. None of the above

    A B C D E

14. WHICH STATEMENT ABOUT THE PERITONEAL ATTACHMENTS OF THE LIVER IS FALSE?
    A. The falciform ligament is attached to the anterior aspect of the liver
    B. The left triangular ligament connects the left lobe with the diaphragm
    C. The right triangular ligament connects the right lobe to the diaphragm
    D. The lesser omentum is attached to the inferior surface of the liver
    E. The falciform ligament contains the ligamentum venosum

    A B C D E

15. WHICH STATEMENT ABOUT THE RIGHT KIDNEY IS FALSE?
    A. Related to branches of the anterior primary rami of the first lumbar nerve
    B. Related posteriorly to the iliacus muscle
    C. Covered by several fascial sheaths
    D. Lower than the left kidney
    E. A retroperitoneal organ

    A B C D E

16.

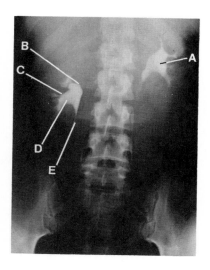

IN THIS INTRAVENOUS PYELO-
GRAM THE PELVIS OF THE
KIDNEY IS INDICATED BY ____.

A B C D E

------------------ ANSWERS, NOTES AND EXPLANATIONS ---------------------

1. D. The fissure for the ligamentum venosum forms the posterior half of the left limb of an H-shaped arrangement of fissures, grooves, and depressions seen on the visceral surface of the liver. The anterior part of the left limb is formed by the fissure for the ligamentum teres (E). The horizontal bar joining the right and left limbs represents the porta hepatis of the liver

(C). The broad right limb, in its anterior half, indicates the fossa for the gall bladder (A) and its posterior half, the groove for the inferior vena cava (B).

2. E. The caudate lobe is rectangular in shape and is limited on the right by the groove for the inferior vena cava and on the left by the fissure for the ligamentum venosum. On the basis of true anatomical division, the caudate lobe belongs to the left half (lobe) of the liver. Posterior to the caudate lobe is the superior recess of the omental bursa (lesser sac).

3. B. The bile duct traverses the head of the pancreas but not the body. After leaving the free margin of the lesser omentum, the bile duct passes behind the first part of the duodenum, where it is related medially to the gastroduodenal artery and lies anterior to the portal vein. Then the bile duct traverses the head of the pancreas to reach the posteromedial wall of the second part of the duodenum, entering at the junction of the foregut and midgut. Behind the head of the pancreas, the superior pancreaticoduodenal artery forms a medial relation to the bile duct.

4. A. The second part of the duodenum overlies the hilum of the right kidney. The quadrate lobe is related to the transverse colon. The right colic vessels cross the inferior pole of the kidney to supply the hepatic flexure of the colon, which is related to the lower lateral part of the anterior surface of the kidney. The gall bladder is located at a higher level and is not related to the kidney.

5. E. The left suprarenal (adrenal) gland is not related to the spleen. The left kidney and the diaphragm form the posterior relations of the left suprarenal gland. Its anterior relations are the pancreas and the stomach.

6. D. The visceral surface of the spleen is not in contact with the left suprarenal (adrenal) gland. The visceral surface of the spleen, however, is in relation to the stomach, left kidney, left colic flexure, and tail of the pancreas.

7. D. The obliterated left umbilical vein is represented in the adult by the ligamentum teres. This vessel returns oxygenated blood from the placenta to the fetus during prenatal life. The ligamentum teres is located in the free border of the falciform ligament.

8. E. The extrabiliary system (apparatus) consists of the: gall bladder, cystic duct, hepatic ducts and bile duct. The gall bladder is described as having a fundus, body, and a neck that leads into the cystic duct. Inflammation of the gall bladder (cholycystitis) occurs commonly in the older obese persons, especially females. Gallstones are frequently found in the gall bladder. The cystic artery, a branch of the right hepatic, is usually the sole blood supply of the gall bladder.

9. B. The bare area of the liver is triangular in shape and is located on its posterior surface. The base of the bare area is the inferior vena cava (A) and the other two sides are formed by the superior and inferior layers of the coronary ligament (not labelled). The apex of the triangle is formed by the right triangular ligament. The rectangular area of the liver, between the inferior vena cava and the fissure for the ligamentum venosum (D), is the caudate lobe (C). The left triangular ligament is indicated by E.

10. D. The anterior and posterior layers of the renal fascia, constituting the outermost covering of the kidney, fail to fuse inferiorly, thereby permitting abnormal movement of the kidney in some instances. This condition is known as

"floating kidney" or nephroptosis. The anterior and posterior layers of the renal fascia fuse at the lateral border of the kidney; superiorly, they fuse at the upper pole enclosing the suprarenal gland; and medially, the two layers blend with the adventitial coverings of the blood vessels. A septum passes between the kidney and the suprarenal gland, thereby enclosing this gland within a separate compartment. For this reason, in the clinical condition of "floating kidney", the suprarenal gland does not follow the kidney.

11. **C.** The right colic flexure is in contact with the inferior surface of the liver, lateral to the gall bladder but not with the quadrate lobe. The quadrate lobe is the visceral surface that lies between the fissure for the ligamentum teres and the fossa for the gall bladder. The quadrate lobe is related to the pylorus of the stomach, the first part of the duodenum, and the transverse colon. The blood supply to the quadrate lobe comes from the left branch of the portal vein and the hepatic artery proper.

12. **A.** The renal pelvis is the upper dilated portion of the ureter that lies enclosed within the renal sinus. The pelvis of the kidney divides into two major calyces (B) and they are subdivided into minor calyces. Each minor calyx receives two or more conical projections of the renal pyramids (C). A coronal section of the kidney exhibits an outer cortex made up of glomeruli and convoluted tubules, and an inner medulla composed primarily of collecting tubules. The cortical tissue separating the pyramids is known as a renal column (D). The beginning of the abdominal ureter is indicated (E).

13. **E.** The suprarenal gland has a rich vascular supply, receiving arteries from three sources: the superior suprarenal from the inferior phrenic; the middle suprarenal directly from the abdominal aorta; and the inferior suprarenal from the renal artery. Each of these arteries gives off several twigs which enter the surface of the suprarenal gland.

14. **E.** The free border of the falciform ligament encloses the ligamentum teres (the obliterated left umbilical vein), but not the ligamentum venosum. The ligamentum venosum lies in a fissure on the posterior surface of the liver and forms the left boundary of the caudate lobe. The left umbilical vein carries oxygenated blood from the placenta to the left branch of the portal vein in the fetus. The sickle-shaped, falciform ligament is attached to the anterior and superior surfaces of the liver. The right and left triangular ligaments connect the right and left lobes respectively to the diaphragm. The lesser omentum extends between the porta hepatis of the liver and the lesser curvature of the stomach.

15. **B.** The right kidney lies anterior to the psoas major, quadratus lumborum, and transversus abdominis muscles, but not the ilacus muscle. The iliohypogastric and ilioinguinal, branches of the first lumbar nerve, lie between the muscles and the kidney. The kidney is a retroperitoneal organ with several fascial coverings. The right kidney is usually lower than the left one because of the large size of the right lobe of the liver.

16. **A.** The pelvis of the kidney is the upper dilated end of the ureter (A). The pelvis usually divides two major calyces, an upper B and a lower (D). Each major calyx divides into several minor calyces. Each minor calyx (C) receives one to three renal pyramids, each of which contains collecting tubules and parts of the secretory tubules. The papilla, or apex of each pyramid, fits into the cup-shaped indentation of a minor calyx.

MULTI-COMPLETION QUESTIONS

DIRECTIONS: In each of the following questions or incomplete statements, ONE OR MORE of the completions given is correct. At the lower right of each question, underline A if 1, 2, and 3 are correct; B if 1 and 3 are correct; C if 2 and 4 are correct; D if only 4 is correct; and E if all are correct.

1. THE QUADRATE LOBE OF THE LIVER IS:
   1. Limited laterally by the fossa for the gall bladder
   2. Located posterior to the porta hepatis of the liver
   3. Related to the transverse colon
   4. Limited medially by the fissure for ligamentum venosum

   A B C D E

2. THE COMMON BILE DUCT IS:
   1. 4-6 cm long
   2. Located in the lesser omentum
   3. Formed by the union of the common hepatic and cystic ducts
   4. Enters the third part of the duodenum

   A B C D E

3. THE LIVER IS:
   1. The largest gland in the body
   2. Completely covered by peritoneum
   3. In the right hypochondrium
   4. Palpable in the healthy adult

   A B C D E

4. WHICH STATEMENT(S) ABOUT THE KIDNEYS IS (ARE) TRUE?
   1. Lie alongside the vertebral column against the psoas major muscle
   2. Three successive sets of excretory organs develop in the embryo
   3. Major renal vessels enter and leave the hilus of the kidney
   4. Nephrons develop from the metanephric mass of mesoderm

   A B C D E

5. THE RIGHT SUPRARENAL GLAND IS:
   1. Semilunar in shape
   2. Related anteriorly to the inferior vena cava
   3. Drained by a vein that joins the right renal vein
   4. Related posteriorly to the diaphragm

   A B C D E

6. CONCERNING THE KIDNEYS AND URETERS:
   1. Doubling of the upper end of the ureter is not uncommon
   2. Persons with a horseshoe kidney are uncommon
   3. Variations in the number of renal arteries are common
   4. Persons with supernumerary kidneys are common

   A B C D E

7. THE SPLEEN:
   1. Is connected with the stomach by the gastrosplenic ligament
   2. Lies in relation to the 9th to 11th ribs on the left side
   3. Is often enlarged in leukemias
   4. Is supplied by the artery of the midgut

   A B C D E

8. THE HEAD OF THE PANCREAS:
   1. Is related posteriorly to the inferior vena cava
   2. Is covered anteriorly by the transverse colon
   3. Lies in the concavity of the duodenum
   4. Encloses the bile duct

   A B C D E

A	B	C	D	E
1,2,3	1,3	2,4	only 4	all correct

9. THE POSTERIOR RELATIONS OF THE RIGHT KIDNEY ARE THE:
   1. Quadratus lumborum
   2. Inferior vena cava
   3. Diaphragm
   4. Iliac crest

   A B C D E

10. THE TRUE ANATOMICAL DIVISION OF THE LIVER IS BASED ON THE:
    1. Attachment of the ligamentum teres to its inferior surface
    2. Attachment of the ligamentum venosum to its posterior surface
    3. Attachment of the falciform ligament to its anterior and superior surfaces
    4. Line drawn between the fossa for the gall bladder and the inferior vena cava

    A B C D E

11. BILE IS:
    1. Secreted only by liver cells
    2. Responsible for the digestion of carbohydrates
    3. Concentrated in the gall bladder
    4. More concentrated when flow is directly into the duodenum

    A B C D E

12. IN DISEASES OF THE RIGHT KIDNEY, PAIN IS USUALLY:
    1. Felt at the tip of the left shoulder
    2. Radiated to the external genitalia
    3. Felt in the right hypochondrium
    4. Felt in the region of the right renal angle

    A B C D E

13. THE BODY OF THE PANCREAS HAS:
    1. An attachment to the transverse mesocolon
    2. An anterior surface that forms part of the stomach bed
    3. An upper border related to the splenic artery
    4. An inferior surface related to the splenic vein

    A B C D E

14. THE BILE DUCT:
    1. Begins at the union of the right and left hepatic ducts
    2. Is related medially to the hepatic artery proper
    3. Opens into the first part of the duodenum
    4. Is related posteriorly to the portal vein

    A B C D E

15. THE RIGHT LATERAL SURFACE OF THE LIVER:
    1. Is called the base
    2. Is related to ribs 7-11 in the right midaxillary line
    3. Accounts for liver dullness in percussion of the right chest wall
    4. For biopsy it is reached through the 7th intercostal space

    A B C D E

16. BASED ON THE TRUE ANATOMICAL DIVISION, THE RIGHT HALF OF THE LIVER:
    1. Includes the quadrate lobe of the liver
    2. Receives the right branch of the portal vein
    3. Is related to the middle of the transverse colon
    4. Is supplied by the right branch of the hepatic artery proper

    A B C D E

17. THE GALL BLADDER:
    1. Is related to the inferior surface of the liver
    2. Is completely enclosed by peritoneum
    3. Has an average capacity of 30 ml
    4. Opens into the bile duct

    A B C D E

A	B	C	D	E
1,2,3	1,3	2,4	only 4	all correct

18. THE LIGAMENTUM TERES OF THE LIVER:
    1. Ascends from the umbilicus to the inferior surface of the liver
    2. Is located in the free edge of the falciform ligament
    3. Represents the obliterated left umbilical vein
    4. Connects the liver to the diaphragm

    A B C D E

---------------------- ANSWERS, NOTES AND EXPLANATIONS ----------------------

1. **B.** <u>1 and 3 are correct</u>. The quadrate lobe is located on the inferior surface of the liver. Its boundaries are: laterally, the fossa for the gall bladder; medially, the fissure for the ligamentum teres; posteriorly, the porta hepatis; and anteriorly, the free margin of the liver. The quadrate lobe is related from anterior to posterior to the transverse colon, the pylorus of the stomach, and the first part of the duodenum. The fissure for the ligamentum venosum forms the left lateral boundary of the caudate lobe of the liver. The quadrate lobe lies anterior to the porta hepatis.

2. **A.** <u>1, 2, and 3 are correct</u>. The common bile duct is formed by the union of the common hepatic and cystic ducts. It is approximately 4-6 cm long and is located together with the portal vein and hepatic artery in the free border of the lesser omentum. The common bile duct enters the second part of the duodenum.

3. **B.** <u>1 and 3 are correct</u>. The liver is the largest gland in the body and accounts for about one-fortieth of the body weight in the adult. It is not palpable in the healthy adult. Though it is primarily located in the right hypochondrium, considerable portions of the liver are related to the epigastrium and left hypochondrium. The peritoneum does not cover the entire liver and is absent over the following areas: the fossa for the gall bladder, the bare area of the liver, and the fissures for various peritoneal ligaments.

4. **E.** <u>All are correct</u>. Although three sets of excretory organs develop, the first one (pronephros) is a transitory, nonfunctional structure. The mesonephroi and the metanephroi (permanent kidneys) function during fetal life. The mesonephroi degenerate during early fetal life, except for their ducts and a few tubules which persist as genital ducts in males and vestigial remnants in females. The ureteric bud, an outgrowth of the mesonephric duct, gives rise to the ureter, pelvis, calyces, and collecting tubules in both sexes. The nephrons develop from the metanephric mass of mesoderm.

5. **C.** <u>2 and 4 are correct</u>. The right suprarenal (adrenal) gland is shaped like a top-hat; it is not semilunar in shape. The right suprarenal gland is related anteriorly to the liver and the inferior vena cava. Posteriorly the gland is related to the kidney and the diaphragm. A single vein passes from the hilum of the gland to the inferior vena cava. The left suprarenal vein drains into the left renal vein.

6. **B.** <u>1 and 3 are correct</u>. Abnormalities of the kidney and ureter occur in three to four per cent of the population, and include variations in blood supply, abnormal positions, and upper urinary tract duplications. As the kidneys ascend during fetal life they are supplied by successively higher vessels. The lower vessels normally degenerate; however, if they persist, vascular variations result, e.g., multiple renal arteries. Horseshoe kidney is present in about 1 in 600 persons, whereas supernumerary kidneys are relatively uncommon.

7. **A. 1, 2, and 3 are correct.** The spleen is the largest ductless gland that lies in the left hypochondrium in relation to the 9th, 10th and 11th ribs. It is completely covered by peritoneum and is connected to the stomach by the gastrosplenic ligament. Splenic enlargements are often seen in leukemias and in tropical diseases like malaria and kala-azar.

8. **E. All are correct.** The expanded and flattened head of the pancreas lies within the C-shaped concavity of the duodenum. The largest structure related anteriorly to the head of the pancreas is the transverse colon. The superior mesenteric vessels cross the lower left part of the head of the pancreas, known as the uncinate process. Posteriorly, the inferior vena cava and the aorta form the largest structures related to the head. The terminal part of the bile duct is enclosed within a groove at the back of the head of the pancreas, near its lateral margin. For this reason, tumors of the head of the pancreas invariably obstruct the flow of the bile and produce jaundice.

9. **B. 1 and 3 are correct.** The quadratus lumborum muscle, along with the psoas major, transversus abdominis and diaphragm form the posterior relations of the right kidney. The iliac crest is at a lower level and is not related to the kidney. The inferior vena cava is medial to the right kidney, overlapping a portion of the right suprarenal gland.

10. **D. Only 4 is correct.** The line drawn between the fossa for the gall bladder and the inferior vena cava is termed the main lobar fissure; it divides the liver into two nearly equal lobes or halves. Each lobe or half receives, respectively, the right and left branches of hepatic artery proper and of the portal vein. Each lobe gives rise to right and left hepatic ducts. The attachment of the falciform ligament to the anterior and superior surfaces of the liver; the ligamentum teres to the inferior surface of the liver; and the ligamentum venosum to the posterior surface of the liver have been used in the past to divide the liver into right and left lobes. However, the present method of dividing it into equal halves is functionally and anatomically more precise.

11. **B. 1 and 3 are correct.** Bile is secreted by liver cells (hepatocytes) and it is transported via the cystic duct to the gall bladder, where it is concentrated and stored. Thus, bile flowing directly from the liver to the duodenum is less concentrated. Bile is involved in the digestion of lipids, but not carbohydrates.

12. **C. 2 and 4 are correct.** Pain in diseases of kidney is felt in the renal angle and is also radiated to the external genitalia and the skin over the gluteal region. The renal angle is the angle subtended between the outer border of the erector spinae (sacrospinalis) and the lower border of the 12th rib. Pain felt at the tip of the left shoulder, in the absence of local trauma, may be referred from an inflamed spleen. Pain in the right hypochondrium is generally due to diseases of the gall bladder and liver.

13. **A. 1, 2, and 3 are correct.** The body of the pancreas is prismatic and has three surfaces and two borders. The upper border is related to the tortuous splenic artery, whereas the lower border gives attachment to the transverse mesocolon. The anterior surface forms part of the stomach bed with the omental bursa intervening. The inferior surface is in contact with the duodenojejunal flexure, coils of jejunum, and the transverse colon, but not with the splenic vein. The splenic vein is intimately related to the posterior surface of the pancreas and separates it from several structures on the posterior abdominal wall.

14. **C, 2 and 4 are correct.** The bile duct begins where the cystic duct joins the common hepatic duct and terminates by entering the second part of the duodenum. This point of entrance represents the junction of the embryonic foregut and midgut. As it lies in the free margin of the lesser omentum, the bile duct is related medially to the hepatic artery proper and posteriorly to the portal vein.

15. **A, 1, 2, and 3 are correct.** The right lateral surface of the liver constitutes its base and is related to ribs 6-11 in the right midaxillary line. On percussion of the right chest in the midaxillary plane, the upper limit of liver dullness is usually encountered at about the level of the 6th rib. Below this level the usual resonance of the lung tissue is replaced by liver dullness. Liver biopsy, for diagnostic purposes, is usually done in the midaxillary line below the level of the eighth rib to avoid the inferior part of the lung. It should be recalled that the inferior margin of the lung reaches the 8th rib in the midaxillary line.

16. **C, 2 and 4 are correct.** The right half of the liver includes the liver substance that is to the right of the plane passing between the gallbladder and the inferior vena cava. It receives the right branch of the hepatic artery proper and the portal vein. This portion of the liver does not include the quadrate lobe and consequently is not related to the middle of the transverse colon but to the right colic flexure.

17. **B, 1 and 3 are correct.** The gall bladder is related to the visceral surface of the liver and is covered by peritoneum on its inferior surface, except for the fundus which projects beyond the free margin of the liver. The gall bladder has a capacity of about 30 ml; however, its size is variable. The gall bladder opens into the cystic duct, not the bile duct. The cystic duct usually joins the common hepatic duct to form the common bile duct, but variations are common.

18. **A, 1, 2, and 3 are correct.** The ligamentum teres forms after birth from the remnant of the left umbilical vein. There is no right umbilical vein in the fetus because it degenerates during the embryonic period. The umbilical vein remains patent for some time and may be used for exchange transfusion in the newborn. The fissure for the ligamentum teres is located on the inferior surface of the liver. The ligamentum teres is found in the lower border of the falciform ligament which connects the liver to the diaphragm and the anterior abdominal wall.

FIVE-CHOICE ASSOCIATION QUESTIONS

DIRECTIONS: Each group of questions below consists of a numbered list of descriptive words or phrases accompanied by a diagram with certain parts indicated by letters, or by a list of lettered headings. For each numbered word or phrase, SELECT THE LETTERED PART OR HEADING that matches it correctly. Then insert the letter in the space to the right of the appropriate number. Sometimes more than one numbered word or phrase may be correctly matched to the same lettered part or heading.

ASSOCIATION QUESTIONS

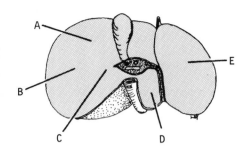

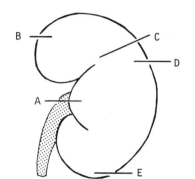

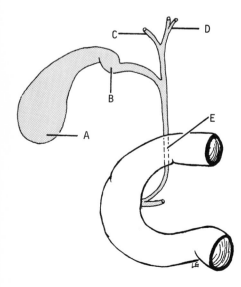

1. ___ An area related to a retroperitoneal solid organ

2. ___ The portion of the liver that forms an anterior relation to the superior recess of the lesser sac

3. ___ A surface related to a hollow viscus located principally in the left hypochondriac region

4. ___ An area related to a part of the small intestine supplied by the foregut and midgut arteries

5. ___ An area in contact with a part of the large gut

6. ___ An area related to a highly vascular, contractile organ

7. ___ Related to an endocrine gland

8. ___ Related to a gland having both exocrine and endocrine functions

9. ___ In contact with an organ supplied by the artery of the hindgut

10. ___ Separated from a portion of the alimentary tract by a peritoneal space

11. ___ Duct draining bile from the quadrate lobe of the liver

12. ___ Region of the gall bladder where gallstones become impacted

13. ___ Bile duct that lies in relation to the gastroduodenal artery

14. ___ Portion of the gall bladder that is in the transpyloric plane

15. ___ Part of the extrabiliary passage that is related to the right hepatic artery

ASSOCIATION QUESTIONS

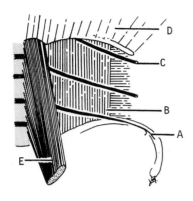

16. ___ A muscle receiving motor innervation from a cervical nerve(s)

17. ___ A muscle enclosed between two layers of lumbar fascia

18. ___ The location of nerves that provide motor innervation to the extensor and adductor compartments of the thigh

19. ___ A nerve that passes through the inguinal canal in both sexes

20. ___ Anterior primary ramus of the last thoracic nerve

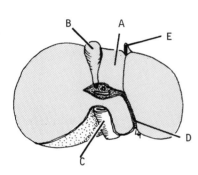

21. ___ A structure that carried oxygenated blood from the placenta to the left branch of the portal vein

22. ___ Encloses a venous channel present in the fetus

23. ___ A structure that usually lies at the level of the ninth right costal cartilage

24. ___ An important venous channel draining into the right atrium

25. ___ A structure in contact with the pylorus of the stomach

-------------------- ANSWERS, NOTES AND EXPLANATIONS --------------------

1. B. This part of the inferior surface is in contact with the right kidney.

2. D. The caudate lobe of the liver forms an anterior relation to the superior recess of the lesser sac (omental bursa).

3. E. The inferior surface of this half of the liver is in contact with the stomach.

4. C. The surface of the liver, immediately lateral to the neck of the gall bladder and continuous with the caudate process of the liver, is related to the second part of the duodenum which receives its blood supply from the celiac and superior mesenteric arteries.

5. A. The anterior half of the inferior surface of the right half of the liver is in contact with the right colic flexure.

6. D. The upper lateral margin of the left kidney is in contact with the visceral surface of the spleen.

7. B. The superomedial aspect of the left kidney is capped by the left suprarenal gland.

8. A. The body of the pancreas, having both endocrine and exocrine functions, lies across the middle of the kidney.

9. E. The lower lateral margin of the left kidney is related to the descending colon, supplied by the inferior mesenteric (hindgut) artery.

10. C. The triangular area on the left kidney is separated by the lesser sac from the posterior surface of the stomach. It is bounded superiorly by the suprarenal gland; laterally by the spleen; and inferiorly by the body of the pancreas.

11. D. The left hepatic duct conveys bile from the quadrate lobe of the liver.

12. B. The neck of the gall bladder is the common site where gallstones become lodged.

13. E. The portion of the common bile duct that is retroduodenal is accompanied by the gastroduodenal artery.

14. A. The fundus of the gall bladder is usually in the transpyloric plane in most instances and is covered by peritoneum on its entire aspect.

15. C. The right hepatic duct drains the right lobe of the liver and is related to the right hepatic artery.

16. D. The diaphragm receives motor innervation from the phrenic nerve, containing primarily fibers of the fourth cervical with contributions from the third and the fifth.

17. B. The quadratus lumborum muscle extends between the iliac crest and the twelfth rib and is enclosed between the anterior and middle layers of the lumbar fascia.

18. E. The lumbar plexus of nerves is located within the substance of the psoas major muscle. Two major nerves that arise from the plexus are the femoral and obturator nerves, supplying the extensor and adductor compartments of the thigh.

19. A. The ilioinguinal nerve passes laterally on the quadratus lumborum muscle and enters the inguinal canal to supply the external genitalia.

20. C. The subcostal nerve is the anterior primary ramus of the twelfth thoracic nerve and passes laterally on the quadratus lumborum muscle immediately below the twelfth rib. Inflammation of the kidney presumably irritates the subcostal, iliohypogastric, and ilioinguinal nerves, causing radiation of pain to the gluteal region and the external genitalia.

21. E. The ligamentum teres is the obliterated left umbilical vein that carried oxygenated blood from the placenta to the left branch of the portal vein in the fetus.

22. D. The fissure for the ligamentum venosum encloses a channel called the ductus venosus in the fetus. This bypass connected the left branch of the portal

vein to the inferior vena cava. It permitted oxygenated blood brought by the left umbilical vein from the placenta to the left branch of the portal vein to be carried to the inferior vena cava without going through the liver.

23. B. The fundus of the gall bladder usually lies at the level of the ninth costal cartilage in the transpyloric plane. The right lateral plane also meets the ninth costal cartilage.

24. C. The inferior vena cava draining the caudal half of the body is lodged in a deep groove on the posterior surface of the liver. This groove has no peritoneal lining.

25. A. The quadrate lobe of the liver is in relation from behind forwards to the pylorus of the stomach, the first part of the duodenum, and the transverse colon.

BLOOD VESSELS, NERVES AND LYMPHATICS

FIVE-CHOICE COMPLETION QUESTIONS

DIRECTIONS: Each of the following questions or incomplete statements is followed by five suggested answers or completions. SELECT THE ONE BEST ANSWER in each case and then underline the appropriate letter at the lower right of each question.

1. WHICH STATEMENT ABOUT THE INFERIOR MESENTERIC ARTERY IS FALSE?
   A. An unpaired branch of the abdominal aorta
   B. Arises at the level of the third lumbar vertebra
   C. Gives rise to the sigmoid arteries
   D. The artery of the hindgut
   E. None of the above   A B C D E

2. THE INFERIOR VENA CAVA IS CROSSED ANTERIORLY BY ALL OF THE FOLLOWING STRUCTURES EXCEPT THE _____.
   A. Third part of the duodenum   D. Root of the mesentery
   B. Superior mesenteric artery   E. Right gonadal artery
   C. Right suprarenal gland   A B C D E

3. WHICH OF THE FOLLOWING IS NOT A BRANCH OF THE SUPERIOR MESENTERIC ARTERY?
   A. Jejunal   D. Right colic
   B. Ileocolic   E. Middle colic
   C. Left colic   A B C D E

4. THE ABDOMINAL AORTA BIFURCATES AT THE LEVEL OF THE _____ LUMBAR VERTEBRA.
   A. First   D. Fourth
   B. Second   E. Fifth
   C. Third   A B C D E

5. WHICH OF THE FOLLOWING IS NOT AN ANTERIOR RELATION OF THE ABDOMINAL AORTA?
   A. Third part of the duodenum   D. Splenic vein
   B. Left lumbar vein   E. Pancreas
   C. Left renal vein   A B C D E

SELECT THE ONE BEST ANSWER

6. WHICH STATEMENT(S) ABOUT THE THORACIC DUCT IS (ARE) CORRECT:
   A. Receives the intestinal trunk of preaortic lymph nodes
   B. Receives the right and left lumbar lymph trunks
   C. Is the main collecting channel of lymph
   D. Begins as the cisterna chyli
   E. All of the above

   A B C D E

7. ALL OF THE FOLLOWING ARE TRIBUTARIES OF THE INFERIOR VENA CAVA EXCEPT THE:
   A. Right inferior phrenic   D. Fourth lumbar
   B. Right suprarenal   E. Renal
   C. Left gonadal

   A B C D E

------------------- ANSWERS, NOTES AND EXPLANATIONS ---------------------

1. E. The inferior mesenteric artery is the caudal unpaired branch of the abdominal aorta. It arises at the level of the third lumbar vertebra and gives off the left colic and sigmoid arteries. It continues as the superior rectal artery after it crosses the site of bifurcation of the left common iliac artery into the external and internal iliacs.

2. C. The medial portion of the anterior surface of the right suprarenal (adrenal) gland forms the posterior relation of the inferior vena cava. Numerous structures are anterior to the inferior vena cava; they include: the root of the mesentery enclosing the superior mesenteric artery and vein; the third part of the duodenum; the right gonadal artery; and the head of the pancreas.

3. C. The left colic artery is not a branch of the superior mesenteric; it arises from the inferior mesenteric. The superior mesenteric artery, arising from the abdominal aorta at the level of the first lumbar vertebra, gives off numerous jejunal and ileal branches from its left convex margin. From its concave right margin, the superior mesenteric artery gives origin to the middle colic, right colic, and ileocolic arteries.

4. D. The aorta passes into the abdomen through the aortic hiatus of the diaphragm and descends anterior to the vertebral column. It bifurcates at the level of the fourth lumbar vertebra to form the common iliac arteries. The short celiac trunk arises from the aorta at the level of the upper part of the first lumbar vertebra. The superior and inferior mesenteric arteries arise at the levels of the lower part of the first and of the third lumbar vertebrae, respectively. These three vessels are the unpaired visceral branches of the aorta supplying the digestive tract.

5. B. The four left lumbar veins run posterior to the abdominal aorta to enter the inferior vena cava. The pancreas and the splenic vein are anterior to the aorta; the superior mesenteric artery lies between them. The third part of the duodenum is anterior to the aorta. The left renal vein crosses the midline, in front of the aorta to join the inferior vena cava.

6. E. All the statements are correct. The thoracic duct is the main collecting channel of body lymph. Its efferents include the right and left lumbar and intestinal lymph trunks. The intestinal trunk usually joins the left lumbar trunk anywhere between the eleventh thoracic and the second lumbar vertebrae. If the union occurs low, the dilated ductal system is known as the cisterna chyli. The thoracic duct ascends into the thoracic cavity through the aortic orifice of the diaphragm.

7. **C.** The left gonadal vein empties into the left renal vein, not the inferior vena cava. The right inferior phrenic, renal, and third and fourth lumbar veins on each side drain into the inferior vena cava. The right gonadal and suprarenal veins drain into the inferior vena cava, whereas the left gonadal and suprarenal veins join the left renal vein.

MULTI-COMPLETION QUESTIONS

DIRECTIONS: In each of the following questions or incomplete statements, ONE OR MORE of the completions given is correct. At the lower right of each question, underline A if 1, 2, and 3 are correct; B if 1 and 3 are correct; C if 2 and 4 are correct; D if only 4 is correct; and E if all are correct.

1. THE PAIRED LATERAL BRANCHES OF THE ABDOMINAL AORTA SUPPLY STRUCTURES DERIVED FROM EMBRYONIC:
    1. Ectoderm
    2. Endoderm
    3. Ectoderm and endoderm
    4. Intermediate cell mass of mesoderm   A B C D E

2. THE FEATURE(S) OF THE LUMBAR PLEXUS OF NERVES IS(ARE):
    1. Located within the psoas major muscle
    2. Formed by the anterior primary rami of L1-L5
    3. Gives rise to the femoral and obturator nerves
    4. Contributes to the lumbosacral trunk   A B C D E

3. THE INFERIOR VENA CAVA IS:
    1. Formed by the union of the common iliac veins
    2. Located on the posterior abdominal wall
    3. Related medially to the abdominal aorta
    4. Anterior to the right psoas major muscle   A B C D E

4. THE BARE AREA OF THE LIVER IS:
    1. Limited medially by the groove for the inferior vena cava
    2. Located principally on the superior surface of the liver
    3. An area not covered by peritoneum
    4. Limited laterally by the left triangular ligament   A B C D E

5. THE LEFT RENAL ARTERY IS:
    1. Anterior to the left psoas major muscle
    2. Anterior to the ureter at the hilum
    3. Posterior to the left renal vein
    4. Longer than the right renal artery   A B C D E

6. THE ARRANGEMENT OF LYMPH NODES AROUND THE AORTA IS CLASSIFIED AS:
    1. Right lumbar chain   3. Retroaortic
    2. Left lumbar chain   4. Preaortic   A B C D E

7. VEINS CONTRIBUTING TO THE FORMATION OF THE PORTAL SYSTEM INCLUDE THE:
    1. Superior mesenteric   3. Splenic
    2. Inferior mesenteric   4. Hepatic   A B C D E

8. WHICH VEINS(S) DRAIN(S) INTO THE INFERIOR VENA CAVA?
    1. Right suprarenal   3. Right gonadal
    2. Left suprarenal   4. Left gonadal   A B C D E

A	B	C	D	E
1,2,3	1,3	2,4	only 4	all correct

9. THE CELIAC GANGLIA ARE:
   1. Two in number
   2. Situated on the crura of the diaphragm
   3. Located in relation to the celiac artery
   4. Supplied by preganglionic fibers from the third and fourth lumbar sympathetic ganglia   A B C D E

10. THE LEFT OVARIAN ARTERY:
    1. Originates at the level of the third lumbar vertebra
    2. Is anterior to the left psoas major muscle
    3. Has the inferior vena cava as a posterior relation
    4. Crosses the external iliac artery   A B C D E

11. THE ABDOMINAL AORTA:
    1. Terminates at the level of the fourth lumbar vertebra
    2. Begins at the aortic hiatus of the diaphragm
    3. Divides into the common iliac vessels
    4. Lies between the sympathetic trunks   A B C D E

12. THE BRANCH(ES) OF THE CELIAC ARTERIAL TRUNK IS (ARE):
    1. Common hepatic   3. Splenic
    2. Right gastric   4. Left gastroepiploic   A B C D E

13. THE PREAORTIC LYMPH NODES ARE LOCATED IN RELATION TO THE:
    1. Anterior surface of the aorta
    2. Inferior mesenteric artery
    3. Superior mesenteric artery
    4. Celiac artery   A B C D E

14. THE LEFT LUMBAR SYMPATHETIC TRUNK:
    1. May be operated on to relieve vasospastic disease of the left lower limb
    2. Runs along the medial border of the left psoas major muscle
    3. Lies on the left side of the abdominal aorta
    4. Is related posteriorly to the left renal vessels   A B C D E

15. THE RIGHT RENAL ARTERY:
    1. Arises at the level of the first lumbar vertebra
    2. Passes posterior to the inferior vena cava
    3. Divides into freely anastomosing branches
    4. Lies anterior to the right ureter   A B C D E

16. THE CAPUT MEDUSAE IS:
    1. Often associated with collection of fluid in the peritoneal cavity
    2. An external sign of portal venous obstruction
    3. Caused by dilatation of the epigastric veins
    4. Located in the region of the umbilicus   A B C D E

17. THE MAIN BRANCHES OF THE CELIAC ARTERIAL TRUNK ARE THE:
    1. Splenic   3. Left gastric
    2. Common hepatic   4. Right gastric   A B C D E

18. THE UNPAIRED VISCERAL BRANCH(ES) OF THE ABDOMINAL AORTA IS (ARE):
    1. Inferior mesenteric   3. Celiac
    2. Superior mesenteric   4. Median sacral   A B C D E

A	B	C	D	E
1,2,3	1,3	2,4	only 4	all correct

19. THE VENOUS CHANNELS IN THE LOWER END OF THE ESOPHAGUS:
    1. Include esophageal veins that drain into the portal veins
    2. Become dilated in portal obstruction forming esophageal varices
    3. Include stomach veins that drain into the azygos veins
    4. May rupture and result in vomiting of blood   A B C D E

20. THE LUMBAR SYMPATHETIC TRUNK:
    1. Runs along the anterior border of the psoas major muscle
    2. Is covered by the inferior vena cava on the right
    3. Begins beneath the medial arcuate ligament
    4. Has only four ganglia   A B C D E

21. THE CELIAC LYMPH NODES:
    1. Drain lymph from the superior gastric nodes
    2. Receive channels from hepatic lymph nodes
    3. Are related to the artery of the foregut
    4. Receive lymph from the descending colon   A B C D E

22. WHICH OF THE FOLLOWING VESSELS IS (ARE) NOT DERIVED FROM THE SPLENIC ARTERY?
    1. Pancreatic branches   3. Left gastroepiploic
    2. Short gastric   4. Right gastroepiploic   A B C D E

23. THE INFERIOR VENA CAVA RECEIVES THE:
    1. Superior mesenteric vein
    2. Hepatic veins
    3. First right lumbar vein
    4. Renal veins at the level of the second lumbar vertebra   A B C D E

------------------- ANSWERS, NOTES AND EXPLANATIONS ---------------------

1. **D.** <u>Only 4 is correct</u>. The paired lateral branches of the abdominal aorta supply the suprarenal glands, kidneys, and gonads which are derived from the intermediate cell mass of the mesoderm. The adrenal medulla develops from neural crest cells. The paired arteries are: the suprarenal, renal, and gonadal.

2. **B.** <u>1 and 3 are correct</u>. The lumbar plexus of nerves is formed in the substance of the psoas major muscle by the anterior primary rami of L1-3, a part of L4, and a contribution from T12. The two major motor nerves that arise from the plexus are the femoral (dorsal divisions of L2,3, and 4) and the obturator (ventral divisions of L2,3, and 4), which supply the extensor and adductor compartments of the thigh, respectively. The lumbosacral trunk is formed by the remaining parts of L4 and L5; it is not part of the lumbar plexus. The lumbosacral trunk enters the pelvis and contributes to the formation of the sacral plexus.

3. **E.** <u>All are correct</u>. The inferior vena cava commences at the level of the fifth lumbar vertebra about 2 cm to the right of the median plane. The inferior vena cava ascends on the posterior abdominal wall to the right of the abdominal aorta in the major part of its course. As it ascends, it lies anterior to the right psoas major muscle. The inferior vena cava drains blood directly from the lower limbs, a large portion of the anterior abdominal wall, and the urogenital structures. Blood from the gut enters the inferior vena cava after circulating through the hepatic portal system.

4. **B.** <u>1 and 3 are correct</u>. The bare area of the liver is located principally on its posterior surface and to a lesser extent on its inferior surface. It is limited medially by the groove for the inferior vena cava and laterally by the right triangular ligament, but not by the left triangular ligament. It is limited superiorly and inferiorly by the superior and inferior layers of the coronary ligaments. The bare area, as the name implies, is devoid of a peritoneal covering. In the liver, the hepatic veins draining into the systemic venous system communicate with the portal veins of the portal venous system. In portal hypertension, these veins dilate, providing an alternate route of venous return.

5. **A.** <u>1, 2, and 3 are correct</u>. The left renal artery arises from the left side of the aorta at the level of the second lumbar vertebra. This artery runs laterally across the left psoas major muscle passing posterior to the left renal vein. In the hilum of the kidney, the ureter lies posterior to the renal artery. The left renal artery is shorter than the right because the aorta is located to the left of the median plane.

6. **E.** <u>All are correct</u>. The lymph nodes draining the adult derivatives of the embryonic gut are all located in relation to the abdominal aorta. The preaortic group is anterior to the aorta and includes the celiac, superior mesenteric, and inferior mesenteric nodes. The left lumbar (paraaortic) chain of nodes (5-10) is located along the left side of the aorta. They are continuous with the common iliac nodes below and drain into the thoracic duct above. The right lumbar chain is more complex, as the nodes are disposed around the inferior vena cava, and is located to the right of the aorta. The retroaortic lymph nodes are not true regional nodes; they drain either the preaortic or the left lumbar (paraaortic) nodes.

7. **A.** <u>1, 2, and 3 are correct</u>. The portal system of veins drains the abdominal part of the gut, the spleen, the pancreas, and the gall bladder. Its tributaries are: splenic, inferior mesenteric, and superior mesenteric veins. The right and left hepatic veins drain into the inferior vena cava. The superior mesenteric vein joins the splenic posterior to the neck of the pancreas; the inferior mesenteric vein enters the splenic behind the body of the pancreas.

8. **B.** <u>1 and 3 are correct</u>. The inferior vena cava is formed by the union of the common iliac veins and ascends through the abdomen on the right side of the aorta. It receives the right suprarenal and right gonadal (ovarian or testicular) veins, the lumbar, the renals, the hepatic, and the right inferior phrenic veins. The left suprarenal, left gonadal (testicular or ovarian), and left inferior phrenic veins drain into the left renal vein.

9. **A.** <u>1, 2, and 3 are correct</u>. The two celiac ganglia are located on the crura of the diaphragm, and are connected by filaments which pass around the celiac artery. Each celiac ganglion receives preganglionic sympathetic fibers from the thoracic spinal cord via the greater and lesser splanchnic nerves. The postganglionic fibers form paired plexuses as follows: phrenic, adrenal, renal, testicular, and unpaired as hepatic, splenic, gastric, and so forth. These are collectively called the abdominal aortic plexuses. The celiac ganglia do not receive fibers from the third and fourth lumbar sympathetic ganglia. Sometimes the first lumbar sympathetic ganglion provides preganglionic fibers to the celiac ganglia. A second major abdominal ganglion is the superior hypogastric plexus located at the bifurcation of the aorta.

10. **C.** <u>2 and 4 are correct</u>. The ovarian arteries arise from the abdominal aorta at the level of the second lumbar artery. The left ovarian artery passes downwards and laterally, anterior to the left psoas major muscle and the left ureter. It enters the true pelvis by crossing the external iliac artery. The

right ovarian artery and not the left ovarian crosses anterior to the inferior vena cava.

11. **E.** <u>All are correct</u>. The abdominal aorta commences at the aortic hiatus of the diaphragm and runs downwards on the lumbar vertebrae and intervertebral discs. It terminates on the fourth lumbar vertebra, slightly to the left of the median plane, by dividing into the common iliac vessels. During its course, the abdominal aorta lies superiorly, between the crura of the diaphragm; and inferiorly, between the two sympathetic trunks.

12. **B.** <u>1 and 3 are correct</u>. The celiac trunk is the first unpaired branch of the abdominal aorta. It gives rise to three branches: the hepatic, splenic, and left gastric arteries. The left gastroepiploic artery is a branch of the splenic; the right gastric is a branch of the common hepatic artery.

13. **E.** <u>All are correct</u>. The preaortic lymph nodes are situated anterior to the aorta and are generally grouped around the unpaired visceral arteries arising from the aorta. Therefore, the lymph nodes are found in relation to the celiac, superior mesenteric, and inferior mesenteric arteries and drain the derivatives of the fore-, mid-, and hindgut.

14. **A.** <u>1, 2, and 3 are correct</u>. The left lumbar sympathetic trunk is located in the groove between the aorta and the left psoas major muscle. In vasospastic disease of the left lower limb (e.g., thromboangiitis obliterans or Buerger's disease) lumbar sympathectomy is performed. The surgical exposure of this trunk is through a lateral extraperitoneal approach. It is generally recommended that the second and third lumbar ganglia be removed for relief of symptoms. Removal of the first lumbar ganglion is not advocated in young men lest the power of ejaculation be lost. No advantage is gained in removing the fourth ganglion, which is beset with technical difficulties. The left renal vessels lie anterior to the left sympathetic trunk.

15. **C.** <u>2 and 4 are correct.</u> The renal arteries arise at the level of the second lumbar vertebra. The right renal artery passes laterally, posterior to the inferior vena cava and the right renal vein. In the hilum of the kidney, the renal artery divides into five branches: four anterior and one posterior, which are segmental in distribution. These segmental arteries have few anastomoses with one another; thus blockage of a segmental artery results in necrosis of the part of the kidney supplied by it.

16. **E.** <u>All are correct</u>. The caput medusae is the term used to describe the dilated veins radiating from the umbilicus because of portal venous obstruction. The veins in this area, enclosed in the falciform ligament, drain into the portal venous system and communicate with the epigastric veins draining into the systemic venous system. A patient with portal venous obstruction has an accumulation of fluid in the peritoneal cavity (ascites). This condition is frequently encountered in patients with cirrhosis of liver resulting from alcoholism.

17. **A.** <u>1, 2, and 3 are correct</u>. The celiac trunk arises from the abdominal aorta at the level of the first lumbar vertebra. It gives rise to three main branches: splenic, common hepatic and left gastric arteries. The right gastric is a branch of the common hepatic artery.

18. **A.** <u>1, 2, and 3 are correct</u>. The three unpaired visceral branches of the abdominal aorta are the celiac, the foregut artery; the superior mesenteric, the midgut artery; and the inferior mesenteric, the hindgut artery. The median sacral artery is a small vessel, representing the caudal part of the embryonic dorsal aorta. The median sacral artery descends anterior to the sacrum to end

at the coccyx.

19. **C.** <u>2 and 4 are correct</u>. In portal obstruction, the veins in the lower end of the esophagus become dilated, forming esophageal varices; their rupture results in vomiting of blood (hematemesis). The esophageal veins communicate with the azygos system of veins, whereas the veins of the stomach drain into the portal system.

20. **E.** <u>All are correct</u>. The lumbar sympathetic trunk on each side runs along the anterior border of the psoas major muscle, descending in front of the lumbar vertebrae and intervertebral discs. Each sympathetic trunk is a continuation of the thoracic sympathetic chain and enters the abdomen beneath the medial arcuate ligament. It then enters the pelvis behind the common iliac vessels. The right sympathetic chain is overlapped by the inferior vena cava; the left lies between the aorta and the medial margin of the left psoas major muscle. The lumbar sympathetic trunk has only four ganglia. Because there are five lumbar nerves, one would expect a fifth ganglion; however, this is not the case because fusion has occurred between two ganglia.

21. **A.** <u>1, 2, and 3 are correct</u>. The celiac lymph nodes belong to the preaortic group of retroperitoneal nodes, situated around the origin of the celiac artery, the artery of the foregut in the embryo. The celiac lymph nodes receive efferents from the hepatic, pancreaticolienal, superior gastric, and superior and inferior mesenteric lymph nodes. The lymphatic drainage of the celiac lymph nodes parallels the distribution of the celiac artery. The lymphatics from the descending colon drain into the inferior mesenteric nodes.

22. **D.** <u>Only 4 is correct</u>. The splenic artery is one of the main branches of the celiac trunk. It is a tortuous vessel and courses along the upper border of the pancreas to supply the spleen via a large number of splenic branches. As it runs along the upper of the pancreas, the splenic artery supplies the pancreas by a series of pancreatic branches. The splenic artery also gives rise to the short gastric branches and the left gastroepiploic artery which supply the fundus and the region of the greater curvature of the stomach. The right gastroepiploic artery is derived from the gastroduodenal branch of the hepatic artery.

23. **C.** <u>2 and 4 are correct</u>. The inferior vena cava receives the right and left renal veins at about the level of the second lumbar vertebra. The superior mesenteric vein, draining the major portion of the small gut and a part of the large intestine, joins the splenic vein to form the portal vein. The first right lumbar vein joins the right subcostal vein to form the azygos vein. The three hepatic veins join the inferior vena cava at the upper border of the liver.

FIVE-CHOICE ASSOCIATION QUESTIONS

DIRECTIONS: Each group of questions below consists of a numbered list of descriptive words or phrases accompanied by a diagram with certain parts indicated by letters, or by a list of lettered headings. For each numbered word or phrase, SELECT THE LETTERED PART OR HEADING that matches it correctly. Then insert the letter in the space to the right of the appropriate number. Sometimes more than one numbered word or phrase may be correctly matched to the same lettered part or heading.

ASSOCIATION QUESTIONS

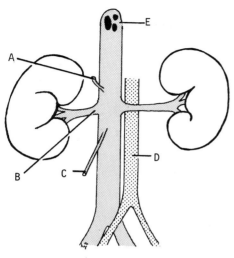

1. ___ A vessel lying to the right of the caudate lobe of the liver

2. ___ A vein draining a gland partly derived from neural crest cells

3. ___ A vein draining a gland that was abdominal in position during embryonic life

4. ___ A portion of the inferior vena cava lying posterior to a retroperitoneal organ having exocrine and endocrine functions

5. ___ A structure that divides at the level of the fourth lumbar vertebra

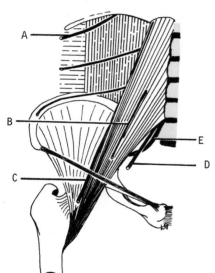

6. ___ A nerve supplying the adductor muscles of the thigh

7. ___ A nerve supplying the skin over the femoral triangle and the cremaster muscle

8. ___ A nerve lying posterior to the kidney

9. ___ A structure innervating muscles of the front of the thigh

10. ___ A structure contributing to the sacral plexus

------------------- ANSWERS, NOTES AND EXPLANATIONS ---------------------

1. **E.** The terminal subdiaphragmatic portion of the inferior vena cava lies in a deep groove located to the right of the caudate lobe of the liver.

2. **A.** The right suprarenal vein drains into the inferior vena cava, whereas the left empties into the left renal vein. The cortex of the suprarenal gland is derived from mesoderm and the medulla from neural crest cells.

3. **C.** The right gonadal vein drains into the inferior vena cava, whereas the left empties into the left renal vein. In embryonic life, the gonads were abdominal in position and descended into the pelvis in the female. In the

219

male, the testis usually passes through the inguinal canal into the scrotum during the perinatal period.

4. B. The portion of the inferior vena cava where the renal veins join is located posterior to the head of the pancreas, an endocrine and exocrine gland.

5. D. The abdominal aorta divides at the level of the fourth lumbar vertebra into the common iliac arteries. Aneurysms (localized dilatations) of the abdominal aorta may occur in the lower half of its course.

6. D. The obturator nerve (ventral divisions of L2, 3, and 4) is one of the two major motor nerves of the lumbar plexus. It emerges from the medial border of the psoas major muscle and supplies the muscles of the adductor compartment of the thigh.

7. B. The genitofemoral nerve is the only nerve of the lumbar plexus that emerges from the anterior surface of the psoas major muscle. It divides into a femoral branch that provides sensory innervation for the skin over the femoral triangle, and a genital branch that innervates the cremaster muscle. Stroking of the skin over the medial thigh in a male causes contraction of the cremaster muscle and elevation of the testis of the corresponding side. This is known as the cremasteric reflex and is dependent on the integrity of the genitofemoral nerve.

8. A. The subcostal nerve is the anterior primary ramus of the 12th thoracic nerve. As it courses laterally and downwards, it passes behind the kidney. Irritation of the subcostal, iliohypogastric, and ilioinguinal nerves in diseases of the kidney results in pain being radiated to the gluteal region and the external genitalia. This is an example of referred pain.

9. C. The femoral nerve (dorsal divisions of L2, 3, and 4) emerges at the lateral border of the psoas major muscle and is one of the major branches of the lumbar plexus. The femoral nerve innervates the iliacus muscle in the abdomen. In the thigh, in addition to supplying the four muscles comprising the quadriceps femoris, it innervates the pectineus muscle.

10. E. The lumbosacral trunk, composed of the anterior primary rami of L4 (partly) and L5, descends over the ala of the sacrum to enter the pelvis. The fourth lumbar nerve is the one ventral ramus that is common to both the lumbar and the lumbosacral plexus.

6. THE PELVIS AND PERINEUM

OBJECTIVES

Pelvic Walls

Be Able To:

* Describe the bones, the joints, the ligaments, the fascia, and the walls of the pelvis.

* Differentiate the female bony pelvis from that of the male.

* Recognize the bony parts of the pelvis in a radiograph.

* Describe and illustrate the muscles of the pelvic diaphragm.

Pelvic Peritoneum and Viscera

Be Able To:

* Describe the pelvic viscera in the male and female, giving their position and peritoneal reflections.

* Discuss briefly the structure, the blood supply, the lymphatic drainage, and the nerve supply of the following: uterus, ovaries, uterine tubes, prostate gland, seminal vesicle, rectum, ureter, and urinary bladder.

* Discuss the anatomy and clinical significance of the anal canal.

* Describe the pelvic vessels (internal iliac arteries and their branches; internal iliac veins and their tributaries), and the nerves of the pelvis (sacral plexus, coccygeal plexus, and the pelvic part of the autonomic nervous system).

Perineum

Be Able To:

* Outline with diagrams the boundaries and subdivisions of the perineum.

* Describe the contents of the anal and urogenital triangles in the male and the female.

* Write short notes on the following, illustrating your answers with labelled sketches: perineal body; sphincter urethrae, ischiorectal fossa; structure and blood supply of the penis; the mechanism of erection; the spermatic cord; testis; epididymis; the greater vestibular glands; the blood and nerve supply of the labia majora and minora; and clitoris.

PELVIC WALLS

FIVE-CHOICE COMPLETION QUESTIONS

DIRECTIONS: Each of the following questions or incomplete statements is followed by five suggested answers or completions. SELECT THE ONE BEST ANSWER in each case and then underline the appropriate letter at the lower right of each question.

1. IN THE ANATOMICAL POSITION, WHICH OF THE FOLLOWING STATEMENTS CONCERNING THE PELVIS IS FALSE?
   A. Supports the weight of the body
   B. Pelvic inclination is 20-30° with the horizontal
   C. Pelvic surface of the symphysis pubis faces upwards and backwards
   D. Concavity of the sacrum is directed downwards and forwards
   E. None of the above   A B C D E

2. WHICH OF THE FOLLOWING STATEMENTS REGARDING THE GREATER PELVIC CAVITY IS FALSE?
   A. Located above the level of the arcuate lines
   B. Is part of the abdominal cavity
   C. Lies between the iliac fossae
   D. Is the axis of the birth canal
   E. None of the above   A B C D E

3. WHICH OF THE FOLLOWING JOINTS OF THE PELVIS IS SYNOVIAL?
   A. Pubic symphysis   D. Sacrococcygeal
   B. Lumbosacral   E. None of the above
   C. Sacroiliac   A B C D E

4. COMPARED TO THE MALE, THE BONES OF THE FEMALE PELVIS ARE USUALLY:
   A. Thicker   D. Larger
   B. Heavier   E. None of the above
   C. Marked by more prominent muscular ridges   A B C D E

5. WHICH OF THE FOLLOWING STATEMENTS REGARDING THE OBTURATOR INTERNUS MUSCLE IS FALSE?
   A. Forms the lateral wall of the pelvis
   B. Originates from the obturator membrane
   C. Its tendon leaves the pelvis via the obturator canal
   D. Inserts on the greater trochanter of the femur
   E. None of the above   A B C D E

6. WHICH OF THE FOLLOWING PELVIC MEASUREMENTS IS THE SHORTEST?
   A. True conjugate   D. Obstetrical conjugate
   B. Oblique diameter   E. Diagonal conjugate
   C. Transverse diameter   A B C D E

7. WHICH OF THE FOLLOWING STATEMENTS REGARDING THE FEMALE PELVIS IS FALSE?
   A. Pelvic inlet is usually oval
   B. Pelvic outlet is larger than in the male
   C. False pelvis is deeper than in the male
   D. Greater sciatic notch is wider than in the male
   E. Subpubic angle is wide   A B C D E

SELECT THE ONE BEST ANSWER

8.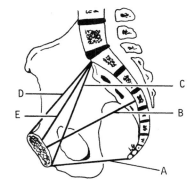

THE TRUE CONJUGATE OF THE PELVIS
IS LABELLED _____ IN THE DIAGRAM.   A B C D E

9. THE TRUE CONJUGATE OF THE PELVIS IS USUALLY DETERMINED FROM WHICH
   OF THE FOLLOWING?
   A. Transverse conjugate   D. Oblique diameter
   B. Obstetrical conjugate   E. Pelvic outlet
   C. Diagonal conjugate
                                                              A B C D E

---------------------- ANSWERS, NOTES AND EXPLANATIONS ----------------------

1. **B.** In the anatomical position, the pelvis supports the weight of the body transmitted via the vertebrae and sacroiliac joints to the hip bones and the femur. The inclination of the pelvis (i.e., the plane of the pelvic inlet in relation to the horizontal) is 50-60°. The pelvic surface of the symphysis is directed upwards and backwards, whereas the concavity of the sacrum faces downwards and forwards.

2. **D.** The curved axis of the birth canal passes through the lesser or true pelvic cavity which is located below the plane demarcated by the arcuate lines. The arcuate line forms part of the terminal line. The greater or false pelvis lies above the plane of the terminal lines, between the iliac fossae. It is considered to be the lower part of the abdominal cavity.

3. **C.** The sacroiliac joint is of the synovial type; it is formed by the articulation between the auricular surfaces of the sacrum and the ilium. The auricular surfaces are covered by hyaline cartilage and the auricular capsule of the joint is lined by synovial membrane. The stability of the sacroiliac joint depends on the roughened interlocking auricular surfaces of the articulating bones and the strong ventral and dorsal interosseous sacroiliac ligaments. As a result of hormonal influences during pregnancy, marked changes occur in this joint, as well as in the pubic symphysis.

4. **E.** The bones of the female pelvis are usually smaller, lighter, and thinner compared to the male. In addition, the muscle markings on the bones are less prominent than in the male. These differences between the bones of the male and female pelvis are usually related to the heavier build and stronger muscles of the male.

5. **C.** The tendon of the obturator internus muscle leaves the pelvis via the lesser sciatic foramen to insert into the greater trochanter of the femur. This muscle has an extensive origin from the internal aspect of the obturator membrane and the adjacent bone, thereby contributing to the formation of the upper part of the lateral pelvic wall. This lateral rotator and abductor of the thigh is innervated by the nerve to the obturator internus (L5, S1 and 2).

6. **D.** The obstetrical conjugate is measured from the posterior aspect of the symphysis pubis to the sacral promontory. It is the shortest of all conjugate diameters and is approximately 10.5 cm long, i.e., slightly less than the true conjugate (11 cm).

7. **C.** The differences between the male and female pelvis are related to its role during parturition (childbirth). The inlet of the female pelvis is usually oval and the outlet is larger because of the everted ischial tuberosities. In the female, the false pelvis is deeper, the greater sciatic notch wider, and the subpubic angle greater than in the male.

8. **D.** The dimensions of the pelvic inlet and outlet are of considerable practical importance in obstetrical practice. For assessment of the female pelvis, several diameters are usually determined. The true conjugate of the inlet (D) is the most important. It extends anteroposteriorly from the upper inner border of the symphysis pubis to the sacral promontory and usually measures 11 cm or more. The obstetrical conjugate (E) extends from the sacral promontory to the inner surface of the pubic symphysis. The true conjugate is difficult to measure but can be estimated from the diagonal conjugate (C). It extends from the lower border of the sacral promontory to the lower border of the symphysis pubis. The anteroposterior diameter of the greatest pelvic dimension (B) and of the pelvic outlet (A) are also indicated in the diagram.

9. **C.** The true conjugate of the pelvic inlet extends from the upper border of the symphysis pubis to the sacral promontory. Although it is the most important of all obstetrical measurements, the true conjugate is difficult to determine directly and is estimated from the diagonal conjugate. The diagonal conjugate (usually 12.5 cm) is the distance from the lower border of the pubic symphysis to the promontory and can be readily determined during vaginal examination. The true conjugate is 1-2 cm less than the diagonal conjugate.

MULTI-COMPLETION QUESTIONS

DIRECTIONS: In each of the following questions or incomplete statements, ONE OR MORE of the completions given is correct. At the lower right of each question, underline A if 1, 2 and 3 are correct; B if 1 and 3 are correct; C if 2 and 4 are correct; D if only 4 is correct; and E if all are correct.

1. THE BONY PELVIS IS FORMED BY THE:
   1. Ilium
   2. Pubis
   3. Ischium
   4. Sacrum   A B C D E

2. WHICH OF THE FOLLOWING JOINTS IS (ARE) PRESENT IN THE PELVIS?
   1. Pubic symphysis
   2. Sacrococcygeal
   3. Sacroiliac
   4. Lumbosacral   A B C D E

3. THE PIRIFORMIS MUSCLE:
   1. Leaves the pelvis via the lesser sciatic foramen
   2. Inserts on the lesser trochanter of the femur
   3. Originates from the obturator membrane
   4. Rotates the thigh laterally   A B C D E

4. THE COCCYGEUS MUSCLE:
   1. Is attached to the sacrospinous ligament
   2. Inserts into both the sacrum and coccyx
   3. Is innervated by only sacral nerves
   4. Arises from the ischial spine   A B C D E

A	B	C	D	E
1,2,3	1,3	2,4	only 4	all correct

5. THE LEVATOR ANI MUSCLE:
   1. Separates the pelvis from the ischiorectal fossa
   2. Is innervated by the sacral plexus
   3. Has free anteromedial borders
   4. Consists of several parts   A B C D E

6. WHICH OF THE FOLLOWING BONES CONTRIBUTE(S) TO THE FORMATION OF THE ACETABULUM?
   1. Ilium   3. Ischium
   2. Pubis   4. Sacrum   A B C D E

7. CONCERNING THE TRUE PELVIC CAVITY:
   1. Is funnel-shaped
   2. Forms part of the birth canal
   3. Its cavity is longer at the back than in front
   4. Lower aperture lies at the plane of the terminal lines   A B C D E

8. FEMALE PELVES MAY BE CLASSIFIED AS FOLLOWS:
   1. Platepelloid   3. Anthropoid
   2. Android   4. Gynecoid   A B C D E

9. THE TERMINAL LINE CONSISTS OF WHICH OF THE FOLLOWING:
   1. Pubic crest   3. Ala of sacrum
   2. Promontory   4. Pectineal line   A B C D E

10. WHICH OF THE FOLLOWING IS (ARE) TRUE FOR THE PELVIC DIAPHRAGM?
    1. Resists intra-abdominal pressure
    2. Lies deep to the ischiorectal fossa
    3. Formed in part by the puborectus muscle
    4. Lies anterior and superficial to the urogenital diaphragm   A B C D E

11. WHICH OF THE FOLLOWING JOINTS PERMIT(S) MOVEMENT DURING PREGNANCY?
    1. Pubic symphysis   3. Sacrococcygeal
    2. Sacroiliac   4. Lumbosacral   A B C D E

12. WHICH OF THE FOLLOWING MUSCLES CONTRIBUTE(S) TO THE FORMATION OF THE LATERAL PELVIC WALL?
    1. Obturator externus   3. Psoas major
    2. Obturator internus   4. Piriformis   A B C D E

13. CONCERNING DIFFERENCES BETWEEN THE MALE AND FEMALE PELVIS:
    1. Subpubic angle is usually more acute in the male
    2. Obturator foramen is usually oval in the female
    3. Ischial tuberosities are everted in the female
    4. Pelvic inlet is oval in the female   A B C D E

14. THE PELVIC DIAPHRAGM CONSISTS OF WHICH OF THE FOLLOWING MUSCLES?
    1. Piriformis   3. Obturator internus
    2. Levator ani   4. Coccygeus   A B C D E

-------------------- ANSWERS, NOTES AND EXPLANATIONS --------------------

1. **E.** <u>All are correct</u>. The ilium, ischium and pubis are parts of the hip bone. The bony pelvis (pelvic girdle) is formed by both hip bones, the sacrum and the coccyx. Anteriorly, the hip bones articulate to form the symphysis pubis; posteriorly, they articulate with the sacrum at the sacroiliac joint.

2. **E.** <u>All are correct</u>. The bones of the pelvis articulate to form the following joints: pubic symphysis, lumbosacral, sacroiliac, and sacrococcygeal. The pubic symphysis, a secondary cartilaginous joint, is formed by the bodies of the pubic bones with an intervening fibrocartilaginous disc. The auricular surfaces of the sacrum and ilium articulate to form the synovial sacroiliac joint. The lumbosacral and sacrococcygeal joints represent cartilaginous articulations between the body of the fifth lumbar vertebra and the sacrum, and between the sacrum and coccyx, respectively. A fibrocartilaginous intervertebral disc intervenes between the articulating surfaces of the bones in each joint.

3. **D.** <u>Only 4 is correct</u>. The piriformis and obturator internus muscles are related to the posterior and lateral walls of the pelvis. The bulky piriformis muscle originates from the internal aspect of the sacrum, the greater sciatic notch and the sacrotuberous ligament, and inserts into the superior aspect of the greater trochanter of the femur. It is an abductor and lateral rotator of the thigh. Branches from the sacral plexus (S1 and 2) innervate the piriformis muscle.

4. **E.** <u>All are correct</u>. The coccygeus and levator ani muscles, together with their fascial coverings, form a muscular sling, the pelvic diaphragm, across the pelvic cavity. The coccygeus muscle is located posterior to the more extensive levator ani. It originates from the ischial spine and the sacrospinous ligament and inserts into the sides of the lower sacrum and upper coccyx. The coccygeus muscle is innervated by sacral nerves 4 and 5.

5. **E.** <u>All are correct</u>. The hammock-shaped levator ani is a complex muscle consisting of three parts: pubococcygeus, puborectalis, and iliococcygeus. Because the anteromedial borders of the levator ani muscle are free, the pelvic diaphragm has an anterior muscular gap, closed by the urogenital diaphragm. The muscular floor, formed by the levator ani, separates the true pelvic cavity from the ischiorectal fossa. Branches from S3, 4, and 5 innervate the levator ani.

6. **A.** <u>1, 2, and 3 are correct</u>. The acetabulum is a large, cup-shaped, incomplete cavity in the hip bone which articulates with the head of the femur. It is formed by the ilium, superiorly; the ischium, posteroinferiorly; and the pubis, anteriorly. The three parts of the hip bone develop independently and fusion occurs in the acetabulum by the age of 18 years.

7. **A.** <u>1, 2, and 3 are correct</u>. The true pelvic cavity is funnel-shaped and located between the lower parts of the ilium. The upper pelvic aperture (pelvic inlet) is in the plane of the terminal lines at an angle of approximately 48° with the horizontal. The diamond-shaped lower pelvic aperture (pelvic outlet) is bounded by the arcuate ligament, inferior rami of the pubis, the coccyx, the ischial tuberosities, and the sacrotuberous ligaments. The cavity of the true pelvis is curved and forms part of the birth canal; its posterior wall is longer than its anterior wall.

8. **E.** <u>All are correct</u>. The shape and size of the female pelvis are of considerable importance during parturition. These are best determined by radiological techniques. Female pelves are classified into four main types, depending on the shape of the inlet, as follows: gynecoid (rounded); android (heart-shaped); anthropoid (oval); and platepelloid (flat). The most common type (50 per cent) is the gynecoid, but often a female pelvis may show characteristics of several types.

9. **E.** <u>All are correct.</u> The terminal line consists of the pubic crest, pectineal line, arcuate line of the ilium, ala of the sacrum, and the sacral promontory.

The terminal lines lie in a plane that separates the false or greater pelvic cavity (above) from the true or lesser pelvic cavity (below). The false pelvis is considered to be the lower part of the abdominal cavity.

10. **D. Only 4 is correct.** The pelvic diaphragm lies posterior and deep to the urogenital diaphragm; medial and deep to the ischiorectal fossa. The pelvic diaphragm consists of the levatores ani and coccygei muscles. The levatores ani, the most important component, are divided into pubococcygeus, puborectalis, and iliococcygeus parts. The functions of the pelvic diaphragm are: (1) supports pelvic viscera; (2) resists increases in intra-abdominal pressure; (3) aids the intrathoracic-abdominal diaphragm in increasing pressure; (4) helps control micturition; (5) flexes the anorectal canal during defecation; and (6) helps direct the fetal head towards the lower part of the birth canal during parturition.

11. **A. 1, 2, and 3 are correct.** Female hormones cause relaxation of the ligaments supporting the pubic symphysis, sacroiliac, and sacrococcygeal joints. The hormone also softens the interpubic disc of the pubic symphysis. The resulting slight increases in pelvic dimensions are important for facilitating the passage of the fetal head through the birth canal during parturition.

12. **C. 2 and 4 are correct.** The piriformis and obturator internus muscles line the internal aspect of the dorsal and lateral walls of the pelvis. Both muscles are covered by pelvic fascia, a continuation of the transversalis fascia. The obturator internus muscle arises from the internal aspect of the obturator membrane and adjacent bone. It leaves the pelvis via the lesser sciatic foramen and inserts on the greater trochanter of the femur. The piriformis muscle originates from the ventral surface of the sacrum and exits through the greater sciatic foramen to insert, like the obturator internus muscle, on the greater trochanter of the femur. The psoas major and obturator externus muscles are not part of the pelvic walls.

13. **E. All are correct.** The female pelvis differs from the male because it forms part of the birth canal. The pelvic inlet is usually oval, whereas in the male it is heart-shaped. The subpubic angle is also greater in the female (about 90°). Because the ischial tuberosities are everted in the female, the distance between them is greater; the pelvic outlet is therefore larger. The obturator foramen is oval in the female, but round in the male.

14. **C. 2 and 4 are correct.** The pelvic diaphragm is the incomplete muscular sling forming the floor of the true pelvic cavity and supporting the pelvic viscera. It consists of the levator ani and coccygeus muscles together with their superior and inferior fasciae. The piriformis and obturator internus muscles are located in the lateral walls of the pelvic cavity.

FIVE-CHOICE ASSOCIATION QUESTIONS

DIRECTIONS: Each group of questions below consists of a numbered list of descriptive words or phrases accompanied by a diagram with certain parts indicated by letters, or by a list of lettered headings. For each numbered word or phrase, SELECT THE LETTERED PART OR HEADING that matches it correctly. Then insert the letter in the space to the right of the appropriate number. Sometimes more than one numbered word or phrase may be directly matched to the same lettered part or heading.

ASSOCIATION QUESTIONS

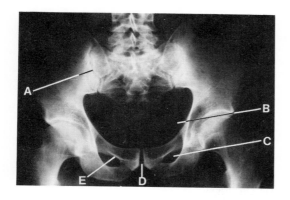

1. ___ Pectineal line
2. ___ Synovial joint
3. ___ Fibrocartilaginous joint
4. ___ Part of the terminal line
5. ___ Attachment of a triangular ligament
6. ___ Covered by a membrane
7. ___ Contributes to the formation of the greater sciatic foramen

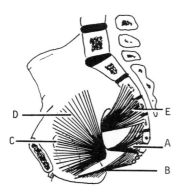

8. ___ Obturator internus muscle
9. ___ Sacrospinous ligament
10. ___ Piriformis muscle
11. ___ Transmits the obturator nerve
12. ___ Gives rise to the falciform process
13. ___ Fascia covering a muscle giving origin to the levator ani muscle

------------------------ ANSWERS, NOTES AND EXPLANATIONS ------------------------

1. E. The pectineal line of the pubic bone or pecten pubis is a well-defined ridge, extending laterally from the pubic tubercle. It forms part of the boundary demarcating the true from the false pelvis. This line also serves for the attachment of the following: the lacunar ligament, the pectineal ligament, the conjoint tendon, and the reflected part of the inguinal ligament.

2. A. The sacroiliac joint is a true synovial articulation between the hyaline cartilage-covered auricular surfaces of the ilium and sacrum. It is strengthened by strong anterior and posterior sacroiliac ligaments. The posterior sacroiliac ligament helps to transmit the weight of the trunk to the legs. Movement is slight in this joint but it increases in females during pregnancy under the influence of female hormones.

3. **D.** The pubic symphysis is a fibrocartilaginous joint between the bodies of the pubic bones. The articular surfaces are united by a thick intermediate fibrocartilaginous interpubic disc. In some cases, a slit-like cavity is present in the disc, but it is not lined by a synovial membrane. The superior and arcuate fibrous ligaments strengthen this joint. These ligaments and the interpubic disc soften during pregnancy because of female hormones, thus permitting a small amount of movement that facilitates childbirth.

4. **E.** The terminal line separates the true (below) from the false (above) pelvis. It consists of the pubic crest, the pectineal line (E), the arcuate line of the ilium, and the ala and promontory of the sacrum.

5. **B.** The triangular sacrospinous ligament is attached to the ischial spine (B) and to the sides of the lower sacral and upper coccygeal vertebrae. It lies anterior to the sacrotuberous ligament which extends from the ischial tuberosity to the sacrum and coccyx.

6. **C.** The obturator foramen (C) is almost completely closed by the obturator membrane, except for a small gap superiorly. Through this opening (obturator canal), the obturator nerve and vessels pass from the pelvis into the thigh. The obturator muscle is attached to the obturator membrane and the margin of the obturator foramen.

7. **B.** The greater sciatic foramen is formed by the sacrospinous and sacrotuberous ligaments and the greater sciatic notch. The sacrospinous ligament extends from the ischial spine (B) to the sacrum and coccyx. The sacrotuberous ligament has a more extensive attachment and passes behind the sacrospinous ligament from the ischial tuberosity to the sacrum and coccyx. Because of the attachments of these ligaments, the greater and lesser sciatic notches become converted into the greater and lesser sciatic foramina, respectively.

8. **D.** The fan-shaped obturator internus muscle contributes to the formation of the lateral wall of the pelvis. It originates from the internal aspect of the obturator membrane, the surrounding margins of the obturator foramen, and the inferior part of the ilium. The tendon of this muscle exits through the lesser sciatic foramen and inserts into the greater trochanter of the femur. The obturator internus muscle is principally a lateral rotator of the thigh but it is also an abductor. It is innervated by the nerve to the obturator internus muscle, a branch of the sacral plexus.

9. **A.** The sacrospinous ligament extends from the ischial spine to the lateral border of the lower sacrum and the upper part of the coccyx. The sacrospinous (A) and the sacrotuberous (B) ligaments convert the greater and lesser sciatic notches of the hipbone into the greater and lesser sciatic foramina, respectively.

10. **E.** The large piriformis muscle, together with the more extensive obturator internus muscle (D), is related to the posterior and lateral walls of the pelvis. The piriformis muscle arises from the ventral surface of the sacrum and inserts into the upper border of the greater trochanter of the femur, after leaving the pelvis via the greater sciatic notch. It is a lateral rotator and abductor of the thigh. Branches from the sacral plexus supply the piriformis muscle.

11. **C.** The obturator nerve, a branch of the lumbar plexus, is formed within the substance of the psoas major muscle, passes through its substance, emerges from its medial border, and passes on the lateral wall of the pelvic cavity. It leaves the pelvis via the obturator canal (C), a small deficiency in the anterosuperior portion of the obturator membrane which is bounded above by the

obturator groove. The obturator nerve is accompanied by the obturator artery, a branch of the internal iliac.

12. B. The sacrotuberous ligament arises from the posterior superior iliac spine and the lateral borders of the lower sacrum and upper coccyx. It inserts into the ischial tuberosity. The falciform process, derived from this ligament, passes to the lower margin of the ramus of the ischium. The pelvic fascia joins the falciform ligament inferiorly.

13. D. The fascia lining the walls of the pelvis is a continuation of the transversalis fascia of the anterior abdominal wall and the fascia iliaca of the posterior abdominal wall. Part of the pelvic fascia covers the piriformis muscle, but the obturator fascia covering the obturator internus muscle is more prominent. Most of the levator ani muscle arises from this fascia. Below the origin of the levator ani, the pelvic fascia continues and lines the lateral walls of the ischiorectal fossae. It eventually splits to form the pudendal (Alcock's) canal, containing the pudendal nerve and the internal pudendal vessels.

PELVIC PERITONEUM AND VISCERA

FIVE-CHOICE COMPLETION QUESTIONS

DIRECTIONS: Each of the following questions or incomplete statements is followed by five suggested answers or completions. SELECT THE ONE BEST ANSWER in each case and then underline the appropriate letter at the lower right of each question.

1. WHICH OF THE FOLLOWING SURFACES OF THE FEMALE URINARY BLADDER IS(ARE) COVERED BY PERITONEUM?
   A. Anterior
   B. Posterior
   C. Lateral
   D. Superior
   E. All of the above

   A B C D E

2. THE URINARY BLADDER IS INNERVATED BY WHICH OF THE FOLLOWING?
   A. Autonomic
   B. Somatic efferent
   C. Afferent
   D. All of the above
   E. None of the above

   A B C D E

3. WHICH OF THE FOLLOWING ARTERIES SUPPLY(IES) THE FEMALE URETER?
   A. Common iliac
   B. Internal iliac
   C. Uterine
   D. Vesical
   E. All of the above

   A B C D E

4. WHICH OF THE FOLLOWING STATEMENTS REGARDING THE FEMALE URETER IS CORRECT?
   A. Enters the true pelvis anterior to the sacroiliac joint
   B. Is related to the posterior border of the ovarian fossa
   C. Lies medial to the obturator vessels and nerve
   D. Crossed anteriorly by the uterine arteries
   E. All of the above

   A B C D E

5. THE UTERUS IS DRAINED BY WHICH OF THE FOLLOWING GROUPS OF LYMPH NODES?
   A. Inguinal
   B. Iliac
   C. Sacral
   D. Lumbar
   E. All of the above

   A B C D E

SELECT THE ONE BEST ANSWER

6. THE BROAD LIGAMENT CONTAINS ALL THE FOLLOWING STRUCTURES, EXCEPT THE:
   A. Uterine tube
   B. Ovary
   C. Uterine artery
   D. Ovarian ligament
   E. Part of the ureter

   A B C D E

7. THE NARROWEST SEGMENT OF THE UTERINE TUBE IS THE:
   A. Infundibulum
   B. Ampulla
   C. Isthmus
   D. Uterine part
   E. None of the above

   A B C D E

8. THE RECTUM IS DRAINED BY WHICH OF THE FOLLOWING GROUPS OF LYMPH NODES?
   A. Inferior mesenteric
   B. Sacral
   C. Internal iliac
   D. Common iliac
   E. All of the above

   A B C D E

9. WHICH OF THE FOLLOWING STATEMENTS REGARDING THE OVARY IS FALSE?
   A. The ovarian ligament contains the ovarian vessels and nerves
   B. Supplied by the ovarian artery
   C. The left ovarian vein drains into the renal vein
   D. Innervated by the abdominal aortic plexus of nerves
   E. Supplied by the uterine artery

   A B C D E

10. WHICH OF THE FOLLOWING STATEMENTS CONCERNING THE RECTUM IS FALSE?
    A. Middle third has no peritoneum
    B. Separated from the uterus by the rectouterine pouch
    C. Transverse folds project into its lumen
    D. Has no mesentery
    E. Taenia coli are absent

    A B C D E

11. WHICH OF THE FOLLOWING STATEMENTS CONCERNING THE PROSTATE GLAND IS FALSE?
    A. Is shaped like a cone
    B. Consists of two lobes
    C. Related inferiorly to the urogenital diaphragm
    D. Innervated by the inferior hypogastric plexus
    E. Can be palpated rectally

    A B C D E

---------------------- ANSWERS, NOTES AND EXPLANATIONS ----------------------

1. D. The superior surface of the urinary bladder is covered by peritoneum in both males and females. The pelvic peritoneum is reflected from the anterior abdominal wall directly onto the upper surface of the bladder. In the female, the peritoneum passes from the bladder to cover the anterior aspect of the uterus, except the cervix; the fundus and posterior aspect of the uterus; and the upper portion of the vagina. The peritoneum then becomes reflected onto the anterior wall of the rectum. The peritoneal space between the bladder and the uterus is the uterovesical pouch; the space between the uterus and the rectum is known as the rectouterine pouch (of Douglas) and is homologous to the rectovesical pouch of the male. The rectouterine pouch is the lowest part of the peritoneal cavity. Blood or pus accumulating here may be drained through the posterior fornix of the vagina.

2. D. The urinary bladder is innervated by the following via the inferior hypogastric and vesical plexuses: autonomic (sympathetic and parasympathetic);

somatic efferent derived from S2, 3, and 4, to the detrusor muscle and the external sphincter; and afferent, travelling with the sympathetic and parasympathetic fibers. When the bladder is distended, stretch receptors in its wall are stimulated and impulses are transmitted via the afferent fibers to the 2nd, 3rd, and 4th sacral segments of the spinal cord. Efferent impulses from the same segments cause contraction of the vesical musculature and relaxation of the external and internal sphincters, resulting in the desire to micturate. Spinal cord injuries affect the nervous control of micturition.

3. E. A large number of arteries supply the ureter, forming a rich anastomosis. The renal, ovarian, common iliac, internal iliac, uterine, vaginal and vesical arteries supply the female ureter. In the male, the testicular arteries replace the ovarian and uterine arteries.

4. E. As in the male, the female ureter enters the true pelvis anterior to the sacroiliac joint. However, its relations to the pelvic viscera differ in its subsequent course. The female ureter passes through the uterosacral and broad ligaments, then along the lateral walls of the cervix and fornices of the vagina to enter the base (fundus or posterior surface) of the urinary bladder, anterior to the cervix. The ureter forms the posterior border of the ovarian fossa and lies medial to the inferior vesical artery and the obturator vessels and nerve. The ureter is obliquely crossed by the uterine artery as the vessel courses towards the uterus. In surgery of the female pelvis, the ureter should be identified and isolated before clamping or ligating of the uterine artery.

5. E. The lymphatic drainage of the uterus is as follows: the fundus and the upper part of the body drain into the lumbar (aortic) and superficial inguinal groups of lymph nodes; the lower part of the body drains into the external iliac nodes; and the cervix into the external iliac, internal iliac, and sacral group of nodes.

6. B. The ovary is not within the two layers of the broad ligament, but is attached to its posterior surface by the mesovarium. The broad ligament contains the following structures: uterine tube, uterine vessels, ovarian ligament, branches of the ovarian artery, uterovaginal plexus of nerves, lymphatics vessels, and the ureter as it passes forwards to enter the bladder.

7. C. The isthmus of the uterine tube is thick-walled and is the narrowest. The other parts of the uterine tube are: the funnel-shaped infundibulum which opens into the peritoneal cavity; the long and wide ampulla; and the uterine (interstitial) part within the wall of the uterus. Implantation may occur in the tube when transport of the zygote is delayed, e.g., in inflammation of the tubes (salpingitis). Such tubal (ectopic) pregnancies frequently lead to rupture of the uterine tube.

8. E. The lymphatic drainage of the rectum is via the pararectal nodes to the inferior mesenteric, sacral, internal iliac and common iliac groups of nodes.

9. A. The ovarian vessels and plexus of nerves are contained within the suspensory (infundibulopelvic) ligament of the ovary; the ovarian ligament contains connective tissue and smooth muscle fibers. The ovary is supplied principally by the ovarian artery, but it also receives a branch from the uterine artery. The right ovarian vein drains into the inferior vena cava, the left into the left renal vein. The lymphatic vessels of the ovary drain into the lumbar group of nodes. Vasomotor fibers from the ovarian plexus of nerves, derived from the abdominal aortic plexus, innervate the ovary.

10. **A.** The middle third of the rectum is covered by peritoneum anteriorly; its upper third has a peritoneal covering anteriorly and laterally; but its lower third has no peritoneum. The uterus and upper part of the vagina are separated from the rectum by the rectouterine pouch. The dilatation of the external os of the cervical canal during childbirth can be assessed by rectal digital examination. Three prominent folds (Houston's valves) project horizontally from the wall of the rectum into its lumen; their functional importance is uncertain. The rectum has neither mesentery nor haustra. The outer horizontal layer of smooth muscle is complete, thus there are no taeniae coli as in other parts of the colon.

11. **B.** The prostate gland is shaped like an inverted and compressed cone, with an apex, base, and inferolateral, anterior and posterior surfaces. It consists of three lobes, two lateral and one median. The base of the prostate gland is related superiorly to the neck of the bladder; its apex, the lowermost part of the gland, lies on the urogenital diaphragm. The prostate gland is innervated by the prostatic plexus, formed by fibers from the inferior hypogastric plexus. The posterior surface of the gland can be palpated by digital rectal examination; it is recognizable by the presence of a relatively prominent median groove in the median lobe.

MULTI-COMPLETION QUESTIONS

DIRECTIONS: In each of the following questions or incomplete statements, ONE OR MORE of the completions given is correct. At the lower right of each question, underline A if 1, 2 and 3 are correct; B if 1 and 3 are correct; C if 2 and 4 are correct; D if only 4 is correct; and E if all are correct.

1. THE OVARY IS:
    1. Partly covered by peritoneum
    2. Enclosed in the broad ligament
    3. Attached to the uterine tube by the ovarian ligament
    4. Located in the ovarian fossa

    A B C D E

2. THE UTERINE TUBE IS:
    1. Supplied by the uterine artery
    2. Innervated by the hypogastric plexus of nerves
    3. Drained by the lumbar lymph nodes
    4. Supplied by the ovarian artery

    A B C D E

3. THE ADULT UTERUS IS:
    1. Partly covered by peritoneum
    2. Anteverted
    3. Anteflexed
    4. 12-15 cm long

    A B C D E

4. THE UTERUS IS SUPPORTED BY WHICH OF THE FOLLOWING LIGAMENTS?
    1. Transverse cervical
    2. Sacrocervical
    3. Pubocervical
    4. Broad

    A B C D E

5. WHICH OF THE FOLLOWING ARTERIES SUPPLY(IES) THE UTERUS?
    1. Umbilical
    2. Ovarian
    3. Internal pudendal
    4. Uterine

    A B C D E

6. THE FEMALE URINARY BLADDER IS SUPPLIED BY WHICH OF THE FOLLOWING ARTERIES?
    1. Superior vesical
    2. Uterine
    3. Vaginal
    4. Inferior vesical

    A B C D E

A	B	C	D	E
1,2,3	1,3	2,4	only 4	all correct

7. THE URINARY BLADDER:
   1. Is round in shape when empty
   2. Has a capacity of 150-200 ml
   3. Is lined by cuboidal epithelium
   4. Has a trigone at its base

   A B C D E

8. CONCERNING THE URETER:
   1. Less than 30 cm long
   2. Abdominal part is twice as long as the pelvic part
   3. Intramural part is the narrowest
   4. Crosses the gonadal vessels anteriorly

   A B C D E

9. CONCERNING THE MALE URINARY BLADDER:
   1. Separated from the pubic symphysis by a definite space
   2. Its posterior surface is related to the rectum
   3. The neck is relatively immobile
   4. Its anterior surface is covered by peritoneum

   A B C D E

10. WHICH OF THE FOLLOWING STATEMENTS ABOUT THE NERVE PLEXUSES IN THE PELVIS IS(ARE) CORRECT?
    1. The sacral plexus is formed by the lumbosacral trunk and the upper four sacral nerves
    2. The coccygeal plexus is formed by the sacral and coccygeal nerves
    3. The coccygeal plexus contains sensory and sympathetic fibers
    4. The sacral plexus contains sensory, somatic motor and sympathetic fibers

    A B C D E

11. PARASYMPATHETIC FIBERS ARE CONVEYED TO THE PELVIC VISCERA BY WHICH OF THE FOLLOWING SPLANCHNIC NERVES?
    1. Sacral
    2. Lumbar
    3. Greater
    4. Pelvic

    A B C D E

12. THE PROSTATE GLAND IS:
    1. Lobated
    2. Traversed by the urethra
    3. Drained by internal iliac lymph nodes
    4. Supplied by the superior vesical artery

    A B C D E

13. THE SEMINAL VESICLES ARE:
    1. Sacculated tubes
    2. About 5 cm long
    3. Supplied by the ductus deferens artery
    4. Palpable rectally

    A B C D E

14. THE ANAL CANAL:
    1. Is 2-4 cm long
    2. Begins below the pelvic diaphragm
    3. Is surrounded by the levator ani muscle
    4. Has a voluntary internal sphincter

    A B C D E

15. THE RECTUM:
    1. Is 20-25 cm long
    2. Begins about the level of the 3rd sacral vertebra
    3. Is narrowest at the ampulla
    4. Has several flexures

    A B C D E

A	B	C	D	E
1,2,3	1,3	2,4	only 4	all correct

16. WHICH OF THE FOLLOWING ARTERIES SUPPLY(IES) THE RECTUM AND ANAL CANAL?
    1. Superior rectal
    2. Middle rectal
    3. Inferior rectal
    4. Median sacral

    A B C D E

17. WHICH OF THE FOLLOWING STATEMENTS ABOUT THE INFERIOR HYPO-GASTRIC PLEXUS OF NERVES IS(ARE) CORRECT?
    1. Is a fan-like continuation of the hypogastric nerve
    2. Contains sensory and autonomic nerve fibers
    3. Receives contributions from the pelvic and sacral splanchnic nerves
    4. Has several subsidiary plexuses of nerves

    A B C D E

------------------------ ANSWERS, NOTES AND EXPLANATIONS ------------------------

1. **D.** <u>Only 4 is correct.</u> The ovary is not covered by peritoneum and lies outside the broad ligament in the ovarian fossa. This fossa is bounded by a peritoneal fold covering the external iliac artery anteriorly, and the ureter and internal iliac artery posteriorly. The ovarian ligament extends from the ovary to the uterus, just below the site of junction of the uterine tube with the uterus, and it is continuous with the round ligament of the uterus.

2. **E.** <u>All are correct.</u> The uterine tube is supplied by branches from both the uterine and ovarian arteries. It is innervated by sensory and autonomic fibers from the ovarian and hypogastric plexuses of nerves. The lumbar (aortic) lymph nodes receive lymph from the uterine tube.

3. **A.** <u>1, 2, and 3 are correct.</u> The adult uterus is about 8 cm long and is divided into the following parts: fundus, body, and cervix. Part of the anterior wall, the fundus and the vesical surface of the body are covered by peritoneum. The uterus is normally anteverted (long axis of the uterus is at right angles to the long axis of the vagina) and anteflexed on itself. Because the uterus is anteflexed, an instrument inserted into the cervical canal may inadvertently penetrate the posterior wall of the uterus and enter the peritoneal cavity (rectouterine pouch).

4. **E.** <u>All are correct.</u> The uterus is held in position by the transverse cervical, sacrocervical, pubocervical, and broad ligaments, together with the levator ani muscle and the round ligament of the uterus. In addition, the uterus is supported by its attachment to the vagina inferiorly, and indirectly by the urinary bladder anteriorly and the rectum posteriorly. All the ligaments of the uterus are formed by condensation of pelvic fascia. When the structures in the pelvic floor and these supporting ligaments are stretched considerably, e.g., during childbirth, prolapse of the uterus and vagina may occur.

5. **C.** <u>2 and 4 are correct.</u> The uterus is supplied principally by the uterine artery, and also by the ovarian artery. The uterine artery, arising from the internal iliac artery, passes forwards and medially in the base of the broad ligament to reach the lateral border of the uterus. The artery lies at first on the lateral side of the ureter and then crosses it on its way to the uterus. The uterine artery gives branches to the cervix and the upper part of the vagina, then ascends between the two layers of the broad ligament. This ascending branch supplies the body and the fundus of the uterus before anastomosing with the ovarian artery.

6. **E.** <u>All are correct.</u> The female urinary bladder receives blood from the following arteries: superior and inferior vesicals, uterine, and vaginal. These vessels all arise from the internal iliac artery. The superior and inferior vesical arteries provide the main blood supply. In the male, the artery of the ductus deferens (middle vesical artery) also supplies the bladder. The veins follow the pattern of the arteries and drain into the internal iliac vein.

7. **D.** <u>Only 4 is correct.</u> When empty the urinary bladder is pyramidal in shape, with an apex, a base, a superior and two inferolateral surfaces. A fibrous cord, the median umbilical ligament, extends from the apex of the bladder to the umbilicus along the anterior abdominal wall. It represents the remnants of the urachus. The normal capacity of the bladder is approximately 500 ml. It is lined by transitional epithelium. The triangular trigone is located on the internal surface of the base (posterior wall). It is formed by the internal urethral opening (apex) and the two ureteric openings (base). Each side of the trigone is approximately 2.5 cm long.

8. **B.** <u>1 and 3 are correct.</u> The ureter begins at the renal pelvis and terminates in the bladder. It is approximately 25 cm long, half in the abdomen and half in the pelvis. The ureter is constricted at three sites: at its origin, where it crosses the common iliac arteries, and at its junction with the bladder. The obliquely directed intramural part (2 cm) in the wall of the bladder is the narrowest. Lodgement of renal calculi may occur at any of these sites of constrictions. In its course, the ureter is crossed anteriorly by the gonadal (ovarian or testicular) artery.

9. **A.** <u>1, 2, and 3 are correct.</u> The male urinary bladder is located in the anteroposterior region of the pelvis. The anterior wall of the bladder, not covered by peritoneum, is separated from the pubic symphysis and the abdominal wall by the retropubic space which contains fat and connective tissue. The posterior surface of the bladder is related to the rectum from which it is separated by the rectovesical pouch, the lowest part of the peritoneal cavity in the male. The neck of the bladder is fixed inferiorly to the pelvic diaphragm and is continuous with the prostate gland. Because of these relationships, the size of the prostate gland can be assessed by rectal (digital) examination.

10. **E.** <u>All are correct.</u> The sacral plexus of nerves is formed on the posterior wall of the pelvis by the ventral rami of the fourth (part of) and the fifth lumbar nerves, forming the lumbosacral trunk, and also by the ventral rami of the upper three sacral nerves with a small contribution from the fourth sacral nerve. The sacral plexus has many sensory and motor branches, but its principal nerve is the sciatic. The coccygeal plexus of nerves is formed by a part of the fourth sacral, the fifth sacral, and the coccygeal nerves. As all spinal nerves receive gray rami communicantes, both the sacral and coccygeal plexuses of nerves contain postganglionic sympathetic fibers. The coccygeal plexus of nerves also contains sensory fibers to the skin in the region of the coccyx.

11. **D.** <u>Only 4 is correct.</u> The pelvic splanchnic nerve is the only nerve carrying parasympathetic fibers; all other splanchnic nerves are sympathetic. The pelvic splanchnic nerves, arising from the sacral segments of the spinal cord (S 2-4), run on each side of the rectum and contribute to formation of the inferior hypogastric plexus of nerves. Through this plexus, preganglionic parasympathetic fibers reach the viscera of the pelvis and perineum. In addition, they supply the left one-third of the transverse colon, the descending colon, and the sigmoid colon by nerves which do not join with the perivascular sympathetic plexus of nerves of the inferior mesenteric artery.

12. **A. 1, 2, and 3 are correct.** The fibromuscular prostate gland consists of three lobes, two lateral and one median. The internal urethral orifice leads into the prostatic part of the urethra traversing the prostate gland. Hypertrophy of the gland, in particular its median lobe, may cause urinary obstruction. Lymphatic vessels from the prostate gland terminate in the internal iliac, external iliac, and sacral nodes. The prostate gland is supplied chiefly by branches from the inferior vesical artery, with contributions from the superior and middle rectal arteries. The prostatic plexus of veins joins the vesical plexus and together they drain into the internal iliac vein.

13. **E. All are correct.** The paired lobulated seminal vesicles are coiled sacculated tubes, approximately 5 cm long. The seminal vesicles produce a major component of seminal fluid; the bulbourethral and prostate glands also contribute to this fluid. The duct of the seminal vesicle joins the ampulla of the ductus deferens to form the ejaculatory duct, which passes through the prostate gland to open into the prostatic part of the urethra. The seminal vesicles are supplied by branches from the artery of the ductus deferens and the inferior vesical artery. The seminal vesicles are located behind the fundus of the urinary bladder and are separated posteriorly from the rectum by the rectovesical pouch. When the bladder is distended, the seminal vesicles are vertical in position and can be palpated rectally.

14. **B. 1 and 3 are correct.** The anal canal is 2-4 cm long. It begins at the anorectal junction, just above the pelvic diaphragm, and ends at the anus. The anal canal passes through the pelvic diaphragm and is surrounded by the levator ani muscle. The internal anal sphincter represents a thickening of the smooth muscle fibers in the muscle wall of the anal canal and is separated from the external anal sphincter, inferiorly, by Hilton's white line.

15. **C. 2 and 4 are correct.** The rectum is 12-15 cm long. It begins at the rectosigmoid junction at the level of the 3rd sacral vertebra and ends at the ampulla just above the pelvic diaphragm. The rectosigmoid junction is the narrowest part of the rectum; its widest part is the ampulla. The lower end of the ampulla suddenly narrows at the anorectal ring. The rectum has two anteroposterior (sacral and perineal) flexures and three lateral. The perineal flexure is formed where the puborectalis muscle slings around the rectum at its junction with the anal canal; it is convex anteriorly. The sacral flexure lies within the curvature of the sacrum and is convex posteriorly.

16. **E. All are correct.** The superior rectal artery arises from the inferior mesenteric and divides into two main branches which descend on the right and left sides of the rectum. Branches from these vessels penetrate the rectal wall and pass downwards in the mucosa of the anal columns to the level of the internal sphincter, where they anastomose with other vessels. The middle rectal arteries, branches of the internal iliac, supply the lower part of the rectum and the upper part of the anal canal. The inferior rectal arteries arising from the internal pudendal, give rise to several branches which traverse the ischiorectal fossa to supply the lower part of the anal canal and the surrounding muscle and skin. Branches from the median sacral artery supply the posterior aspect of the rectum. The rectum is supplied by all these arteries, which form an extensive anastomosis.

17. **E. All are correct.** The superior hypogastric plexus of nerves of the abdomen continues into the pelvis as the hypogastric nerves. These nerves ramify on each side of the rectum, the prostate, and the seminal vesicles (or the uterine cervix and lateral vaginal fornices), and continue in a fan-like fashion as the inferior hypogastric plexus of nerves. This plexus receives

parasympathetic contributions from the pelvic splanchnic nerves. The inferior hypogastric plexus contains sensory, sympathetic, and parasympathetic fibers. This plexus supplies the viscera in the pelvis and the perineum. The subsidiary plexuses of the inferior hypogastric plexus are the middle rectal, vesical, deferential, and prostatic in the male, and the uterovaginal plexus in the female.

FIVE-CHOICE ASSOCIATION QUESTIONS

DIRECTIONS: Each group of questions below consists of a numbered list of descriptive words or phrases accompanied by a diagram with certain parts indicated by letters, or by a list of lettered headings. For each numbered word or phrase, SELECT THE LETTERED PART OR HEADING that matches it correctly. Then insert the letter in the space to the right of the appropriate number. Sometimes more than one numbered word or phrase may be directly matched to the same lettered part or heading.

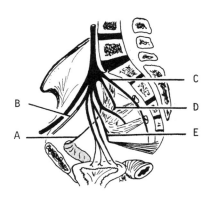

1. ___ Superior gluteal artery
2. ___ Enters the ischiorectal fossa
3. ___ Passes through the obturator foramen
4. ___ Supplies the spinal cord and meninges
5. ___ Supplies the clitoris
6. ___ Formerly an umbilical artery
7. ___ Gives rise to the inferior rectal artery

-------------------------- ANSWERS, NOTES AND EXPLANATIONS ----------------------

1. **D.** The superior gluteal artery, the largest branch of the internal iliac, arises from its posterior trunk. It leaves the pelvis via the greater sciatic foramen above the piriformis muscle to enter the gluteal region.

2. **E.** The internal pudendal artery, a terminal branch of the internal iliac, leaves the pelvic cavity via the greater sciatic foramen below the piriformis muscle. Accompanied by the pudendal nerve (S2-4), it crosses the ischial spine and enters the ischiorectal fossa via the lesser sciatic foramen.

3. **A.** The obturator artery arises from the anterior trunk of the internal iliac artery. It passes along the lateral pelvic wall on the obturator muscle and leaves the pelvic cavity via the obturator foramen, accompanied by the obturator nerve and vein. The obturator artery supplies the adductor region of the thigh.

4. **C.** The superior and inferior lateral sacral arteries arise from the posterior trunk of the internal iliac artery. Branches of the lateral sacral arteries enter the sacral foramina and supply the spinal cord, meninges, and

spinal nerves.

5. E. Both the clitoris and penis are supplied by the deep and dorsal arteries, derived from the internal pudendal artery.

6. B. The superior vesical artery, formerly the proximal part of the umbilical, is a branch of the anterior trunk of the internal iliac. It gives off the artery to the ductus deferens and then continues along the lateral pelvic wall to supply the lower end of the ureter, the seminal vesicle, and the upper part of the bladder. The distal end of the umbilical artery obliterates and becomes a solid cord, the medial umbilical ligament (formerly called the lateral umbilical ligament). It runs up the posterior surface of the anterior abdominal wall to the umbilicus. It is not important in the fixation of the bladder.

7. E. The internal pudendal artery passes through the pudendal canal, giving rise to the inferior rectal artery, which crosses the ischiorectal fossa and supplies the lower end of the rectum and anal canal. The external iliac and the smaller internal iliac arteries are terminal branches of the common iliac at the level of the lumbosacral joint. The internal iliac artery divides into anterior and posterior trunks which give rise to several named but not always constant branches. The following arteries arise from the anterior trunk: superior vesical, umbilical, obturator, uterine, vaginal, middle rectal, inferior vesical, inferior gluteal and internal pudendal. The iliolumbar, lateral sacral, and superior gluteal arteries arise from the posterior trunk of the internal iliac artery.

PERINEUM

FIVE-CHOICE COMPLETION QUESTIONS

DIRECTIONS: Each of the following questions or incomplete statements is followed by five suggested answers or completions. SELECT THE ONE BEST ANSWER in each case and then underline the appropriate letter at the lower right of each question.

1. WHICH OF THE FOLLOWING STATEMENTS REGARDING THE EXTERNAL ANAL SPHINCTER IS FALSE?
   A. Located below the pelvic diaphragm
   B. Composed of striated muscle fibers
   C. Innervated by inferior rectal nerves
   D. Its subcutaneous part inserts in the central tendon of the perineum
   E. Is under voluntary control

   A B C D E

2. WHICH OF THE FOLLOWING STATEMENTS CONCERNING THE VENOUS DRAINAGE OF THE ANAL CANAL IS FALSE?
   A. Internal are more common than external hemorrhoids
   B. The inferior rectal veins drain below the pectinate line
   C. The venous channels draining the upper and lower parts communicate
   D. Internal hemorrhoids involve the inferior rectal vessels
   E. The upper half is drained only by the superior rectal veins

   A B C D E

SELECT THE ONE BEST ANSWER

3. THE CREMASTERIC FASCIA OF THE SPERMATIC CORD IS A CONTINUATION OF THE _____ OF THE ANTEROLATERAL ABDOMINAL WALL.
   A. Peritoneum
   B. Transversalis fascia
   C. External oblique muscle
   D. Transversus abdominus muscle
   E. None of the above

4. ALL OF THE FOLLOWING CAN USUALLY BE PALPATED DURING A RECTAL EXAMINATION, EXCEPT THE:
   A. Prostate gland in the male
   B. Uterine tubes in the female
   C. Posterior vaginal wall in the female
   D. Enlarged seminal vesicles in the male
   E. Hollow of the sacrum in both sexes

5. THE GREATER VESTIBULAR GLANDS OF THE FEMALE ARE HOMOLOGOUS TO THE _____ GLAND(S) IN THE MALE.
   A. Urethral
   B. Prostate
   C. Bulbourethral
   D. Seminal vesicles
   E. None of the above

6. TUMORS OF THE TESTIS TEND TO SPREAD (METASTASIZE) TO THE _____ LYMPHATIC NODES.
   A. Lumbar
   B. Sacral
   C. Internal iliac
   D. External iliac
   E. Common iliac

7. WHICH OF THE FOLLOWING NERVES SUPPLY(IES) THE FEMALE EXTERNAL GENITALIA?
   A. Ilioinguinal
   B. Pudendal
   C. Uterovaginal plexus
   D. All of the above
   E. None of the above

8. THE FEMALE URETHRA IS _____ CM LONG.
   A. 1-2
   B. 3-4
   C. 5-6
   D. 6-8
   E. 8-10

9. THE NARROWEST PART OF THE MALE URETHRA IS THE:
   A. Prostatic part
   B. Membranous part
   C. Internal urethral orifice
   D. External urethral orifice
   E. None of the above

SELECT THE ONE BEST ANSWER

10. WHICH OF THE FOLLOWING STATEMENTS CONCERNING THE VAGINA IS FALSE?
    A. The posterior wall is longer than the anterior
    B. Its axis is directed downwards and forwards
    C. The cervix projects into its posterior wall
    D. Its posterior fornix is deeper than the anterior
    E. The lower half is located below the pelvic diaphragm

    A B C D E

11. WHICH OF THE FOLLOWING MUSCLES IS NOT ATTACHED TO THE PERINEAL BODY?
    A. Superficial transverse perineal
    B. Deep transverse perineal
    C. Ischiocavernosus
    D. Levator ani
    E. Bulbospongiosus

    A B C D E

12. WHICH OF THE FOLLOWING IS NOT PRESENT IN THE DEEP PERINEAL POUCH?
    A. Deep transverse perineal muscle
    B. Bulbospongiosus muscle
    C. Sphincter urethrae
    D. Bulbourethral gland
    E. Perineal nerve

    A B C D E

13. THE SKIN OF THE PENIS IS MAINLY SUPPLIED BY THE _____ NERVES.
    A. Ilioinguinal
    B. Deep perineal
    C. Pudendal
    D. Cavernous
    E. None of the above

    A B C D E

------------------------ ANSWERS, NOTES AND EXPLANATIONS ------------------------

1. D. The external anal sphincter consists of striated muscle surrounding the lower part of the anal canal. It is located below the pelvic diaphragm and consists of three parts: subcutaneous, superficial and deep. The subcutaneous part encircles the anal orifice, around which the fibers decussate. The superficial or main part arises from the tip of the coccyx, encircles the anal canal, and inserts into the central tendon of the perineum. The upper anal canal is surrounded by the deep part of the muscle. The external anal sphincter is under voluntary control; it is innervated by the inferior rectal nerves and a branch from the fourth sacral nerve.

2. D. The upper part of the anal canal, above the pectinate or mucocutaneous line, is drained by the superior rectal veins, tributaries of the inferior mesenteric. The lower part of the anal canal, below the pectinate line, drains into the internal iliac vein via the inferior rectal and internal pudendal veins. These two venous channels communicate and represent a site of anastomosis between the portal and systemic circulations. This collateral circulation may become functional in cases of portal hypertension. Varicosities of the superior rectal plexus of veins give rise to internal hemorrhoids; these occur more frequently than external hemorrhoids which involve the inferior rectal veins, located below the pectinate line.

3. **E.** The cremasteric fascia of the spermatic cord is a continuation of the internal oblique muscle of the anterolateral abdominal wall. The cremasteric fascial layer is superficial to the internal spermatic fascia, a continuation of the transversalis fascia. The third layer, superficial to the cremasteric layer, is the external spermatic fascia, which is a continuation of the external oblique. The tunica vaginalis, a double serous membrane, covering the front and sides of the testis and epididymis, is derived from the abdominal peritoneum.

4. **B.** During a digital examination the following structures can usually be felt in the female: vagina, cervix and external os of the uterus, the body of the uterus, and the rectouterine fossa. The following may be palpated in the male: the membranous urethra (when catheterized), the prostate gland, the rectovesical fossa, the seminal vesicles (when distended), and the bladder (when full).

5. **C.** The greater vestibular glands of the female are homologous to the bulbourethral glands in the male. These vestibular glands, located behind the bulb of the vestibule, secrete mucus which lubricates the vestibule of the vagina. The seminal vesicles, prostate, and bulbourethral, considered accessory genital organs in the male, produce the bulk of the seminal fluid. The epididymis also adds a mucoid secretion to the seminal fluid containing spermatozoa produced by the testis.

6. **A.** The lymphatic vessels draining the testis and epididymis pass upwards with the testicular vessels and drain into the lumbar (aortic) nodes. Most of the pelvic structures drain via four groups of lymph nodes, named according to the arteries with which they are associated. They are the sacral and internal, external, and common iliac lymph nodes. In the female, the uterine tube and the upper part of the body of the uterus drain into the lumbar lymph nodes.

7. **D.** The external female genitalia are innervated as follows: the labia majora and minora by the anterior and posterior labial nerves from the ilio-inguinal and pudendal nerves, respectively; the bulb of the vestibule by the uterovaginal plexus; and the clitoris by the uterovaginal plexus and the dorsal nerve of the clitoris. These nerves carry sensory and autonomic fibers.

8. **B.** The female urethra is 3-4 cm long, much shorter than in the male (18-20 cm). For this reason, the spread of infection from the vagina and vulva to the bladder is common. The female urethra runs from the bladder, through the sphincter urethrae and opens into the vestibule in front of the vagina orifice about 2.5 cm behind the clitoris.

9. **D.** The male urethra is 18-20 cm long and extends from the internal urethral orifice in the wall of the bladder (apex of the trigone) to the external urethral orifice. It is divided into three parts: prostatic, membranous, and penile (spongy). The external urethral orifice is the narrowest part. The prostatic urethra (about 3 cm long) passes through the prostate gland and is the widest and most dilatable part. It receives the two ejaculatory ducts and many prostatic ducts. The membranous urethra (about 2 cm long) passes through the urogenital diaphragm and is surrounded by the external sphincter urethrae muscle, which is under voluntary control. The penile or spongy urethra is the longest part (about 15 cm long); it lies within the corpus spongiosum of the penis.

10. **C.** The posterior wall of the vagina is about 2 cm longer than the anterior wall, but is highly dilatable. The vagina extends from the vestibule to the uterus. The uterine cervix projects into the upper part of the vagina and is

surrounded by the anterior, posterior and lateral fornices. The posterior fornix is deeper than the anterior and is related to the rectouterine pouch (of Douglas). The vagina is directed downwards and forwards. The upper half (approximately 4 cm) of the vagina is located above the pelvic diaphragm, the lower half in the perineum.

11. **C.** The perineal body or central tendon of the perineum is a fibromuscular mass located in the midline between the anal canal and the vagina in the female, and the bulb of the penis in the male. The following muscles are attached to the perineal body: superficial and deep transverse perineal, bulbospongiosus, external anal sphincter, and levator ani. Because the perineal body may be damaged during childbirth, the posterior vaginal wall and the perineal body are usually incised (posterior episiotomy) to facilitate delivery of the infant. Sometimes a lateral episiotomy is performed.

12. **B.** The deep perineal pouch or space is located between the superior and inferior layers of fascia of the urogenital diaphragm. It contains the following: deep transverse perineal muscle, sphincter urethrae, membranous part of the urethra, bulbourethral glands, branches of the internal pudendal vessels, and perineal nerve.

13. **C.** The skin of the penis, especially on the glans penis, is supplied by branches of the pudendal nerve. The ilioinguinal nerve supplies a small area of skin near the root of the penis; the deep perineal nerves supply the urethra. The cavernous nerves supply the erectile tissue of the bulb, crura, corpus spongiosum and corpora cavernosa. These nerves contain large numbers of sensory fibers including: pain, special receptor, sympathetic, and parasympathetic.

MULTI-COMPLETION QUESTIONS

DIRECTIONS: In each of the following questions or incomplete statements, ONE OR MORE of the completions given is correct. At the lower right of each question, underline A if 1, 2 and 3 are correct; B if 1 and 3 are correct; C if 2 and 4 are correct; D if only 4 is correct; and E if all are correct.

1. WHICH OF THE FOLLOWING CONTRIBUTE(S) TO FORMATION OF THE PERINEUM?
   1. Sacrospinous ligament
   2. Arcuate pubic ligament
   3. Symphysis pubis
   4. Tip of the coccyx   A B C D E

2. THE PERINEAL BODY IS:
   1. Located anterior to the anus
   2. A fibromuscular mass
   3. A midline structure
   4. Clinically important   A B C D E

3. WHICH OF THE FOLLOWING GROUPS OF LYMPH NODES DRAIN(S) THE ANAL CANAL?
   1. Internal iliac
   2. External iliac
   3. Superficial inguinal
   4. Inferior mesenteric   A B C D E

A	B	C	D	E
1,2,3	1,3	2,4	only 4	all correct

4. WHICH OF THE FOLLOWING STATEMENTS ABOUT THE PUDENDAL NERVE IS(ARE) CORRECT?
   1. Has anterior scrotal or labial branches
   2. Lies in the deep perineal space
   3. Arises from L2, 3, and 4
   4. Enters the perineum through the lesser sciatic foramen

   A B C D E

5. CONCERNING THE SPHINCTER URETHRAE MUSCLE IN THE FEMALE:
   1. Arises from the pubic bone
   2. Inserts into the vaginal wall
   3. Fibers form a sphincter around the urethra
   4. Damage to the perineal nerve results in incontinence

   A B C D E

6. THE VULVA INCLUDES THE:
   1. Labia majora
   2. Vaginal vestibule
   3. Clitoris
   4. Greater vestibular glands

   A B C D E

7. THE VAGINA IS SUPPORTED BY WHICH OF THE FOLLOWING:
   1. Levator ani muscles
   2. Transverse cervical ligament
   3. Perineal body
   4. Urogenital diaphragm

   A B C D E

8. THE VAGINA IS:
   1. Supplied by the vaginal artery
   2. Partly drained by the superficial inguinal lymph nodes
   3. Innervated by the pelvic plexus of nerves
   4. Supplied by the uterine artery

   A B C D E

9. THE MUSCLE(S) MAKING UP THE UROGENITAL DIAPHRAGM IS(ARE) THE:
   1. Bulbospongiosus
   2. Deep transverse perineal
   3. Ischiocavernosus
   4. Sphincter urethrae

   A B C D E

10. THE DEEP TRANSVERSE PERINEAL MUSCLE IN THE MALE:
    1. Originates from the ischium
    2. Inserts into the perineal body
    3. Is innervated by the perineal nerve
    4. Assists in fixing the central tendon of the perineum

    A B C D E

11. THE INFERIOR RECTAL NERVE:
    1. Is derived from the pudendal nerve
    2. Innervates the upper part of the anal canal
    3. Contains sensory fibers
    4. Provides motor innervation to the internal anal canal

    A B C D E

12. THE LABIA MAJORA:
    1. Are homologous to the scrotum in the male
    2. Contain smooth muscle fibers
    3. Are supplied by the pudendal arteries
    4. Are drained by the iliac lymph nodes

    A B C D E

A	B	C	D	E
1,2,3	1,3	2,4	only 4	all correct

13. THE ANTERIOR BOUNDARY OF THE ISCHIORECTAL FOSSA IS COMPOSED OF THE:
    1. Perineal membrane
    2. Superficial transverse perineal muscle
    3. Deep transverse perineal membrane
    4. Obturator fascia   A B C D E

14. THE ISCHIORECTAL FOSSA IS DESCRIBED AS HAVING A:
    1. Medial wall formed by the rectum
    2. Posterior wall contributed by the gluteus maximus muscle
    3. Lateral wall formed by the ischium
    4. Floor formed by the skin over the anal triangle   A B C D E

15. THE BULBOSPONGIOSUS MUSCLE:
    1. Is located in the superficial perineal pouch
    2. Originates from the perineal body
    3. Is innervated by the perineal nerve
    4. Is pierced by the vagina   A B C D E

16. THE ISCHIORECTAL FOSSA IS:
    1. Located behind an imaginary line joining the ischial tuberosities
    2. Frequently a site of infection
    3. A space containing fat
    4. Of clinical importance   A B C D E

---------------------- ANSWERS, NOTES AND EXPLANATIONS ------------------------

1. **C.** <u>2 and 4 are correct</u>. The perineum is a diamond-shaped area of the trunk that is enclosed between the medial aspects of the thighs. It is bounded anteriorly by the arcuate pubic ligament and the inferior rami of the pubis and ischium; posteriorly by the tip of the coccyx; and laterally by the sacrotuberous ligament and the ischial tuberosities. These boundaries also define the outlet of the pelvis. The perineal region is divided into an anterior urogenital triangle and a posterior anal triangle by a line passing between the ischial tuberosities, in front of the anal orifice.

2. **E.** <u>All are correct</u>. The median perineal body, the tendinous center of the perineum, is a midline structure consisting of smooth and skeletal muscle and elastic and collagenous fibers. It lies anterior to the anal canal and fuses with the urogenital diaphragm. This body serves as a site of attachment for many muscles of the perineal region. This structure is often surgically incised during parturition by an incision passing through the posterior wall of the vagina and the perineal body. This procedure, called posterior episiotomy, is performed to prevent possible laceration of the perineum during passage of the fetal head through the pelvic outlet.

3. **B.** <u>1 and 3 are correct</u>. Above the pectinate line, lymph from the anal canal drains into the internal iliac nodes. The superficial inguinal nodes receive lymph from the anal canal below the pectinate line, as well as from the scrotum, and perineum. The reason the lymphatic drainage above the pectinate line differs from that below it is that the upper part of the anal canal is derived from the hindgut, whereas the lower part originates from the proctodeum (anal pit).

4. **D.** <u>Only 4 is correct</u>. The pudendal nerve, supplying most of the perineum,

originates from spinal segments S2, 3, and 4. It passes through the greater sciatic notch, below the piriformis muscle, and runs along the medial aspect of the ischial spine, medial to the internal pudendal artery, with which it enters the perineum through the lesser sciatic foramen. The pudendal nerve is enclosed in the pudendal canal in the lateral wall of the ischiorectal fossa. It gives off the inferior rectal nerve and then divides into the perineal and dorsal nerves of the penis (or clitoris).

5. **A.** <u>1,2 and 3 are correct.</u> The sphincter urethrae muscle in the female arises from the inner aspect of the inferior ramus of the pubis and inserts largely into the lateral wall of the vagina. A few fibers pass anterior to the urethra and between the urethra and vagina. The perineal nerve supplies the sphincter urethrae and damage of it does not lead to incontinence. In the male, this muscle also forms a sphincter around the urethra and is innervated by the perineal nerve.

6. **E.** <u>All are correct.</u> The vulva or female external genitalia consists of the following: the mons pubis, labia majora, labia minora, vaginal vestibule, clitoris, bulb of the vestibule, greater vestibular glands, and vaginal orifice. If you excluded the greater vestibular glands from the vulva, you were wrong. The male external genital organs are the penis and the scrotum containing the testes.

7. **E.** <u>All are correct.</u> The support of the vagina is clinically important because it helps to maintain the uterus in its position. Thus, uterine prolapse is often associated with prolapse of the vagina. The vagina is supported by the: pelvic diaphragm, particularly the levator ani muscles; the transverse cervical (cardinal); pubocervical; and sacrocervical ligaments; the urogenital diaphragm; and the perineal body in its lowermost part.

8. **E.** <u>All are correct.</u> The vagina is supplied by branches from the uterine artery and by the vaginal artery, arising from the internal iliac. The veins of the vagina form a venus plexus which drains into the internal iliac vein. Most of the vagina is innervated by the uterovaginal (pelvic) plexus of nerves; the lowest part is supplied by the pudendal nerve. Lymph from the upper third of the vagina drains into the external and internal lymph nodes; the middle third into the internal iliac and sacral nodes; and the lowest third into the superficial inguinal nodes.

9. **C.** <u>2 and 4 are correct.</u> The urogenital diaphragm consists of the deep transverse perineal and the sphincter urethrae muscles, invested by the superior and inferior perineal fascia. The urogenital diaphragm is pierced by the urethra in the male and by the urethra and vagina in the female.

10. **E.** <u>All are correct.</u> In both sexes, the deep transverse perineal muscle arises from the inner surface of the ramus of the ischium and inserts into the central tendon of the perineum (perineal body), assisting in its fixation. The perineal nerve supplies the deep transverse perineal muscle. The muscles of the deep perineal space in the female are less developed than in the male.

11. **B.** <u>1 and 3 are correct.</u> The rectum and anal canal are supplied by nerves derived from the superior and middle rectal plexuses, and from the inferior rectal nerves derived from the pudendal. The inferior rectal nerve innervates the lower part of the anal canal below the pectinate line; the upper part, above the pectinate line, innervated by the superior rectal nerve, is relatively insensitive. The inferior rectal nerve contains motor fibers,

which innervate the external anal sphincter muscle; vasomotor fibers; and general sensory fibers.

12. **A. 1, 2, and 3 are correct.** The labia majora are homologous to the scrotum in the male and contain smooth muscle fibers, nerves, blood vessels and lymphatics. The round ligaments of the uterus terminate in these folds. Branches from the external and internal pudendal arteries supply both the labia majora and minora. The lymphatic drainage of the labia majora, as well as from other external genital organs, is to the superficial inguinal nodes.

13. **A. 1, 2, and 3 are correct.** The anterior boundary of the ischiorectal fossa is composed of the superficial transverse perineal muscle, the perineal membrane, and the deep transverse perineal muscle. The obturator fascia covering the obturator internus muscle constitutes the lateral wall of the ischiorectal fossa. The splitting of the obturator fascia results in the pudendal (Alcock's) canal, enclosing the internal pudendal vessels and the pudendal nerve and its terminal branches.

14. **C. 2 and 4 are correct.** The ischiorectal fossa is located on both sides of the anal orifice in the anal triangle. The posterior wall of this fossa is formed by the gluteus maximus muscle and the underlying sacrotuberous ligament. Its floor is the skin over the anal triangle. Although the fossa is termed the ischiorectal fossa, neither the ischium nor the rectum is an immediate relation. The ischium is covered by the obturator internus muscle and its fascia. In the medial wall of the fossa, the levator ani prevents the rectum from being a direct relation. The sphincter ani externus muscle contributes to the lower portion of the medial boundary of the ischiorectal fossa.

15. **E. All are correct.** The bulbospongiosus muscle arises from the perineal body (central tendon of the perineum) and the fibrous raphe of the bulb of the penis. It inserts in the inferior fascia of the urogenital diaphragm, the corpus spongiosum which it covers, and the deep fascia of the penis. The bulbospongiosus muscle is innervated by the perineal branch of the pudendal nerve. Contraction of the bulbospongiosus and the ischiocavernous muscles constricts the corpus spongiosum, containing the urethra, and the crura, thereby squeezing out the last few drops of urine or semen from the urethra. In the female, these muscles are less developed; contraction of them causes constriction of the vagina (sphincter vaginae) because the vagina traverses the bulbospongiosus.

16. **E. All are correct.** The ischiorectal fossa is located in the anal triangle, on each side of tha anal orifice. The anal triangle is the posterior part of the perineum that lies behind a transverse line drawn between the ischial tuberosities, passing anterior to the anal orifice. The contents of the ischiorectal fossa include fat, the inferior rectal (hemorrhoidal) nerve, artery, and vein. An infection in this area often results in an ischiorectal abscess. If untreated, such an abscess may perforate into the anal canal, forming an anal fistula. The ischiorectal fossa is somewhat wedge-shaped, with its base directed to the surface of the perineum and its thin edge at the line of meeting of the obturator internus and levator ani, covered by the obturator fascia and inferior fascia of the pelvic diaphragm.

FIVE-CHOICE ASSOCIATION QUESTIONS

DIRECTIONS: Each group of questions below consists of a numbered list of descriptive words or phrases accompanied by a diagram with certain parts indicated by letters, or by a list of lettered headings. For each numbered word or phrase, SELECT THE LETTERED PART OR HEADING that matches it correctly. Then insert the letter in the space to the right of the appropriate number. Sometimes more than one numbered word or phrase may be directly matched to the same lettered part or heading.

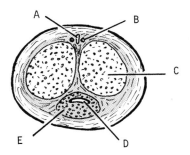

1. ___ Deep dorsal vein of the penis
2. ___ Corpus cavernosum
3. ___ A terminal branch of the internal pudendal artery
4. ___ Corpus spongiosum

------------------------ ANSWERS, NOTES AND EXPLANATIONS ------------------------

1. **A.** The unpaired deep dorsal vein of the penis is located in the midline deep to the deep fascia enclosing the body of the penis. It passes through the suspensory ligament of the penis below the infrapubic ligament and drains into the prostatic and pudendal (pelvic) plexuses of veins.

2. **C.** The corpora cavernosa are two elongated cylindrical masses consisting of highly vascular erectile tissue. The corpora cavernosa are the continuation of the crura of the penis and are surrounded by a deep heavy fascia (tunica albuginea) which forms an incomplete septum between them. The deep artery of the penis, a terminal branch of the internal pudendal, passes through each corpus cavernosum.

3. **B.** The paired dorsal arteries of the penis are terminal branches of the internal pudendal. They are located outside the deep fascia on each side of the deep dorsal vein (A) of the penis, accompanied by the dorsal nerves. Branches of the dorsal arteries and the deep arteries of the penis, supplying the erectile tissue, anastomose extensively.

4. **E.** The unpaired, elongated corpus spongiosum consists of erectile tissue. It is located ventral to the corpora cavernosa and is traversed by the urethra (D). The distal end of the corpus spongiosum expands to form the glans penis. The corpus spongiosum and the corpora cavernosa are the major components of the penis.

7. THE BACK

OBJECTIVES

<u>Vertebral Column and Joints</u>

Be Able To:

* Compare and contrast the principle regions of the vertebral column, giving the number of vertebrae and the characteristics of a typical vertebra in each region.

* Illustrate the normal curvatures of the vertebral column in a midterm fetus and an adult; and the exaggerated curvatures of kyphosis, lordosis and scoliosis in an adult.

* Describe the intervertebral articulations between adjacent bodies and arches of vertebrae in the cervical, thoracic, and lumbar regions of the vertebral column, giving the ligamentous support and movements allowed in each case.

* Compare the articulating surface, the movements, and the supporting ligaments of the atlantoaxial and atlanto-occipital joints.

* Discuss the anatomy of the lumbosacral junction giving its clinical significance in regard to fractures.

* Recognize the components of the articulated vertebral column in radiographs.

<u>Muscles, Nerves, and Blood Supply of the Back</u>

Be Able To:

* Describe the extrinsic muscles of the back, giving their attachments, innervation and general functions.

* Discuss the functions of the erector spinae muscles.

* Use simple sketches to illustrate and label the boundaries, the roof, the floor, and the contents of the suboccipital triangle, and indicate the innervation of the muscles involved.

* Discuss the segmental nature of the arterial supply and the venous and lymphatic drainage of the deep structures of the back.

<u>Spinal Cord and Meninges</u>

Be Able To:

* Illustrate the shape and the level of origin and termination of the spinal

cord in the newborn and adult.

Discuss the functions and the clinical significance of the spinal meninges and the spaces between them.

Illustrate and label the gross components in a cross section of a typical spinal cord segment, showing its nerves and their connections with the sympathetic ganglia.

Make a sketch showing the segmental cord origin and the emergence of the spinal nerves from the intervertebral foramina, particularly in the lumbar region.

VERTEBRAL COLUMN AND JOINTS

F I V E - C H O I C E   C O M P L E T I O N   Q U E S T I O N S

DIRECTIONS: Each of the following questions or incomplete statements is followed by five suggested answers or completions. SELECT THE ONE BEST ANSWER in each case and then underline the appropriate letter at the lower right of each question.

1. THE UNIQUE FEATURE OF THE FIRST CERVICAL VERTEBRA IS THAT IT HAS NO:
   A. Superior articular facet
   B. Transverse foramen
   C. Body
   D. Transverse processes
   E. Posterior tubercles

   A B C D E

2. WHICH OF THE FOLLOWING IS INCORRECT FOR THORACIC VERTEBRAE?
   A. Larger transverse processes than in the cervical region
   B. Vertebral foramina are larger than the bodies
   C. Bodies decrease in size towards the cranium
   D. Spinous processes project caudally
   E. Have costal facets (foveae

   A B C D E

3.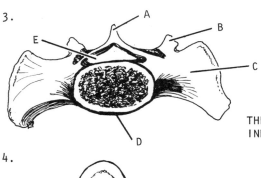

   THE ALA OF THE SACRUM IS INDICATED BY _____.

   A B C D E

4.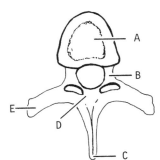

   THE LAMINA OF THE VERTEBRA IS INDICATED BY _____.

   A B C D E

SELECT THE ONE BEST ANSWER

5. WHICH OF THE FOLLOWING IS INCORRECT FOR ONE OF THE SECONDARY CURVATURES OF THE ADULT VERTEBRAL COLUMN?
   A. Opposite in direction to the fetal curvature
   B. Located in the thoracic region
   C. Located in the cervical region
   D. Develops as a child begins to lift his/her head
   E. Develops as the child begins to sit

   A B C D E

6. KYPHOSIS OF THE VERTEBRAL COLUMN CAN BE DEFINED AS A (AN):
   A. Lateral deviation
   B. Accentuated secondary curvature
   C. Deviation at the lumbosacral angle
   D. Accentuated primary curve
   E. None of the above

   A B C D E

-------------------- ANSWERS, NOTES AND EXPLANATIONS --------------------

1. **C.** The first cervical vertebra (atlas) is unique in that it has no body. It consists of anterior and posterior arches that meet two lateral masses bearing superior and inferior articular surfaces. The atlas articulates above with the occipital condyles of the skull, forming the atlanto-occipital joints; and below with the axis forming the atlantoaxial joints.

2. **B.** The vertebral foramina are small and circular in the thoracic vertebrae and increase in size from above downwards. Their spines are long and slender and project caudally. Costal facets (foveae) are present on the sides of the bodies where the heads of the ribs articulate, and on the transverse processes for articulation with the tubercles of the ribs. The eleventh and twelfth vertebrae have no facets on their transverse processes. The facets of the superior articular processes face backwards and laterally, whereas those on the inferior articular processes face forwards and medially.

3. **C.** The ala is the upper surface of the lateral parts of the sacrum (C). The base of the sacrum is the upper portion which articulates with the body of L5. Its anterior edge is marked by the promontory (D) and its posterior edge forms part of the sacral canal (E). The spinous processes of the five fused vertebrae making up the sacrum form the median crest (A). The superior articular process of the sacrum (B) forms part of the lumbosacral joint.

4. **D.** A typical vertebra consists of a posterior vertebral arch and an anterior body (A). These basic components enclose the spinal cord and its meninges within the vertebral foramen (not labelled). The vertebral arch consists of paired pedicles (B) laterally; and paired laminae (D) posteriorly. The vertebral arch gives rise to seven processes: one spinous (C), two transverse (E), and four articular (two superior and two inferior).

5. **B.** The secondary curvatures of the adult vertebral column are located in the cervical and lumbar regions; whereas the primary curvatures are in the thoracic and sacral regions. The secondary curvatures are convex anteriorly, opposite to those of the primary. The secondary curvatures compensate for and counteract the primary curvatures. The cervical curvature develops as the child begins to lift his/her head, whereas the lumbar curvature develops as the child begins to sit and stand erect.

6. **D.** Kyphosis (hunchback) is an abnormal exaggerated primary curvature of the vertebral column; whereas lordosis is an abnormal accentuation of a secondary

curvature. A lateral deviation, right or left, is termed scoliosis. The lumbosacral angle is the angle between the long axis of the lumbar part of the vertebral column and the sacrum.

MULTI-COMPLETION QUESTIONS

DIRECTIONS: In each of the following questions or incomplete statements ONE OR MORE of the completions given is correct. At the lower right of each question, underline A if 1, 2 and 3 are correct; B if 1 and 3 are correct; C if 2 and 4 are correct; D if only 4 is correct; and E if all are correct.

1. WHICH OF THE FOLLOWING STATEMENTS IS (ARE) CORRECT FOR THE ATLANTOAXIAL JOINTS?
   1. Their major movement is rotation
   2. Are synovial
   3. Supported by the accessory atlantoaxial ligament
   4. Are three in number   A B C D E

2. WHICH OF THE FOLLOWING STATEMENTS IS (ARE) CORRECT FOR THE INTERVERTEBRAL DISCS?
   1. Contain spirally arranged collagen bundles
   2. Function in absorbing mechanical stress
   3. Contain mucoid material
   4. Located between C1 and C2   A B C D E

3. CONCERNING THE LUMBOSACRAL JOINT:
   1. Separated by a disc thicker in front than behind
   2. Articular processes at this junction prevent anterior displacement
   3. Sharpest change in direction of the vertebral column
   4. Upper surface of the sacrum forms a 22-33° angle with the horizontal plane   A B C D E

4. WHICH OF THE FOLLOWING IS (ARE) TRUE FOR MOVEMENTS OF THE VERTEBRAL COLUMN?
   1. Atlanto-occipital joint is primarily involved in flexion and extension
   2. In the lumbar region a small amount of rotation is permitted
   3. Rotation in the thoracic region is limited by the sternum
   4. The primary movement at the atlantoaxial joint is rotation   A B C D E

5. THE JOINT TYPE(S) FOUND IN THE VERTEBRAL COLUMN OR AT ITS ARTICULATION WITH THE SKULL IS(ARE):
   1. Pivot   3. Symphysis
   2. Plane   4. Ellipsoid   A B C D E

6. WHICH OF THE FOLLOWING IS (ARE) CORRECT FOR THE LUMBAR VERTEBRAE?
   1. Bodies are wider transversely than anteroposteriorly
   2. Superior vertebral notches are deep
   3. Distinguished by their thin transverse processes
   4. Have short narrow spinous processes   A B C D E

7. THE ATLANTO-OCCIPITAL JOINT IS:
   1. Supported by membranes anteriorly and posteriorly
   2. Involved in rotation
   3. An ellipsoid type
   4. Partitioned by a disk   A B C D E

---------------------- ANSWERS, NOTES AND EXPLANATIONS ----------------------

1. **E.** <u>All are correct</u>. The atlantoaxial joints are synovial and three in number: two plane and one pivot. The lateral plane joints are between the superior and inferior articular facets of the atlas and axis, respectively; whereas the median joint is between the dens of the axis and the anterior arch and transverse ligament of the atlas. The lateral atlantoaxial joint capsules are supported posteriorly by the accessory atlantoaxial ligament. The dens is supported by the cruciform, apical, and alar ligaments, and the membrana tectoria. The major movement allowed at these joints is rotation of the skull and atlas on the axis.

2. **A.** <u>1, 2, and 3 are correct</u>. The intervertebral discs form fibrocartilaginous joints between the bodies of adjacent vertebrae, from below the level of C2. The disc consists of an outer layer of spirally arranged collagen bundles, the anulus fibrosus; and an inner mucoid mass, the nucleus pulposus. Both structures function in the shock-absorbing mechanism and the nucleus pulposus is also involved in equalizing stresses and exchanging of fluid between the disc and capillaries of the vertebrae. The anulus fibrosus binds the vertebral column together, retains the nucleus, and permits small amounts of movement.

3. **A.** <u>1, 2, and 3 are correct</u>. The lumbosacral joint marks the sharpest change in direction of the vertebral column and is the site of greatest weight bearing and leverage. The sacrum forms an angle of some 37-48° with the horizontal and is separated from the body of L5 by a thick intervertebral disc, which is thicker anteriorly than posteriorly. There is a natural tendency for the vertebral column to slip forward on the sacrum. This tendency is checked by the overlapping articular processes of the sacrum and those of L5 (inferior articular). The fifth lumbar vertebra is a common site of abnormal development and also a common site of pathological lesions (spondylolisthesis). Quite often these abnormalities and the stresses placed on the lumbosacral junction by the vertebral column are the causes of low back pain.

4. **E.** <u>All are correct</u>. The atlanto-occipital joint is primarily involved in flexion and extension; the atlantoaxial joint is involved in rotation of the head. In the remainder of the cervical region all movements are extremely free. The thoracic region is capable of all movements at its articulations, but these movements are limited by rib and sternal attachments. For this reason the lower thoracic region is more mobile than the upper region. The lumbar region is relatively free in that flexion and extension are permitted, but because of the locking of the articulating facets between vertebrae, little rotation is allowed.

5. **E.** <u>All are correct</u>. The vertebral column has both cartilaginous and synovial articulations. The fibrocartilaginous (symphysis, intervertebral) joints are located between the bodies of most vertebrae, the exception being between the atlas and axis (atlantoaxial joint). The remaining joints are synovial. The joint between the atlas and axis is a pivot joint; the joint between the atlas and the condyles of the occipital bone is an ellipsoid joint. The articulations between the superior articulating and inferior articulating facets of adjacent vertebrae, from C2 through S1, are plane joints. All these joints are strengthened by many ligaments.

6. **B.** <u>1 and 3 are correct</u>. The five lumbar vertebrae are large and heavy, with bodies that are wider transversely than in the anteroposterior diameter. The superior vertebral notches are shallow, whereas the inferior notches are deep. The spinal processes are short and broad (quadrilateral). Distinguishing features of the lumbar vertebrae are their thin transverse processes and their absence of costal facets and transverse foramina.

7. B. 1 and 3 are correct. The atlanto-occipital joint, ellipsoid in shape, is between the superior articulating facets of the atlas and occipital condyles of the occipital bone. This joint is supported posteriorly and anteriorly by the atlanto-occipital membranes extending from the arches of the atlas to the anterior and posterior margins of the foramen magnum. The movement allowed at this joint is flexion, extension and some lateral flexion of the head.

F I V E - C H O I C E   A S S O C I A T I O N   Q U E S T I O N S

DIRECTIONS: Each group of questions below consists of a numbered list of descriptive words or phrases accompanied by a diagram with certain parts indicated by letters, or by a list of lettered headings. For each numbered word or phrase, SELECT THE LETTERED PART OR HEADING that matches it correctly. Then insert the letter in the space to the right of the appropriate number. Sometimes more than one numbered word or phrase may be correctly matched to the same lettered part or heading.

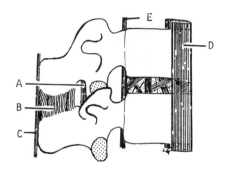

1. ___ Ligament within the vertebral canal
2. ___ Ligamentum flavum
3. ___ Structure pierced during a spinal puncture
4. ___ Ligament checking dorsiflexion of the vertebral column
5. ___ A ligament continuing superiorly with the ligamentum nuchae

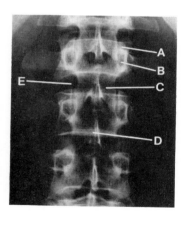

6. ___ Intervertebral disc
7. ___ Superior articulating process
8. ___ A pedicle
9. ___ Spinous process

------------------ ANSWERS, NOTES AND EXPLANATIONS --------------------

1. E. The posterior longitudinal ligament, located within the vertebral canal, supports the articulation between adjacent vertebral bodies. This ligament is broad in the cervical region but becomes narrow over the middle of each vertebral body in the thoracic and lumbar regions; it extends to the sacrum. It supports the posterior aspect of the vertebral column and limits its flexion.

2. A. The ligamentum flavum, the strongest posteriorly placed ligament of the vertebral column, may be pierced during spinal puncture. These paired ligaments run between two adjacent laminae. Because the laminae are separated farther from each other by flexion, this movement stretches the posterior ligaments and gives more room between laminae for the procedure of lumbar or spinal puncture.

3. B. The interspinous ligament passes between the lower border of one spinous process and the upper border of the one below. It blends with the supraspinous ligament (C).

4. D. The anterior longitudinal ligament functions in limiting dorsiflexion (extension) of the vertebral column. This ligament, a broad band, extends between the bodies of vertebrae on their anterior and anterolateral surfaces. This long ligament extends from the base of the skull to the upper part of the sacrum.

5. C. The supraspinous ligament runs over the tips of the spinous processes of the many vertebrae of the vertebral column, and its fibers blend with the interspinous ligament (B). In the cervical region the supraspinous ligament, between the spinous processes of the second and seventh vertebrae, is represented by the ligamentum nuchae. This septum separates the muscles of the back of the neck into halves; and in quadrupeds with heavy heads, this septum is strong and thick, and helps to support the head.

6. C. The intervertebral disc, located between the bodies of two lumbar vertebrae, consists of a central mucoid substance (nucleus pulposus) and a surrounding fibrocartilaginous lamina (anulus fibrosus).

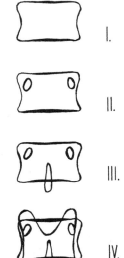

7. A. The superior articular process of one vertebra articulates with the inferior articular process (E) of the vertebra above by a synovial joint. These two processes of the same vertebra can be seen in diagram IV.

8. B. The pedicles of the vertebrae consist of cortical tissue and appear as cylinders on a radiograph as depicted in diagram II.

9. D. The spinous process is indicated in diagram III and the body of the vertebra above is indicated in diagram I. Diagram V depicts all the structures that one would see in a radiograph of an isolated lumbar vertebra.

Modified from Squire, 1964

MUSCLES, NERVES, AND BLOOD SUPPLY OF THE BACK

FIVE-CHOICE COMPLETION QUESTIONS

DIRECTION: Each of the following questions or incomplete statements is followed by five suggested answers or completions. SELECT THE ONE BEST ANSWER in each case and then underline the appropriate letter at the lower right of each question.

1.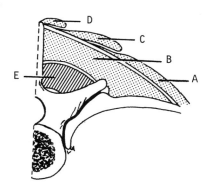

   THE TRANSVERSOSPINALIS GROUP OF MUSCLES IS INDICATED BY _____.

   A B C D E

2. WHICH OF THE FOLLOWING ARTERIES DOES NOT SUPPLY STRUCTURES OF THE BACK?
   A. Vertebral
   B. External iliac
   C. Lumbar
   D. Lateral sacral
   E. Internal iliac

   A B C D E

3. WHICH OF THE FOLLOWING STATEMENTS IS NOT TRUE FOR THE THORACOLUMBAR FASCIA?
   A. Forms the roof of the lumbar triangle
   B. Extends laterally from the vertebral spines
   C. Forms an aponeurotic sheet to which the transversus abdominis muscle attaches
   D. Lies deep to the latissimus dorsi muscle
   E. Attaches to the sacrum and the iliac crest

   A B C D E

4. WHICH OF THE FOLLOWING IS NOT AN EXTRINSIC MUSCLE OF THE BACK?
   A. Trapezius
   B. Longissimus thoracis
   C. Sternocleidomastoid
   D. Levator scapulae
   E. Rhomboid major

   A B C D E

5. WHICH OF THE FOLLOWING IS NOT A MEMBER OF THE TRANSVERSOSPINALIS MUSCLE GROUP OF THE BACK?
   A. Rotatores
   B. Semispinalis
   C. Splenius
   D. Multifidus
   E. All of the above

   A B C D E

---------------------- ANSWERS, NOTES AND EXPLANATIONS ----------------------

1. **E.** The intrinsic muscles of the back are located deep within the "gutters" of the vertebral column. The muscles of the back are divided into two groups: the obliquely running transversospinalis (E); and the longitudinally arranged erector spinalis (B). The serratus posterior (A), latissimus dorsi (C), and trapezius (D) muscles are part of the superficial extrinsic muscles of the back.

2. **B.** The structures of the back receive their blood supply from arteries located in the various regions. The external iliac artery has no branches supplying the back. In the neck region muscular branches are given off from the occipital artery. Spinal and muscular branches arise from the ascending cervical, vertebral, and deep cervical arteries. Posterior intercostal, subcostal, and lumbar arteries give off muscular and spinal branches in the thorax and abdomen. In the pelvis, branches of the internal iliac, such as the lateral sacral and iliolumbar, supply structures located in the back.

3. **D.** The thoracolumbar (lumbar) fascia invests the deep muscles of the back. It extends laterally from the vertebral spines. In the thoracic region, it attaches to the angles of the ribs: in the lumbar region, it forms a number of thick investing layers. This fascia splits laterally to enclose the latissimus dorsi muscle and then forms the roof of the lumbar triangle. The fascial layers invest the quadratus lumborum muscle and then reunite laterally to form an aponeurotic layer to which the fascia of the transversus abdominus and internal oblique muscles attach. Inferiorly, the lumbar fascia attaches to the sacrum and the iliac crests.

4. **B.** The posterior muscles of the back are divided into extrinsic and intrinsic layers. The extrinsic muscles are divided into a superficial group, including the trapezius, latissimus dorsi, and sternocleidomastoid, and a deeper group including the levator scapulae, rhomboids, and serrati posteriores. These extrinsic muscles are primarily involved with movements of the head, upper limbs, and ribs. The intrinsic (true) muscles of the back are deep to these extrinsic muscles and are involved in movements of the vertebral column. There are two groups of intrinsic muscles, the erector spinae and the transversospinalis. The longissimus thoracis muscle belongs to the erector spinae group.

5. **C.** The deep transversospinalis group of back muscles includes the: rotatores, semispinales, and multifidis. These short oblique muscles of the back are primarily rotators of the vertebral column. The splenius muscles are oblique muscles of the neck and are not considered deep muscles of the back.

MULTI-COMPLETION QUESTIONS

DIRECTIONS: In each of the following questions or incomplete statements ONE OR MORE of the completions given is correct. At the lower right of each question, underline A if 1, 2 and 3 are correct; B if 1 and 3 are correct; C if 2 and 4 are correct; D if only 4 is correct; and E if all are correct.

1. THE SPLENIUS MUSCLE:
    1. Consists of two parts
    2. Runs obliquely across the back of the neck
    3. Rotates the head
    4. Is one of the erector spinae group

   A B C D E

2. WHICH OF THE FOLLOWING STATEMENTS IS (ARE) CORRECT FOR THE SPINAL NERVES OF THE BACK?
    1. Dorsal rami communicate with each other
    2. Dorsal rami contain sympathetic fibers
    3. Medial branches supply periosteum
    4. Dorsal rami of C1 has cutaneous branches

   A B C D E

A	B	C	D	E
1,2,3	1,3	2,4	only 4	all correct

3. THE SUBOCCIPITAL NERVE:
   1. Emerges above the posterior arch of the axis
   2. Is derived from the dorsal ramus of C1
   3. Supplies the levator scapulae muscle
   4. Lies inferior to the vertebral artery within the suboccipital triangle

   A B C D E

4. THE ERECTOR SPINAE MUSCLES FUNCTION AS _____ OF THE VERTEBRAL COLUMN.
   1. Rotators
   2. Lateral flexors
   3. Flexors
   4. Extensors

   A B C D E

5. WHICH OF THE FOLLOWING ARE CORRECT WITH RESPECT TO THE SURFACE ANATOMY OF THE BACK?
   1. A horizontal plane between the iliac crests passes through the spinous process of L4
   2. The spinous process of C2 is the common place to begin counting vertebrae
   3. The spinous process of C7 is more prominent when the neck is flexed
   4. The sacral hiatus and coccyx are usually not palpable

   A B C D E

6. WHICH OF THE FOLLOWING STATEMENTS IS (ARE) INCORRECT FOR THE ERECTOR SPINAE GROUP OF MUSCLES?
   1. Form a longitudinal group
   2. Lie in the "gutters" of the vertebral column
   3. Extend from the sacrum to the skull
   4. Supplied by ventral rami of spinal nerves

   A B C D E

7. THE ROOF OF THE SUBOCCIPITAL TRIANGLE IS FORMED BY THE _____ MUSCLE(S).
   1. Rectus capitis posterior minor
   2. Semispinalis capitus
   3. Rectus capitis posterior major
   4. Longissimus capitis

   A B C D E

-------------------- ANSWERS, NOTES AND EXPLANATIONS --------------------

1. **E.** <u>All are correct</u>. The splenius muscle, considered a member of the erector spinae group, has two parts. The splenius capitis arises from the ligamentum nuchae and the spines of C7 and the upper thoracic vertebrae. Its fibers run upwards and laterally to insert into the mastoid process of the temporal bone and the superior nuchal line of the occipital bone. The splenius cervicis, the second part, lies lateral to the splenius capitis and arises from the spines of the upper thoracic vertebrae and inserts into the transverse processes of the upper cervical vertebrae. This muscle functions as a rotator and extensor of the head.

2. **A.** <u>1, 2, and 3 are correct</u>. Dorsal rami of spinal nerves contain motor, sensory, and sympathetic fibers that supply the muscles, vessels, bones, joints, and skin of the back. These rami communicate with each other above and below. Most dorsal rami divide into medial and lateral branches which supply muscles of the back, skin, periosteum, ligaments, and the joints of the articular facets. Dorsal rami contain cutaneous branches except C1.

3. **C.** <u>2 and 4 are correct</u>. The suboccipital nerve, derived from the dorsal ramus of C1, is located within the suboccipital triangle. This nerve emerges

between the posterior arch of the atlas, below, and the vertebral artery, above. It supplies the muscles of the suboccipital triangle and the semispinalis capitis. The levator scapulae muscle is innervated by branches of C3 and 4.

4. **C. 2 and 4 are correct.** The erector spinae muscles acting bilaterally are primarily extensors of the vertebral column. They are antagonistic to the muscles lying in front of the vertebral column, i.e., prevertebral, sternocleidomastoid, rectus abdominus, psoas major, and others. By contracting unilaterally, the erector spinae muscles are involved in lateral (bending) flexing of the vertebral column. These muscles are primarily involved in maintaining erect posture, but not in rotation or flexion of the vertebral column.

5. **B. 1 and 3 are correct.** The spinous processes of C6, 7, and T1 are prominent in an erect individual and become even more prominent when the neck and trunk are flexed. These spines are often used in counting and palpating vertebrae. Thus, the spine of C2 is not a common point to begin counting of vertebrae as it is not very prominent. In the lumbar region a horizontal plane between the iliac crests passes through or just below the spine of L4. This reference point is used in performing a lumbar puncture where a needle may be safely inserted into the subarachnoid space to draw cerebral spinal fluid. The spinal cord of the adult ends at the intervertebral disc between L1 and 2; whereas in the newborn the spinal cord terminates at the level of L3. In the sacral region the sacral hiatus and the coccyx are usually palpable.

6. **D. Only 4 is correct.** The deep muscles of the back lie within the "gutters" of the vertebral column. They form a superficial longitudinal (erector spinae) group and a deeper short oblique (transversospinalis) group. The erector spinae group extends from the sacrum to the skull, but is better developed in the thoracic region. Here the erector spinae can be divided into a medial spinalis group, an intermediate longissimus group, and a lateral iliocostalis group of muscles. Both the erector spinae and transversospinalis groups of muscles are innervated by branches of the dorsal rami of spinal nerves.

7. **C. 2 and 4 are correct.** The roof of the suboccipital triangle is formed by the semispinalis capitis and longissimus capitis muscles. The boundaries of the triangle are formed by the rectus capitis posterior major muscle, medially; the oblique capitis inferior, inferiorly; and the oblique capitis superior, laterally. The floor is formed by the posterior arch of the atlas and the posterior atlanto-occipital membrane, connecting the posterior arch of the atlas to the posterior margin of the foramen magnum. Fluid in the subarachnoid space of the cisterna magna (cerebellomedullary cisterna) can be tapped (cisterna puncture) by inserting a needle through this membrane.

FIVE-CHOICE ASSOCIATION QUESTIONS

DIRECTIONS: Each group of questions below consists of a numbered list of descriptive words or phrases accompanied by a diagram with certain parts indicated by letters, or by a list of lettered headings. For each numbered word or phrase, SELECT THE LETTERED PART OR HEADING that matches it correctly. Then insert the letter in the space to the right of the appropriate number. Sometimes more than one numbered word or phrase may be correctly matched to the same lettered part or heading.

ASSOCIATION QUESTIONS

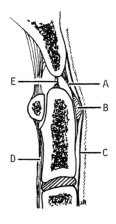

1. ___ Membrana tectoria
2. ___ Anterior longitudinal ligament
3. ___ Apical ligament
4. ___ Transverse component of the cruciform ligament

5. ___ Rectus capitis posterior major muscle
6. ___ Posterior atlanto-occipital membrane
7. ___ Oblique capitis inferior muscle
8. ___ Semispinalis capitis muscle

------------------ ANSWERS, NOTES AND EXPLANATIONS ---------------------

1. **C.** The broad membrana tectoria connects the axis to the occipital bone. It extends from the posterior longitudinal ligament and the body of the axis upwards to the basilar part of the occipital bone, anterior to the foramen magnum. This membrane lies external to the dura mater of the spinal cord (not labelled).

2. **D.** The anterior longitudinal ligament is a broad band extending from the atlas to the pelvic surface of the sacrum. It lies on the anterolateral surface of the vertebral column and limits dorsiflexion (extension).

3. **E.** The apical ligament (of the dens) connects the upper tip of the odontoid process (dens) of the axis to the occipital bone.

4. **B.** The cruciform (cruciate) ligament, so named because it is shaped like a cross, consists of two parts: the transverse ligament (B), and a vertical portion formed by the superior longitudinal band (A) and the inferior longitudinal band (not labelled).

5. C. The rectus capitis posterior major muscle is attached to the spine of the axis and the lateral part of the inferior nuchal line. Thus muscle forms the medial boundary of the suboccipital triangle. The rectus capitis posterior minor muscle (not labelled) lies medial and deep to this muscle.

6. E. The posterior atlanto-occipital membrane lies between the posterior arch of the atlas and posterior margin of the foramen magnum. This membrane forms the floor of the suboccipital triangle and is pierced by the vertebral artery as it ascends to the foramen magnum.

7. D. The oblique capitis inferior muscle runs between the spine of the axis and the posterior surface of the transverse process of the atlas. It forms the inferior boundary of the suboccipital triangle, containing the suboccipital nerve (C1) and the vertebral artery. The lateral boundary of this triangle is formed by the oblique capitis superior muscle (B). The greater occipital nerve (C3) hooks around the inferior border of this muscle to supply the skin of the occipital region.

8. A. The semispinalis capitis muscle (A) and longissimus capitis muscle (not shown) form the roof of the suboccipital triangle. Both these muscles belong to the erector spinae group of back muscles.

SPINAL CORD AND MENINGES

FIVE-CHOICE COMPLETION QUESTIONS

DIRECTIONS: Each of the following questions or incomplete statements is followed by five suggested answers or completions. SELECT THE ONE BEST ANSWER in each case and then underline the appropriate letter at the lower right of each question.

1. THE FILUM TERMINALE IS A CONTINUATION OF THE:
   A. Denticulate ligament   D. Arachnoid
   B. Coccygeal ligament   E. Pia mater
   C. Dura mater

   A B C D E

2. THERE ARE _____ PAIRS OF SPINAL NERVES EMERGING FROM THE SPINAL CORD.
   A. 27   D. 33
   B. 29   E. 35
   C. 31

   A B C D E

3.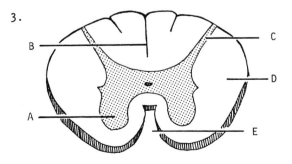

   THE LATERAL COLUMN OF THE SPINAL CORD IS INDICATED BY _____.
   THE LATERAL COLUMN OF THE SPINAL CORD IS INDICATED BY _____.

   A B C D E

SELECT THE ONE BEST ANSWER

4. WHICH OF THE FOLLOWING IS INCORRECT FOR THE LUMBAR SPINAL NERVES?
   A. L5 is larger than L1
   B. L5 exits through a larger foramen than L1
   C. L1 conveys sympathetic fibers
   D. L5 is longer than L1
   E. Contain motor fibers to the lower limb

   A B C D E

5. THE VENTRAL RAMUS OF THE SPINAL NERVE IS INDICATED BY _____.

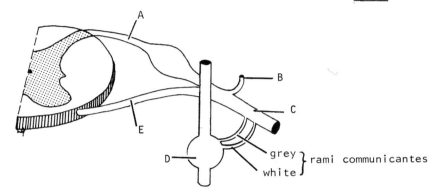

   A B C D E

6. THE INTERNAL VERTEBRAL VENOUS PLEXUSES ARE FOUND IN THE _____.
   A. Epidural space   D. Dural sac
   B. Subdural space   E. None of the above
   C. Subarachnoid space

   A B C D E

------------------ ANSWERS, NOTES AND EXPLANATIONS --------------------

1. E. The filum terminale is a continuation of the pia mater from the tip (conus medullaris) of the spinal cord. It therefore begins where the spinal cord ends, at the level of the disc between L1 and L2. It is 15-20 cm long, lies in the midst of the cauda equina, and ends by attaching to the dorsum of the coccyx.

2. C. There are 31 pairs of spinal nerves emerging from the spinal cord: 8 cervical, 12 thoracic, 5 lumbar, 5 sacral, and 1 coccygeal. The first pair of cervical nerves emerge between the atlas and the skull; cervical nerves two to seven emerge from the vertebral canal above the correspondingly numbered vertebrae; and the eight cervical nerves emerge below the seventh cervical vertebra. The remaining spinal nerves all exit below their correspondingly numbered vertebrae.

3. D. The lateral column (funiculus) is one of three pairs of columns of white matter seen in a transverse section of the cord. The dorsal columns of white matter lie on each side of the dorsal median sulcus (B) whereas the ventral columns lie on each side of the ventral median fissure (E). The grey matter of the cord, the central H-shaped structure, consists of dorsal (C) and ventral (A) horns and an intermediate zone (unlabelled).

4. B. The five lumbar spinal nerves increase in size and length from above downwards; however the intervertebral foramina decrease in diameter. Therefore

the fifth lumbar nerve passes through a smaller foramen than the first; thus any misalignment or protrusion of the nucleus pulposus of the intervertebral disc will compress the fifth nerve. The lumbar nerves are all involved with motor and sensory innervation of the lower limbs, but the upper lumbar nerves are also involved with the sympathetic outflow.

5. **C.** The ventral ramus of a spinal nerve is indicated by C and its dorsal ramus by B. The ventral ramus communicates with the sympathetic ganglion (D) via the ramus communicans. The dorsal root (A) and its ganglion contain sensory fibers, whereas the ventral root (E) contains motor fibers.

6. **A.** The internal vertebral venous plexuses, fat and some connective tissue occupy the space (epidural) between the walls of the vertebral canal and the outer covering (meninx) of the cord, the dura mater. These venous plexuses, one anterior and the other posterior, drain the vertebral column, spinal cord and surrounding soft tissue.

MULTI-COMPLETION QUESTIONS

DIRECTIONS: In each of the following questions or incomplete statements ONE OR MORE of the completions given is correct. At the lower right of each question, underline A if 1, 2 and 3 are correct; B if 1 and 3 are correct; C if 2 and 4 are correct; D if only 4 is correct; and E if all are correct.

1. WHICH STATEMENT(S) IS (ARE) CORRECT FOR THE ADULT SPINAL CORD?
   1. Has a cervical enlargement
   2. Is flattened and cylindrical
   3. Averages 45 cm in length
   4. Most often ends at the disc between L3 and 4   A B C D E

2. WHICH OF THE FOLLOWING STATEMENTS IS (ARE) CORRECT FOR THE SPINAL NERVES?
   1. Emerge below their site of origin in the thoracic region
   2. The cervical roots are the thickest
   3. Length of roots increase from above downwards
   4. Cervical spinal ganglia lie within the intervertebral foramina   A B C D E

3. THE CAUDA EQUINA IS:
   1. A collection of spinal nerve roots
   2. Contained within the subarachnoid space
   3. Located below the conus medullaris
   4. A rigid structure   A B C D E

4. WHICH OF THE FOLLOWING STATEMENTS IS (ARE) CORRECT FOR THE BLOOD VESSELS SUPPLYING THE SPINAL CORD?
   1. Located in the subdural space
   2. Consists of three longitudinal arteries
   3. Do not form adequate anastomoses
   4. Derived from segmental arteries   A B C D E

5. THE DENTICULATE LIGAMENT:
   1. Is composed of collagen
   2. Is fused laterally to the arachnoid and dura mater
   3. Usually has 21 paired lateral processes
   4. Supports the spinal cord   A B C D E

A	B	C	D	E
1,2,3	1,3	2,4	only 4	all correct

6. THE SPINAL DURA MATER:
   1. Attaches to the walls of the vertebral canal
   2. Blends with the perineurium of spinal nerves
   3. Adheres to the arachnoid
   4. Ends blindly

   A B C D E

------------------------ ANSWERS, NOTES AND EXPLANATIONS ------------------------

1. **A. 1, 2, and 3 are correct.** The spinal cord, flattened and cylindrical, begins at the foramen magnum and usually ends at the disc between L1 and 2. It averages about 45 cm in length and has prominent enlargements in the cervical and lumbar regions where the nerves to the upper and lower limbs exit. The conus medullaris, the lower end of the cord, is conical and from it an extension of pia mater, called the filum terminalis, blends with the dura at the apex of the dural sac and attaches to the coccyx.

2. **B. 1 and 3 are correct.** The level of origin of the spinal nerve roots is higher than their level of emergence from the intervertebral foramina in the thoracic, lumbar, and sacral regions. The cervical nerve roots exit at the same level or may even course upwards. The roots of the spinal nerves increase in thickness and length from above downwards. The spinal ganglia and nerves usually lie within the intervertebral foramina, but in the cervical region they lie outside.

3. **A. 1, 2, and 3 are correct.** The cauda equina ("horse's tail") is formed by the lumbosacral nerve roots. It is located within the subarachnoid space, below the level of the conus medullaris (L2). The cauda equina is not a rigid structure but is free to float in the cerebrospinal fluid.

4. **C. 2 and 4 are correct.** The spinal cord receives its blood supply from arteries located in the subarachnoid space. It receives blood from three longitudinal arterial branches: the anterior spinal artery, a continuation of a union between the spinal branches of the vertebral arteries; and two posterior spinal arteries, branches of the vertebral or the posterior inferior cerebellar arteries. These longitudinal arteries run from the medulla oblongata to the conus medullaris and receive branches from the segmental (medullary) arteries. The arterial branches supplying the spinal cord form extensive anastomoses.

5. **E. All are correct.** The denticulate ligament, a collagenous membrane, is a lateral extension of the pia mater. For the most part, this lateral edge of the ligament is free, but it is fused laterally by about 21 processes to the arachnoid and dura mater. The denticulate ligaments help to anchor the spinal cord within the vertebral canal.

6. **D. Only 4 is correct.** The spinal dura mater is the outermost protective layer of the spinal cord; the middle layer is the arachnoid, and the inner is the pia mater. The dura mater is separated from the walls of the vertebral canal by an epidural (extradural) space containing fat and many thin-walled veins. Laterally, the dura mater forms an outer sheath for the nerve roots and blends with the epineurium of spinal nerves. It is separated from the arachnoid by a potential subdural space. The spinal dura mater ends as a blind sac at the level of the second segment of the sacrum.

FIVE-CHOICE ASSOCIATION QUESTIONS

DIRECTIONS: Each group of questions below consists of a numbered list of descriptive words or phrases accompanied by a diagram with certain parts indicated by letters, or by a list of lettered headings. For each numbered word or phrase, SELECT THE LETTERED PART OR HEADING that matches it correctly. Then insert the letter in the space to the right of the appropriate number. Sometimes more than one numbered word or phrase may be correctly matched to the same lettered part or heading.

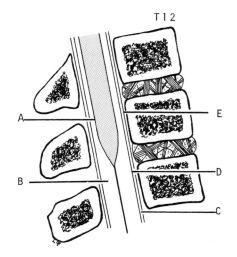

1. ___ Pia mater

2. ___ Middle meningeal layer

3. ___ Space containing cerebrospinal fluid

4. ___ Dura mater

5. ___ Potential space between the arachnoid and dura mater

6. ___ A space ending at the level of the second segment of the sacrum

---------------------- ANSWERS, NOTES AND EXPLANATIONS ----------------------

1. A. The pia mater covering the spinal cord and its nerves continues below the conus medullaris as the filum terminalis.

2. D. The arachnoid is the middle layer of the three layers of meninges covering the spinal cord.

3. B. The subarachnoid space, between the pia mater covering the spinal cord and the arachnoid layer, contains cerebrospinal fluid.

4. E. The dura mater, the outer layer of the meninges of the spinal cord, is separated from the vertebral canal by a space, termed the epidural space. It contains venous plexuses and connective tissue.

5. C. The potential space between the arachnoid and dura mater is called the subdural space.

6. B. The subarachnoid space ends blindly at the level of the second segment of the sacrum. This space is entered (lumbar puncture) between the arches of the third and fourth lumbar vertebrae to introduce air for diagnostic X-ray studies (pneumoencephalography) or to obtain a sample of cerebrospinal fluid. Note that the site of insertion of the needle is well below the termination of the spinal cord (L2).

PART TWO

REVIEW EXAMINATIONS
ON THE SYSTEMS

8. REVIEW EXAMINATION ON THE ARTICULAR, MUSCULAR, AND SKELETAL SYSTEMS

INTRODUCTORY NOTE. The following 50 multiple-choice questions are based on the instructional objectives pertaining to the articular, muscular and skeletal systems in Chapters 1-7 of this Study Guide & Review Manual. You should be able to answer these questions in 35-40 minutes. Before beginning, tear out an answer sheet from the back of the book and read the directions on how to use it. The key to the correct responses is on page 351.

FIVE-CHOICE COMPLETION QUESTIONS

DIRECTIONS: Each of the following questions or incomplete statements is followed by five suggested answers or completions. SELECT THE ONE BEST ANSWER in each case and then blacken the appropriate space on the answer sheet.

1. WHICH MUSCLE IS NOT ATTACHED TO THE PERINEAL BODY?
    A. Superficial transverse perineal
    B. Deep transverse perineal
    C. Ischiocavernosus
    D. Bulbospongiosus
    E. Levator ani

2. WHICH STATEMENT IS INCORRECT FOR ONE OF THE SECONDARY CURVATURES OF THE ADULT VERTEBRAL COLUMN?
    A. Develops as a child begins to lift his/her head
    B. Opposite in direction to the fetal curvature
    C. Located in the thoracic region
    D. Located in the cervical region
    E. Develops as the child begins to sit

3. WHICH OF THE FOLLOWING MUSCLES IS AN ADDUCTOR OF THE SCAPULA?
    A. Infraspinatus
    B. Pectoralis major
    C. Pectoralis minor
    D. Rhomboid major
    E. Serratus anterior

4. THE _____ MUSCLE IS A PLANTAR FLEXOR OF THE FOOT.
    A. Extensor digitorum longus
    B. Tibialis anterior
    C. Peroneus tertius
    D. Peroneus longus
    E. Popliteus

5. WHICH STATEMENT ABOUT THE TRAPEZIUS MUSCLE IS INCORRECT?
    A. Weakness of this muscle results in drooping of the shoulder
    B. Receives blood supply from the transverse cervical artery
    C. Forms a boundary of the posterior triangle of the neck
    D. Inserts into the scapula and the length of the clavicle
    E. Innervated by motor fibers of the accessory nerve

SELECT THE ONE BEST ANSWER

6. THE HEAD OF THE RADIUS ARTICULATES WITH THE:
   A. Capitulum of the humerus
   B. Olecranon fossa of the humerus
   C. Coronoid fossa of the humerus
   D. Head of the ulna
   E. Tuberosity of the ulna

7. THE LATERAL MENISCUS OF THE KNEE JOINT IS:
   A. Semicircular in shape
   B. Grooved by the popliteus tendon
   C. Completely attached to the tibia
   D. Fused with the fibular collateral ligament
   E. More frequently torn than the medial meniscus

8. THE STERNAL ANGLE EXISTS AT THE ARTICULATION BETWEEN THE:
   A. Body of the sternum and and xiphoid process
   B. Manubrium and body of the sternum
   C. Manubrium and xiphoid process
   D. Sternum and clavicle
   E. None of the above

9. WHICH IS NOT AN EXTRINSIC MUSCLE OF THE BACK?
   A. Longissimus thoracis
   B. Sternocleidomastoid
   C. Levator scapulae
   D. Rhomboid major
   E. Trapezius

10. WHICH OF THE FOLLOWING STATEMENTS IS CORRECT FOR THE FLEXOR RETINACULUM? IT PASSES BETWEEN THE:
    A. Lateral malleolus and the lateral surface of the calcaneus
    B. Medial malleolus and the medial surface of the calcaneus
    C. Lateral malleolus and the medial surface of the calcaneus
    D. Medial malleolus and the lateral surface of the calcaneus
    E. Tibia and fibula above the ankle joint

11. THE SELLA TURCICA LIES DIRECTLY ABOVE THE:
    A. Frontal sinus
    B. Foramen ovale
    C. Pons
    D. Sphenoid sinus
    E. Maxillary sinus

12. THE REGION OF THE ABDOMINAL CAVITY DELINEATED BY THE XIPHOID PROCESS AND THE ADJACENT COSTAL MARGINS, THE TWO VERTICAL PLANES AND THE SUBCOSTAL PLANE IS KNOWN AS _____ REGION.
    A. Hypochrondriac
    B. Hypogastric
    C. Suprapubic
    D. Epigastric
    E. Iliac

13. THE CRISTA GALLI SERVES AS AN ATTACHMENT FOR THE:
    A. Diaphragma sellae
    B. Tentorium cerebelli
    C. Falx cerebelli
    D. Falx cerebri
    E. None of the above

14. WHICH OF THE FOLLOWING IS INCORRECT FOR THORACIC VERTEBRAE?
    A. Have costal facets
    B. Spinous processes project caudally
    C. Bodies decrease in size towards the cranium
    D. Vertebral foramina are larger than the bodies
    E. Larger transverse processes than in the cervical region

SELECT THE ONE BEST ANSWER

15. HOW MANY PAIRS OF RIBS ARE USUALLY ATTACHED DIRECTLY TO THE STERNUM BY THEIR COSTAL CARTILAGES?
    A. Five
    B. Seven
    C. Nine
    D. Ten
    E. Twelve

16. THE MAJOR MOTOR SIGN IN PARALYSIS OF THE QUADRICEPS FEMORIS MUSCLE IS LOSS OF _____.
    A. Lateral rotation of the leg
    B. Extension of the thigh
    C. Extension of the leg
    D. Adduction of the thigh
    E. Flexion of the leg

17. THE STRUCTURE INDICATED BY THE ARROW IS THE:
    A. Trochlear notch
    B. Radial notch
    C. Radial tuberosity
    D. Supinator fossa
    E. Coronoid process

18. WHICH OF THE FOLLOWING STATEMENTS REGARDING THE EXTERNAL ANAL SPHINCTER IS FALSE?
    A. Is under voluntary control
    B. Consists of striated muscle fibers
    C. Innervated by inferior rectal nerves
    D. Located below the pelvic diaphragm
    E. Subcutaneous part inserts in the central tendon of the perineum

19. THE PROXIMAL RADIOULNAR JOINT PERMITS WHICH OF THE FOLLOWING MOVEMENTS?
    A. Abduction
    B. Adduction
    C. Extension
    D. Flexion
    E. Rotation

MULTI-COMPLETION QUESTIONS

DIRECTIONS: In each of the following questions or incomplete statements, ONE OR MORE of the completions given is correct. On the answer sheet blacken the space under A if 1, 2, and 3 are correct; B if 1 and 3 are correct; C if 2 and 4 are correct; D if only 4 is correct; and E if all are correct.

20. THE PSOAS MAJOR MUSCLE:
    1. Is a powerful flexor of the thigh
    2. Inserts into the greater trochanter of the femur
    3. Lies posterior to the kidney
    4. Is innervated by the femoral nerve

21. WHICH STATEMENT(S) ABOUT THE THYROID CARTILAGE IS(ARE) CORRECT?
    1. Derived from the cartilage of the fourth branchial arch
    2. Has a synovial articulation with the cricoid cartilage
    3. Produces the laryngeal prominence
    4. Comprises two laminae

A	B	C	D	E
1,2,3	1,3	2,4	only 4	all correct

22. THE ATLANTO-OCCIPITAL JOINT IS:
    1. Supported by membranes anteriorly and posteriorly
    2. Partitioned by a disc
    3. An ellipsoid type
    4. Involved in rotation

23. THE ANTERIOR COMPARTMENT OF THE LEG CONTAINS THE FOLLOWING MUSCLE(S):
    1. Extensor digitorum longus
    2. Sartorius
    3. Peroneus tertius
    4. Peroneus longus

24. WHICH BONE(S) CONTRIBUTE(S) TO THE FORMATION OF THE ACETABULUM?
    1. Ilium
    2. Pubis
    3. Ischium
    4. Sacrum

25. THE ZYMOGATIC ARCH IS FORMED BY THE:
    1. Zygomatic process of the maxilla
    2. Zygomatic process of the temporal bone
    3. Zygomatic process of the frontal bone
    4. Temporal process of the zygomatic bone

26. WHICH OF THE FOLLOWING IS(ARE) TRUE FOR MOVEMENTS OF THE VERTEBRAL COLUMN?
    1. Atlanto-occipital joint is involved in flexion and extension
    2. A small amount of rotation is permitted in the lumbar region
    3. Rotation in the thoracic region is limited by the sternum
    4. Rotation is the primary movement at the atlantoaxial joint

27. THE ACETABULAR LABRUM:
    1. Adds stability to the hip joint
    2. Consists of fibrocartilage
    3. Deepens the socket
    4. Is an incomplete ring

28. THE CORACOHUMERAL LIGAMENT:
    1. Is united to the fibrous capsule
    2. Is attached to the greater tubercle of the humerus
    3. Strengthens the upper part of the capsule of the glenohumeral joint
    4. Holds the long head of biceps in the bicipital groove

29. THE RECTUS ABDOMINIS MUSCLE:
    1. Has a segmental innervation
    2. Originates from the pubic crest
    3. Inserts into the 5th, 6th, and 7th costal cartilages
    4. Is covered anteriorly in its entire length by the aponeurosis of the three abdominal muscles

30. THE TRAPEZIUS MUSCLE ORIGINATES FROM THE:
    1. Spines of thoracic vertebrae
    2. Ligamentum nuchae
    3. Superior nuchal line
    4. Spine of the scapula

31. THE ANTERIOR WALL OF THE AXILLA IS FORMED BY THE _____ MUSCLES.
    1. Pectoralis minor
    2. Deltoid
    3. Subclavius
    4. Coracobrachialis

A	B	C	D	E
1,2,3	1,3	2,4	only 4	all correct

32. WHICH STATEMENT(S) IS(ARE) CORRECT FOR THE ANKLE JOINT?
    1. Supported laterally by the deltoid ligament
    2. Has three supporting ligaments medially
    3. Calcaneus articulates with the fibula
    4. Is a hinge joint

33. THE CORACOCLAVICULAR JOINT:
    1. Is a synovial joint
    2. Is a fibrous joint
    3. Is biaxial
    4. Prevents overriding of the clavicle on the acromion

34. WHICH STATEMENT(S) CONCERNING THE THYROHYOID MUSCLE IS(ARE) CORRECT?
    1. Attached to the hyoid bone lateral to the omohyoid muscle
    2. Is innervated by the ansa cervicalis
    3. Lies lateral to the thyroid notch
    4. Is one of the infrahyoid muscles

35. THE DENTICULATE LIGAMENT:
    1. Is composed of collagen
    2. Supports the spinal cord
    3. Usually has 21 paired lateral processes
    4. Is fused laterally to the arachnoid and dura mater

36. A TYPICAL THORACIC VERTEBRA HAS:
    1. Articular facets on the transverse processes
    2. Articular facets on the sides of the body
    3. A long, slender, downward-sloping spine
    4. A heart-shaped body

37. THE FIBROUS CAPSULE OF THE HIP JOINT:
    1. Is thickened anteriorly by the iliofemoral ligament
    2. Is thicker posterior to the femoral head
    3. Is weak inferiorly
    4. Attaches to the base of the femoral head

38. WHICH STATEMENT(S) IS(ARE) CORRECT FOR THE ATLANTOAXIAL JOINTS?
    1. Supported by the accessory atlantoaxial ligament
    2. Their major movement is rotation
    3. They are three in number
    4. Are synovial joints

39. THE PSOAS MAJOR MUSCLE:
    1. Takes origin from the twelfth rib
    2. Is one of the muscles of the posterior abdominal wall
    3. Enters the thigh within the inguinal ligament
    4. Contains the lumbar plexus within its substance

40. THE BULBOSPONGIOSUS MUSCLE:
    1. Is located in the superficial perineal pouch
    2. Is innervated by the perineal nerve
    3. Originates from the perineal body
    4. Is pierced by the vagina

A	B	C	D	E
1,2,3	1,3	2,4	only 4	all correct

41. THE THENAR GROUP OF MUSCLES IS COMPOSED OF THE:
   1. Opponens pollicis
   2. Flexor pollicis brevis
   3. Abductor pollicis brevis
   4. Adductor pollicis

FIVE-CHOICE ASSOCIATION QUESTIONS

DIRECTIONS: Each group of questions below consists of a numbered list of descriptive words or phrases accompanied by a diagram with certain parts indicated by letters, or by a list of lettered headings. For each numbered word or phrase, SELECT THE LETTERED PART OR HEADING that matches it correctly. Then on the appropriate line of the answer sheet, blacken the space under the letter of that part or heading. Sometimes more than one numbered word or phrase may be correctly matched to the same lettered part or heading.

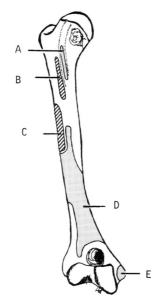

42. ____ The insertion of an abductor of the arm

43. ____ The insertion of an extensor and medial rotator of the arm

44. ____ The insertion of an adductor of the arm

45. ____ The origin of a strong flexor of the forearm

46. ____ A common site of origin for the flexor muscles of the forearm

ASSOCIATION QUESTIONS

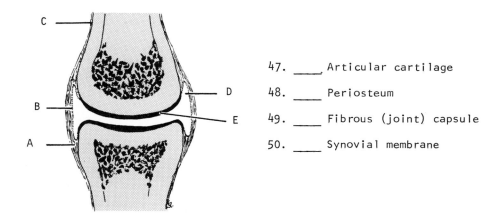

47. _____ Articular cartilage

48. _____ Periosteum

49. _____ Fibrous (joint) capsule

50. _____ Synovial membrane

Interpretation of Your Score

Number of Correct Responses	Level of Performance
43-50	Excellent - Exceptional
38-42	Superior - Very Superior
32-37	Average - Above Average
28-31	Poor - Marginal
27 or less	Very Poor - Failure

9. REVIEW EXAMINATION ON THE CARDIOVASCULAR SYSTEM

INTRODUCTORY NOTE. The following 50 multiple-choice questions are based on the objectives pertaining to the cardiovascular system in Chapters One to Seven of this Study Guide & Review Manual. You should be able to answer these questions in 35-40 minutes. Before beginning, tear out an answer sheet from the back of the book and read the directions on how to use it. The key to the correct responses is on page 351.

FIVE-CHOICE COMPLETION QUESTIONS

DIRECTIONS: Each of the following questions or incomplete statements is followed by five suggested answers or completions. SELECT THE ONE BEST ANSWER in each case and then blacken the appropriate space on the answer sheet.

1. WHICH STATEMENT CONCERNING THE PULMONARY CIRCULATORY SYSTEM IS INCORRECT?
   A. Two pairs of pulmonary veins are involved
   B. One pair of pulmonary arteries are present
   C. Pulmonary arteries contain oxygenated blood
   D. Pulmonary arteries carry deoxygenated blood
   E. Pulmonary veins transport oxygenated blood to the heart

2. THE _____ ARTERY IS THE FIRST LARGE BRANCH OF THE AORTA.
   A. Left common carotid   D. Right common carotid
   B. Right subclavian   E. Left subclavian
   C. Brachiocephalic

3. THE THORACIC DUCT RECEIVES LYMPH FROM THE:
   A. Left side of the thorax   D. Pelvis and perineum
   B. Lower limbs   E. All of the above
   C. Abdomen

4. THE RIGHT ATRIUM RECEIVES ALL THE FOLLOWING VESSELS EXCEPT THE:
   A. Small coronary veins   D. Superior vena cava
   B. Inferior vena cava   E. Coronary sinus
   C. Pulmonary veins

5. THE CORONARY SINUS IS DERIVED FROM THE:
   A. Right anterior cardinal and right common cardinal veins
   B. Right horn of the sinus venosus
   C. Left horn of the sinus venosus
   D. Sinus venosus
   E. None of the above

6. WHICH OF THE FOLLOWING VEINS DRAINS THE FIRST INTERCOSTAL SPACE?
   A. Azygos   D. Superior intercostal
   B. Hemiazygos   E. None of the above
   C. Brachiocephalic

SELECT THE ONE BEST ANSWER

7. ON THE ANTERIOR CHEST WALL, THE MITRAL VALVE IS MOST AUDIBLE OVER THE:
   A. Right junction of the sternal body and the costal margin
   B. Left junction of the sternal body and the costal margin
   C. Right second intercostal space
   D. Left second intercostal space
   E. Left fifth intercostal space

8. WHICH STATEMENT ABOUT THE CONDUCTING SYSTEM OF THE HEART IS FALSE?
   A. Fibers from the right crus of the atrioventricular bundle pass through the moderator band
   B. The atrioventricular bundle is a collection of specialized nerve fibers
   C. The sinuatrial node is located in the wall of the right atrium
   D. The atrioventricular node lies in the interatrial septum
   E. Cardiac impulses are initiated in the sinuatrial node

9. WHICH OF THE FOLLOWING ARTERIES DOES NOT SUPPLY THE ESOPHAGUS?
   A. Direct branches from the aorta
   B. Bronchial
   C. Left gastric
   D. Inferior thyroid
   E. Right gastric

10. WHICH STATEMENT REGARDING THE RIGHT VENTRICLE IS FALSE?
    A. Communicates with the right atrium via the mitral valve
    B. Has papillary muscles attached to its internal wall
    C. Contains the moderator band
    D. Part of its wall is smooth
    E. Is triangular in shape

11. THE LYMPHATIC DRAINAGE OF THE LUNGS IS TO WHICH OF THE FOLLOWING GROUPS OF NODES?
    A. Tracheobronchial
    B. Paratracheal
    C. Pulmonary
    D. Bronchopulmonary
    E. All of the above

12. WHICH OF THE FOLLOWING STATEMENTS IS FALSE?
    A. The right coronary artery is located in the atrioventricular groove
    B. The left coronary artery arises from the posterior aortic sinus
    C. The circumflex branch arises from the right coronary artery
    D. The marginal branch arises from the right coronary artery
    E. The right coronary artery is smaller than the left

13. WHICH STATEMENT(S) ABOUT THE VENOUS DRAINAGE OF THE STOMACH IS(ARE) TRUE?
    A. The left gastric vein drains the upper half of the lesser curvature
    B. The right gastric vein drains into the portal vein
    C. The left gastroepiploic vein drains into the splenic vein
    D. The left gastric veins communicate with the esophageal veins
    E. All of the above

14. WHICH STATEMENT ABOUT THE SUPERIOR MESENTERIC ARTERY IS FALSE?
    A. Originates at the level of the first lumbar vertebra
    B. Supplies the jejunum and the ileum
    C. Gives off the middle colic artery
    D. Gives off the left colic artery
    E. Is the artery of the midgut

SELECT THE ONE BEST ANSWER

15. THE SUPRARENAL GLANDS RECEIVE THEIR ARTERIAL SUPPLY FROM:
    A. The middle suprarenal artery from the aorta
    B. The inferior suprarenal, a branch of the renal artery
    C. The superior suprarenal, a branch of the inferior phrenic artery
    D. All of the above
    E. None of the above

16. THE INFERIOR VENA CAVA IS CROSSED ANTERIORLY BY ALL THE FOLLOWING STRUCTURES EXCEPT THE _____.
    A. Right gonadal artery
    B. Root of the mesentery
    C. Right suprarenal gland
    D. Superior mesenteric artery
    E. Third part of the duodenum

17. WHICH OF THE FOLLOWING IS NOT A BRANCH OF THE SUPERIOR MESENTERIC ARTERY?
    A. Middle colic
    B. Right colic
    C. Left colic
    D. Ileocolic
    E. Jejunal

18. WHICH STATEMENT ABOUT THE INFERIOR MESENTERIC ARTERY IS FALSE?
    A. Arises at the level of the third lumbar vertebra
    B. An unpaired branch of the abdominal aorta
    C. Gives rise to the sigmoid arteries
    D. The artery of the hindgut
    E. None of the above

19. WHICH OF THE FOLLOWING IS NOT AN ANTERIOR RELATION OF THE ABDOMINAL AORTA?
    A. Pancreas
    B. Splenic vein
    C. Left renal vein
    D. Left lumbar vein
    E. Third part of the duodenum

20. ALL THE FOLLOWING ARE TRIBUTARIES OF THE INFERIOR VENA CAVA EXCEPT THE:
    A. Right inferior phrenic
    B. Right suprarenal
    C. Fourth lumbar
    D. Left gonadal
    E. Renal

MULTI-COMPLETION QUESTIONS

DIRECTIONS: In each of the following questions or incomplete statements, ONE OR MORE of the completions given is correct. On the answer sheet blacken the space under A if 1, 2, and 3 are correct; B if 1 and 3 are correct; C if 2 and 4 are correct; D if only 4 is correct; and E if all are correct.

21. THE THORACIC AORTA:
    1. Passes through the diaphragm at the level of the 10th thoracic vertebra
    2. Gives rise to all posterior intercostal arteries
    3. Descends on the right side of the vertebral column
    4. Lies behind the esophagus

22. THE PERICARDIAL SAC:
    1. Fuses with the diaphragm
    2. Lies immediately in front of the esophagus
    3. Extends from the level of the 2nd to 6th costal cartilages
    4. Lies posterior to the phrenic nerves

A	B	C	D	E
1,2,3	1,3	2,4	only 4	all correct

23. WHICH STATEMENT(S) ABOUT THE HEART IS(ARE) CORRECT?
    1. Structural defects of the ventricular septum are rare
    2. The floor of the fossa ovalis represents the persistent portion of the septum primum of the embryonic heart
    3. The inner wall of the right atrium is divided into two parts by the sulcus terminalis
    4. The infundibulum of the right ventricle is derived from the bulbus cordis of the embryonic heart

24. CONCERNING THE FOSSA OVALIS IN THE RIGHT ATRIUM:
    1. The limbus fossae ovalis is derived from the septum secundum
    2. It is bounded by the limbus (annulus) fossae ovalis
    3. Its floor may be recognized in the right atrium
    4. Its floor is derived from the septum primum

25. THE SUPERIOR VENA CAVA:
    1. Is formed by the brachiocephalic veins
    2. Receives blood from the azygos vein
    3. Contains no valves
    4. Is thin-walled

26. THE RIGHT CORONARY ARTERY:
    1. Is located in the coronary sinus
    2. Passes between the right auricle and the pulmonary trunk
    3. Arises from the posterior aortic sinus
    4. Anastomoses with the left coronary artery

27. THE EXTRINSIC INNERVATION OF THE HEART IS VIA WHICH OF THE FOLLOWING:
    1. Thoracic and cervical ganglia
    2. Superficial cardiac plexus
    3. Vagus nerve
    4. Deep cardiac plexus

28. CONCERNING THE LEFT VENTRICLE:
    1. The aortic vestibule is smooth
    2. The mitral valve has only two cusps
    3. Its wall is thicker than the right
    4. Its internal surface is trabeculated

29. CONCERNING THE CORONARY ARTERIES:
    1. Lie deep in the myocardium
    2. Flow of blood through them occurs during diastole
    3. Their branches anastomose with each other
    4. The first branches of the aorta

30. THE CORONARY SINUS:
    1. Is a thin-walled venous channel
    2. Opens directly into the right atrium
    3. Lies in the coronary sulcus
    4. Is 3-5 cm long

31. THE LEFT ATRIUM:
    1. Has smooth and trabeculated portions on its internal wall
    2. Is demarcated from the right atrium by the coronary sulcus
    3. Receives the pulmonary veins
    4. Communicates with the left ventricle via the tricuspid valve

A	B	C	D	E
1,2,3	1,3	2,4	only 4	all correct

32. THE DIAPHRAGM IS SUPPLIED BY WHICH OF THE FOLLOWING ARTERIES?
    1. Musculophrenic
    2. Pericardiacophrenic
    3. Phrenic
    4. Intercostal

33. WHICH STRUCTURE(S) IS(ARE) LOCATED IN THE VENTRICULAR WALLS OF THE HEART?
    1. Chordae tendineae
    2. Musculi pectinati
    3. Trabeculae carneae
    4. Crista terminalis

34. THE TRACHEA IS SUPPLIED BY BRANCHES FROM THE FOLLOWING ARTERIES:
    1. Internal thoracic
    2. Superior thyroid
    3. Bronchial
    4. Inferior thyroid

35. THE ARTERIAL SUPPLY OF THE MAMMARY GLAND IS DERIVED FROM THE _____ ARTERIES.
    1. Intercostal
    2. Internal thoracic
    3. Superior epigastric
    4. Lateral thoracic

36. WHICH OF THE FOLLOWING ARTERIES SUPPLY(IES) THE UTERUS?
    1. Ovarian
    2. Umbilical
    3. Uterine
    4. Internal pudendal

37. THE VENOUS DRAINAGE OF THE LOWER LIMB:
    1. Is aided by numerous valves
    2. Consists of two isolated systems
    3. Is via deep veins during exercise
    4. Depends on smooth muscle in the venous walls

38. THE CAPUT MEDUSAE IS:
    1. Located in the region of the umbilicus
    2. Caused by dilatation of the epigastric veins
    3. An external sign of portal venous obstruction
    4. Often associated with collection of fluid in the peritoneal cavity

39. THE MAIN BRANCHES OF THE CELIAC ARTERIAL TRUNK ARE THE:
    1. Common hepatic
    2. Left gastric
    3. Splenic
    4. Right gastric

40. THE INFERIOR VENA CAVA IS:
    1. Located on the posterior abdominal wall
    2. Related medially to the abdominal aorta
    3. Anterior to the right psoas major muscle
    4. Formed by union of the common iliac veins

FIVE-CHOICE ASSOCIATION QUESTIONS

DIRECTIONS: Each group of questions below consists of a numbered list of descriptive words or phrases accompanied by a diagram with certain parts indicated by letters, or by a list of lettered headings. For each numbered word or phrase, SELECT THE LETTERED PART OR HEADING that matches it correctly. Then on the appropriate line of the answer sheet, blacken the space under the letter of that part or heading. Sometimes more than one numbered word or phrase may be correctly matched to the same lettered part or heading.

ASSOCIATION QUESTIONS

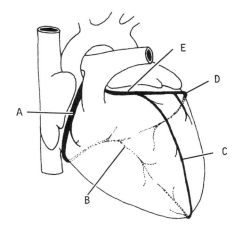

41. ____ Arises from the right aortic sinus
42. ____ Gives rise to the anterior interventricular branch
43. ____ Gives rise to the posterior interventricular branch
44. ____ The circumflex branch

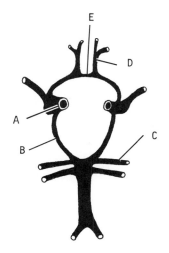

45. ____ Posterior communicating branch
46. ____ Posterior cerebral artery
47. ____ Anterior communicating branch

ASSOCIATION QUESTIONS

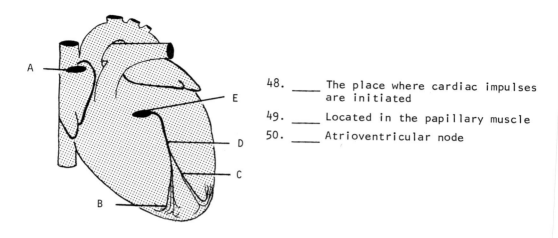

48. _____ The place where cardiac impulses are initiated
49. _____ Located in the papillary muscle
50. _____ Atrioventricular node

Interpretation of Your Score

Number of Correct Responses	Level of Performance
43-50	Excellent - Exceptional
38-42	Superior - Very Superior
32-37	Average - Above Average
28-31	Poor - Marginal
27 or less	Very Poor - Failure

10. REVIEW EXAMINATION ON THE DIGESTIVE SYSTEM

INTRODUCTORY NOTE. The following 50 multiple-choice questions are based on the instructional objectives pertaining to the digestive system in Chapters Three to Six of this Study Guide & Review Manual. You should be able to answer these questions in 35-40 minutes. Before beginning, tear out an answer sheet from the back of the book and read the directions on how to use it. The key to the correct responses is on page 351.

FIVE-CHOICE COMPLETION QUESTIONS

DIRECTIONS: Each of the following questions or incomplete statements is followed by five suggested answers or completions. SELECT THE ONE BEST ANSWER in each case and then blacken the appropriate space on the answer sheet.

1. CONCERNING THE ORAL CAVITY WHICH STATEMENT IS CORRECT?
   A. The oral cavity comprises the vestibule, oral cavity proper and oropharynx
   B. The total number of deciduous teeth is 16 in each jaw
   C. The oral cavity communicates with the pharynx through the faucial isthmus
   D. The oral cavity is bounded anteriorly and laterally by the alveolar arches, teeth, gingivae, and palatine tonsils.
   E. The floor of the mouth is supported by the oral diaphragm

2. WHICH STATEMENT(S) ABOUT THE SMALL INTESTINE IS(ARE) TRUE?
   A. Begins at the pylorus
   B. Terminates at the ileocecal junction
   C. Includes the duodenum which is mostly retroperitoneal
   D. Supplied by the celiac and superior mesenteric arteries
   E. All of the above

3. WHICH STATEMENT CONCERNING THE VENOUS DRAINAGE OF THE ANAL CANAL IS FALSE?
   A. Venous channels draining the upper and lower parts communicate
   B. Internal hemorrhoids involve the inferior rectal vessels
   C. Upper half is drained only by the superior rectal veins
   D. Inferior rectal veins drain below the pectinate line
   E. Internal are more common than external hemorrhoids

4. WHICH STATEMENT ABOUT THE TRANSVERSE COLON IS FALSE?
   A. Suspended by the lesser omentum
   B. The most mobile part of the colon
   C. Gives attachment to the greater omentum
   D. Mainly supplied by the artery of the midgut
   E. Attached to the lower border of the pancreas

SELECT THE ONE BEST ANSWER

5. THE LIGAMENTUM TERES REPRESENTS THE OBLITERATED:
   A. Right umbilical artery
   B. Left umbilical vein
   C. Left umbilical artery
   D. Right umbilical vein
   E. None of the above

6. ALL THE FOLLOWING CAN USUALLY BE PALPATED DURING A RECTAL EXAMINATION, EXCEPT THE:
   A. Uterine tubes in the female
   B. Prostate gland in the male
   C. Enlarged seminal vesicles in the male
   D. Posterior vaginal wall in the female
   E. Hollow of the sacrum in both sexes

7. ALL STATEMENTS ABOUT THE VERMIFORM APPENDIX ARE TRUE EXCEPT:
   A. Subject to considerable variation in position
   B. Opens into the apex of the cecum in fetuses
   C. Retroileal appendices are the most common type
   D. Opens about two cm below the ileocecal junction in adults
   E. When the cecum is full, the free appendix usually hangs in the pelvis

8. THE HEAD OF THE PANCREAS:
   A. Is related posteriorly to the inferior vena cava
   B. Is covered anteriorly by the transverse colon
   C. Lies in the concavity of the duodenum
   D. Encloses the bile duct
   E. All of the above

9. WHICH STATEMENT(S) ABOUT THE GALL BLADDER IS(ARE) TRUE?
   A. Consists of a fundus, body, and neck
   B. Is supplied only by the cystic artery
   C. Forms part of the extrahepatic biliary system
   D. Inflammation of the gall bladder is common in older obese persons
   E. All of the above

10. ALL THESE VEINS CONTRIBUTE TO THE FORMATION OF THE PORTAL SYSTEM EXCEPT THE:
    A. Superior mesenteric
    B. Inferior mesenteric
    C. Splenic
    D. Hepatic
    E. None of the above

11. WHICH STATEMENT ABOUT A MECKEL'S DIVERTICULUM IS INCORRECT?
    A. Occurs in nearly all newborn infants
    B. Occurs at a variable distance from the cecum
    C. May become a leading point for an intussusception
    D. May contain gastric and pancreatic tissues in its wall
    E. Always located on the antimesenteric side of the ileum

12. WHICH STATEMENT REGARDING THE EXTERNAL ANAL SPHINCTER IS FALSE?
    A. Is under voluntary control
    B. Consists of striated muscle
    C. Located below the pelvic diaphragm
    D. Innervated by inferior rectal nerves
    E. Subcutaneous part inserts in the central tendon of the perineum

13. WHICH OF THE FOLLOWING ARTERIES SUPPLY(IES) THE RECTUM AND ANAL CANAL?
    A. Superior rectal
    B. Middle rectal
    C. Inferior rectal
    D. Median sacral
    E. All of the above

SELECT THE ONE BEST ANSWER

14. WHICH OF THE FOLLOWING STATEMENTS ABOUT THE STOMACH IS FALSE?
    A. Is a common site of ulcers
    B. Is divided into cardiac and pyloric portions
    C. Receives its blood supply from the celiac artery
    D. Its fundic portion receives its blood supply from the left gastroepiploic artery
    E. Has the greater omentum attached to its greater curvature

15. THE DUODENUM IS SUPPLIED BY WHICH OF THE FOLLOWING ARTERIES?
    A. Pancreaticoduodenal
    B. Superior mesenteric
    C. Gastroduodenal
    D. Common hepatic
    E. All of the above

16. WHICH STATEMENT ABOUT THE BILE DUCT, AFTER IT LEAVES THE LESSER OMENTUM, IS INCORRECT?
    A. Traverses the medial part of the body of the pancreas
    B. Related to the superior pancreaticoduodenal artery
    C. Related medially to the gastroduodenal artery
    D. Lies behind the first part of the duodenum
    E. Related posteriorly to the portal vein

17. WHICH STATEMENT ABOUT THE FALCIFORM LIGAMENT IS TRUE?
    A. Attached to the anterior and superior aspects of the liver
    B. Contains the ligamentum teres in its free border
    C. Attached to the anterior abdominal wall
    D. A sickle-shaped peritoneal fold
    E. All of the above

18. WHICH STATEMENT CONCERNING THE RECTUM IS FALSE?
    A. Separated from the uterus by the rectouterine pouch
    B. Transverse folds project into its lumen
    C. Middle third has no peritoneum
    D. Taenia coli are absent
    E. Has no mesentery

19. WHICH STATEMENT ABOUT THE SECOND PART OF THE DUODENUM IS INCORRECT?
    A. Related superiorly to the uncinate process of the pancreas
    B. Related laterally to the hepatic flexure of the colon
    C. Related posteriorly to the hilum of the right kidney
    D. Related medially to the head of the pancreas
    E. Related anteriorly to the transverse colon

20. WHICH STATEMENT ABOUT THE DESCENDING COLON IS FALSE?
    A. Covered by peritoneum on its front and sides
    B. Terminates at the inlet of the true pelvis
    C. Related to the lower pole of the spleen
    D. Commences at the left colic flexure
    E. Supplied by the middle colic artery

21. WHICH STATEMENT ABOUT THE EPIPLOIC FORAMEN IS INCORRECT?
    A. Related anteriorly to the gastrohepatic ligament
    B. Related superiorly to the caudate lobe of the liver
    C. Bounded inferiorly by the first part of the duodenum
    D. Related posteriorly to the portal vein
    E. The opening between the greater and lesser peritoneal sacs

MULTI-COMPLETION QUESTIONS

DIRECTIONS: In each of the following questions or incomplete statements, ONE OR MORE of the completions given is correct. On the answer sheet blacken the space under A if 1, 2, and 3 are correct; B if 1 and 3 are correct; C if 2 and 4 are correct; D if only 4 is correct; and E if all are correct.

22. THE FEATURES OF THE STOMACH INCLUDE WHICH OF THE FOLLOWING?
    1. Has a pyloric orifice located at the level of the first lumbar vertebra
    2. Has a lesser omentum containing the gastroepiploic vessels
    3. Receives its blood supply from the artery of the foregut
    4. Its cardiac orifice is located at the level of the eighth thoracic vertebra

23. THE GREATER OMENTUM:
    1. Contains the left and right gastric vessels
    2. Is the term applied to a fold of peritoneum
    3. Is attached to the lower border of the pancreas
    4. Is attached to the greater curvature of the stomach

24. THE TRANSVERSE COLON:
    1. Begins anterior to the right kidney
    2. Terminates immediately below the spleen
    3. Is supplied mainly by the middle colic artery
    4. Is the shortest segment of the colon

25. CONCERNING THE DUODENUM:
    1. Supplied by the celiac trunk and the superior mesenteric artery
    2. The beginning of its first part is called the free part
    3. Derived from the foregut and midgut of the embryo
    4. Is relatively fixed except for its first part

26. THE COMMON BILE DUCT IS:
    1. Formed by the union of the common hepatic and cystic ducts
    2. Located in the lesser omentum
    3. 4-6 cm long
    4. Enters the third part of the duodenum

27. THE CELIAC LYMPH NODES:
    1. Are related to the artery of the foregut
    2. Receive channels from hepatic lymph nodes
    3. Drain lymph from the superior gastric nodes
    4. Receive lymph from the descending colon

28. THE BRANCH(ES) OF THE CELIAC ARTERIAL TRUNK IS(ARE):
    1. Left gastroepiploic   3. Right gastric
    2. Common hepatic   4. Splenic

29. THE VENOUS CHANNELS IN THE LOWER END OF THE ESOPHAGUS:
    1. Include esophageal veins that drain into the portal veins
    2. May rupture and result in vomiting of blood
    3. Include stomach veins that drain into the azygos veins
    4. Become dilated in portal obstruction forming esophageal varices

30. THE RIGHT FLEXURE OF THE COLON IS:
    1. Anterior to the right kidney
    2. Supplied by the right colic artery
    3. Related to the right lobe of the liver
    4. Also called the splenic flexure

A	B	C	D	E
1,2,3	1,3	2,4	only 4	all correct

31. THE ANTERIOR WALL OF THE LESSER SAC IS COMPOSED OF:
    1. The posterior layer of the lesser omentum
    2. Peritoneum on the posterior surface of the liver
    3. Peritoneum covering the posterior surface of the stomach
    4. The posterior peritoneal layer extending downwards from the greater curvature

32. THE FIRST PART OF THE DUODENUM:
    1. Terminates at the neck of the gall bladder
    2. Commences in the transpyloric plane
    3. Is a common site for ulcers
    4. Is the most stable part

33. BILE IS:
    1. Concentrated in the gall bladder
    2. Responsible for the digestion of carbohydrates
    3. Secreted only by liver cells
    4. More concentrated when flow is directly into the duodenum

34. THE RIGHT LATERAL SURFACE OF THE LIVER:
    1. Accounts for liver dullness in percussion of the right chest wall
    2. Is called the base
    3. Is related to ribs 7-11 in the right midaxillary line
    4. For biopsy it is reached through the 7th intercostal space

35. THE ANATOMICAL FEATURES DISTINGUISHING THE JEJUNUM FROM THE ILEUM ARE:
    1. The jejunal mesentery has less fat
    2. The jejunum has a thicker wall
    3. The jejunum has a greater diameter
    4. The jejunal arterial arcades are four or five in number

36. THE PYLORUS OF THE STOMACH:
    1. Has the prepyloric vein on its anterior surface
    2. Is related posteriorly to the hepatic artery
    3. Is the junction between the stomach and duodenum
    4. Contains the pyloric antrum

37. THE GALL BLADDER:
    1. Opens into the bile duct
    2. Is related to the inferior surface of the liver
    3. Is completely enclosed by peritoneum
    4. Has an average capacity of 30 ml

38. CONCERNING MECKEL'S DIVERTICULUM:
    1. May produce gross bleeding from the bowel during infancy
    2. Common congenital malformation of the digestive tract
    3. Occurs in about two per cent of persons
    4. Represents a remnant of the yolk stalk

39. THE LESSER OMENTUM:
    1. Has a free right margin
    2. Is attached inferiorly to the lesser curvature of the stomach
    3. Is attached superiorly to the porta hepatis of the liver
    4. Encloses the inferior vena cava in its free margin

A	B	C	D	E
1,2,3	1,3	2,4	only 4	all correct

40. THE ASCENDING COLON IS:
    1. Related posteriorly to the quadratus lumborum muscle
    2. Related laterally to the right paracolic gutter
    3. Supplied by the ileocolic artery
    4. Covered by peritoneum on its anterior and posterior surfaces

41. THE POSTERIOR SURFACE OF THE STOMACH IS RELATED TO THE:
    1. Left kidney
    2. Left suprarenal
    3. Splenic artery
    4. Head of the pancreas

42. THE BODY OF THE PANCREAS HAS:
    1. An upper border related to the splenic artery
    2. An anterior surface that forms part of the stomach bed
    3. An attachment to the transverse mesocolon
    4. An inferior surface related to the splenic vein

43. THE QUADRATE LOBE OF THE LIVER IS:
    1. Related to the transverse colon
    2. Located posterior to the porta hepatis of the liver
    3. Limited laterally by the fossa for the gall bladder
    4. Limited medially by the fissure for ligamentum venosum

44. THE BILE DUCT:
    1. Begins at the union of the right and left hepatic ducts
    2. Is related posteriorly to the portal vein
    3. Opens into the first part of the duodenum
    4. Is related medially to the hepatic artery proper

45. THE LIVER IS:
    1. The largest gland in the body
    2. In the right hypochondrium
    3. Partly covered by peritoneum
    4. Palpable in the healthy adult

46. WHICH VESSEL(S) IS(ARE) NOT DERIVED FROM THE SPLENIC ARTERY?
    1. Short gastric
    2. Left gastroepiploic
    3. Pancreatic branches
    4. Right gastroepiploic

FIVE-CHOICE ASSOCIATION QUESTIONS

DIRECTIONS: Each group of questions below consists of a numbered list of descriptive words or phrases accompanied by a diagram with certain parts indicated by letters, or by a list of lettered headings. For each numbered word or phrase, SELECT THE LETTERED PART OR HEADING that matches it correctly. Then on the appropriate line of the answer sheet, blacken the space under the letter of that part or heading. Sometimes more than one numbered word or phrase may be correctly matched to the same lettered part or heading.

ASSOCIATION QUESTIONS

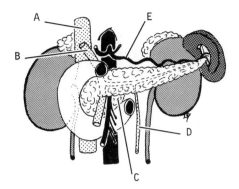

47. _____ Portal Vein
48. _____ Uncinate process of the pancreas
49. _____ Splenic artery
50. _____ Inferior mesenteric vein

Interpretation of Your Score

Number of Correct Responses	Level of Performance
43-50	Excellent - Exceptional
38-42	Superior - Very Superior
32-37	Average - Above Average
28-31	Poor - Marginal
27 or less	Very Poor - Failure

11. REVIEW EXAMINATION ON THE REPRODUCTIVE SYSTEM

INTRODUCTORY NOTE. The following 35 multiple-choice questions are based on the instructional objectives pertaining to the reproductive system in Chapter Six of this Study Guide & Review Manual. You should be able to answer these questions in 25-28 minutes. Before beginning, tear out an answer sheet from the back of the book and read the directions on how to use it. The key to the correct responses is on page 352.

FIVE-CHOICE COMPLETION QUESTIONS

DIRECTIONS: Each of the following questions or incomplete statements is followed by five suggested answers or completions. SELECT THE ONE BEST ANSWER in each case and then blacken the appropriate space on the answer sheet.

1. THE UTERUS IS DRAINED BY WHICH GROUP(S) OF LYMPH NODES?
   A. Inguinal
   B. Lumbar
   C. Sacral
   D. Iliac
   E. All of the above

2. WHICH NERVE(S) SUPPLY(IES) THE EXTERNAL GENITALIA OF FEMALES?
   A. Uterovaginal plexus
   B. Ilioinguinal
   C. Pudendal
   D. All of the above
   E. None of the above

3. THE CREMASTERIC FASCIA OF THE SPERMATIC CORD IS A CONTINUATION OF THE _____ OF THE ANTEROLATERAL ABDOMINAL WALL.
   A. Peritoneum
   B. Transversalis fascia
   C. External oblique muscle
   D. Transversus abdominus muscle
   E. Internal oblique muscle

4. THE WIDEST SEGMENT OF THE UTERINE TUBE IS THE:
   A. Infundibulum
   B. Ampulla
   C. Isthmus
   D. Uterine part
   E. None of the above

5. WHICH OF THE FOLLOWING STATEMENTS CONCERNING THE VAGINA IS FALSE?
   A. The anterior wall is longer than the posterior wall
   B. Its anterior wall is related to the urethra
   C. The cervix projects into its anterior wall
   D. Its posterior fornix is deeper than the anterior fornix
   E. The lower half is located below the pelvic diaphragm

6. THE WIDEST PART OF THE MALE URETHRA IS THE:
   A. Internal urethral orifice
   B. Membranous part
   C. Prostatic part
   D. External urethral orifice
   E. None of the above

SELECT THE ONE BEST ANSWER

7. WHICH OF THE FOLLOWING IS PRESENT IN THE SUPERFICIAL PERINEAL POUCH?
   A. Bulbospongiosus muscle
   B. Deep transverse perineal muscle
   C. Sphincter urethrae
   D. Bulbourethral gland
   E. Membranous urethra

8. THE BROAD LIGAMENT CONTAINS ALL THE FOLLOWING STRUCTURES, EXCEPT THE:
   A. Uterine artery
   B. Ovary
   C. Uterine tube
   D. Ovarian ligament
   E. Part of the ureter

9. WHICH STATEMENT REGARDING THE OVARY IS FALSE?
   A. The ovarian ligament contains the ovarian vessels and nerves
   B. Innervated by the abdominal aortic plexus of nerves
   C. The left ovarian vein drains into the renal vein
   D. Supplied by the ovarian artery
   E. Supplied by the uterine artery

10. THE SKIN OF THE PENIS IS MAINLY SUPPLIED BY THE _____ NERVES.
    A. Deep perineal
    B. Ilioinguinal
    C. Cavernous
    D. Pudendal
    E. None of the above

11. WHICH STATEMENT CONCERNING THE PROSTATE GLAND IS INCORRECT?
    A. Is shaped like a cone
    B. Related inferiorly to the urogenital diaphragm
    C. Contains the second part of the urethra
    D. May become enlarged in older persons
    E. Can be palpated rectally

MULTI-COMPLETION QUESTIONS

DIRECTIONS: In each of the following questions or incomplete statements, ONE OR MORE of the completions given is correct. On the answer sheet blacken the space under A if 1, 2, and 3 are correct; B if 1 and 3 are correct; C if 2 and 4 are correct; D if only 4 is correct; and E if all are correct.

12. THE EPOOPHORON IS A VESTIGIAL STRUCTURE THAT:
    1. Consists of a duct and a few blind tubules
    2. Corresponds to the ductus deferens
    3. Lies in the layers of the broad ligament
    4. Is of no clinical significance

13. THE SEMINAL VESICLES ARE:
    1. About 5 cm long
    2. Located posterior to the prostate
    3. Palpable rectally
    4. Supplied by the superior rectal artery

14. THE GUBERNACULUM IN THE FEMALE EMBRYO BECOMES THE:
    1. Pubocervical ligament
    2. Round ligament
    3. Cardinal ligament
    4. Ovarian ligament

A	B	C	D	E
1,2,3,	1,3	2,4	only 4	all correct

15. THE PROCESSUS VAGINALIS IS:
    1. A peritoneal evagination
    2. The primordium of the vagina
    3. Covered by layers of the abdominal wall
    4. Present only in female embryos

16. THE UTERINE ARTERY:
    1. Is closely related to the ureter
    2. Supplies the uterus and vagina
    3. Is a branch of the posterior division of the internal iliac artery
    4. Runs medially along the upper border of the broad ligament

17. THE PROSTATE GLAND IS:
    1. Lobated
    2. Traversed by the urethra
    3. Drained by the internal iliac nodes
    4. Supplied by the superior vesical artery

18. THE VAGINA IS:
    1. Partly drained by the superficial inguinal lymph nodes
    2. Supplied by the vaginal artery
    3. Supplied by the uterine artery
    4. Innervated by the pelvic plexus of nerves

19. THE MESONEPHRIC DUCTS GIVE RISE TO THE:
    1. Epididymis
    2. Ductus deferens
    3. Ejaculatory duct
    4. Seminal vesicles

20. THE UROGENITAL DIAPHRAGM CONTAINS THE:
    1. Membranous urethra
    2. Deep transverse perineal muscle
    3. Sphincter urethrae muscle
    4. Ischiocavernosus muscle

21. THE VAGINA IS SUPPORTED BY WHICH OF THE FOLLOWING?
    1. Levator ani muscles
    2. Urogenital diaphragm
    3. Perineal body
    4. Transverse cervical ligament

22. THE LABIA MAJORA:
    1. Are homologous to the scrotum in the male
    2. Are supplied by the pudendal arteries
    3. Contain smooth muscle fibers
    4. Are drained by the iliac lymph nodes

23. THE CERVIX OF THE UTERUS:
    1. Is divided into supravaginal and infravaginal portions
    2. Is frequently a site of cancer
    3. Is related laterally to the ureter
    4. Has a soft consistency in the nonpregnant uterus

A	B	C	D	E
1,2,3	1,3	2,4	only 4	all correct

24. WHICH OF THE FOLLOWING ARTERIES SUPPLY(IES) THE UTERUS?
    1. Umbilical
    2. Uterine
    3. Internal pudendal
    4. Ovarian

25. CONCERNING THE UTERUS:
    1. The normal uterus is described as being anteflexed and retroverted
    2. The normal uterus is described as being anteflexed and anteverted
    3. The round ligament of uterus lies in the posterior layer of the broad ligament
    4. The ovaries are related to the posterior layer of the broad ligament

26. THE UTERUS IS SUPPORTED BY WHICH OF THE FOLLOWING LIGAMENTS?
    1. Transverse cervical
    2. Sacrocervical
    3. Pubocervical
    4. Broad

27. CONCERNING THE TESTIS:
    1. In the fetus the testis lies on the posterior abdominal wall
    2. Descent of the testis is caused by intra-abdominal pressure
    3. The serous sac which envelops the testis is known as the tunica vaginalis
    4. Accumulation of fluid within the tunica vaginalis is known as a hernia

28. THE ADULT UTERUS IS:
    1. Partly covered by peritoneum
    2. Anteflexed
    3. Anteverted
    4. 12-15 cm long

29. CONCERNING THE FEMALE PERINEUM:
    1. It is divided into urogenital and anal triangles
    2. The perineal body gives attachment to most muscles of the perineum
    3. A badly repaired perineal tear results in a weak pelvic floor predisposing to prolapse of the viscera
    4. The levator ani muscle is inserted into the perineal body

FIVE-CHOICE ASSOCIATION QUESTIONS

DIRECTIONS: Each group of questions below consists of a numbered list of descriptive words or phrases accompanied by a diagram with certain parts indicated by letters, or by a list of lettered headings. For each numbered word or phrase, SELECT THE LETTERED PART OR HEADING that matches it correctly. Then on the appropriate line of the answer sheet, blacken the space under the letter of that part or heading. Sometimes more than one numbered word or phrase may be correctly matched to the same lettered part or heading.

ASSOCIATION QUESTIONS

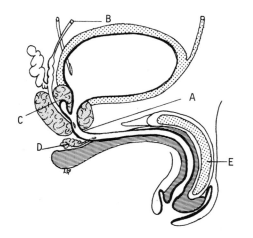

30. ____ Derived from the cranial part of the mesonephric duct
31. ____ Passes through the inguinal canal
32. ____ Enlargement of this structure gives rise to difficulty in micturition
33. ____ Covered by the ischiocavernosus muscle
34. ____ Opens into the shortest portion of the urethra
35. ____ The shortest segment of the urethra

Interpretation of Your Score

Number of Correct Responses	Level of Performance
32-35	Excellent - Exceptional
28-31	Superior - Very Superior
24-27	Average - Above Average
20-23	Poor - Marginal
18 or less	Very Poor - Failure

12. REVIEW EXAMINATION ON THE RESPIRATORY SYSTEM

INTRODUCTORY NOTE. The following 35 multiple-choice questions are based on the instructional objectives pertaining to the respiratory system in Chapters Three and Four of this Study Guide & Review Manual. You should be able to answer these questions in 25-28 minutes. Before beginning, tear out an answer sheet from the back of the book and read the directions on how to use it. The key to the correct responses is on page 352.

FIVE-CHOICE COMPLETION QUESTIONS

DIRECTIONS: Each of the following questions or incomplete statements is followed by five suggested answers or completions. SELECT THE ONE BEST ANSWER in each case and then blacken the appropriate space on the answer sheet.

1. WHICH OF THE FOLLOWING DIAMETERS OF THE THORAX IS(ARE) INCREASED DURING INSPIRATION?
   A. Transverse
   B. Anteroposterior
   C. Vertical
   D. All of the above
   E. None of the above

2. WHICH STATEMENT ABOUT THE DIAPHRAGM IS FALSE?
   A. Involved in respiration
   B. Central part is muscular
   C. Intrapleural pressure is decreased following contraction
   D. Is bilaterally innervated
   E. Contraction promotes venous return

3. WHICH OF THE FOLLOWING DOES NOT INNERVATE THE LUNGS?
   A. Vagus nerve
   B. Phrenic nerve
   C. Thoracic sympathetic
   D. Anterior pulmonary plexus
   E. Posterior pulmonary plexus

4. FETAL BLOOD IS OXYGENATED IN THE:
   A. Fetal lungs
   B. Liver via the ductus venosus
   C. Placenta
   D. Aorta via the ductus arteriosus
   E. Maternal lungs

5. WHICH OF THE FOLLOWING INTRINSIC MUSCLES OF THE LARYNX ABDUCT(S) THE VOCAL FOLDS?
   A. Vocalis
   B. Arytenoid
   C. Posterior cricoarytenoid
   D. Cricothyroid
   E. All of the above

6. THE TRACHEA IS SUPPLIED BY BRANCHES FROM THE FOLLOWING ARTERIES:
   A. Inferior thyroid
   B. Superior thyroid
   C. Bronchial
   D. Internal thoracic
   E. All of the above

SELECT THE ONE BEST ANSWER

7. THE DIAPHRAGM IS SUPPLIED BY WHICH OF THE FOLLOWING ARTERIES?
   A. Phrenic
   B. Pericardiacophrenic
   C. Musculophrenic
   D. Intercostal
   E. All of the above

8. WHICH OF THE FOLLOWING ARTERIES SUPPLY(IES) THE NASAL CAVITY?
   A. Anterior ethmoidal
   B. Sphenopalatine
   C. Superior labial
   D. Greater palatine
   E. All of the above

9. THE MUSCULAR FIBERS OF THE DIAPHRAGM ARISE FROM THE:
   A. Xiphoid process of the sternum
   B. Lumbocostal arches and crura
   C. Lower costal cartilages
   D. Lower ribs
   E. All of the above

10. WHICH OF THE FOLLOWING STATEMENTS ABOUT THE LARYNX IS(ARE) CORRECT?
    A. The aryepiglottic fold contains the corniculate and cuneiform cartilages
    B. The rima vestibuli is the opening between the false vocal cords
    C. The thyroid cartilage articulates with the cricoid cartilage by a synovial joint
    D. All of the above
    E. None of the above

11. ALL THE FOLLOWING STATEMENTS REGARDING THE LEFT LUNG ARE CORRECT, EXCEPT:
    A. Has only two lobes
    B. Has a groove on its mediastinal surface for the azygos vein
    C. The cardiac notch is present in its upper lobe
    D. Is subdivided into eight bronchopulmonary segments
    E. Is supplied by the bronchial arteries

12. WHICH OF THE FOLLOWING STATEMENTS CONCERNING THE RIGHT PRIMARY BRONCHUS IS FALSE? COMPARED TO THE LEFT PRIMARY BRONCHUS IT IS:
    A. Less vertical
    B. Shorter
    C. Wider
    D. Clinically more important
    E. All of the above

13. WHICH STATEMENT(S) CONCERNING THE RIGHT LUNG IS(ARE) CORRECT?
    A. Has both oblique and horizontal fissures
    B. Has ten bronchopulmonary segments
    C. Is divided into three lobes
    D. Is shorter than the left
    E. All of the above

14. WHICH OF THE FOLLOWING STATEMENTS CONCERNING THE TRACHEA IS FALSE?
    A. Is 20-25 cm long
    B. Is extremely mobile
    C. Terminates at the level of the sternal angle
    D. Begins at the level of the 6th cervical vertebra
    E. Rigidity of its wall is due to the presence of imcomplete cartilaginous rings

MULTI-COMPLETION QUESTIONS

DIRECTIONS: In each of the following questions or incomplete statements, ONE OR MORE of the completions given is correct. On the answer sheet blacken the space under A if 1, 2, and 3 are correct; B if 1 and 3 are correct; C if 2 and 4 are correct; D if only 4 is correct; and E if all are correct.

15. THE "PUMP-HANDLE" MOVEMENT OF THE RIBS INVOLVES:
    1. Elevation of the sternum
    2. An increase in the anteroposterior diameter
    3. Depression of the sternum
    4. No movement at the costovertebral joint

16. THE DIAPHRAGM IS SUPPLIED BY:
    1. Branches from the aorta
    2. Lower intercostal nerves
    3. Branches of the internal thoracic artery
    4. Phrenic nerves (C3, 4 and 5)

17. WHICH OF THE FOLLOWING STATEMENTS IS(ARE) CORRECT?
    1. The maxillary air sinus may become infected from an infected tooth
    2. The trigeminal nerve supplies general sensory innervation to the nasal mucosa
    3. The facial nerve supplies parasympathetic innervation to the nasal mucosa
    4. The nasolacrimal duct opens in the anterior part of the middle nasal meatus

18. THE LYMPHATIC DRAINAGE OF THE LUNGS IS TO WHICH OF THE FOLLOWING GROUPS OF NODES?
    1. Paratracheal
    2. Tracheobronchial
    3. Bronchopulmonary
    4. Pulmonary

19. WHICH OF THE FOLLOWING PARANASAL SINUSES OPEN(S) INTO THE MIDDLE NASAL MEATUS?
    1. Frontal
    2. Middle ethmoidal
    3. Anterior ethmoidal
    4. Posterior ethmoidal

20. WHICH OF THE FOLLOWING STATEMENT(S) ABOUT A BRONCHOPULMONARY SEGMENT IS(ARE) FALSE?
    1. The term bronchopulmonary segment is applied to the largest segment within a lobe
    2. Surgical removal of a bronchopulmonary segment in disease is feasible
    3. Bronchopulmonary segments are separated from each other by connective tissue septa
    4. Subdivision of the lobes of the lungs into bronchopulmonary segments is useful anatomically, but is of little clinical importance

21. THE PARIETAL PLEURA IS SUPPLIED BY THE FOLLOWING ARTERIES, EXCEPT THE:
    1. Internal thoracic
    2. Superior phrenic
    3. Posterior intercostal
    4. Bronchial

22. THE LEFT PRIMARY BRONCHUS:
    1. Passes below the aortic arch
    2. Is shorter than the right
    3. Is located in front of the esophagus
    4. Lies above the left pulmonary artery

A	B	C	D	E
1,2,3	1,3	2,4	only 4	all correct

23. WHICH OF THE FOLLOWING STATEMENTS ABOUT THE LARYNX IS(ARE) CORRECT?
    1. The cricothyroid muscle is innervated by the inferior laryngeal nerve
    2. The mucous membrane is supplied by branches of the internal laryngeal nerve
    3. The thyroarytenoid muscle is innervated by the internal laryngeal nerve
    4. The vocal cords are abducted by the posterior cricoarytenoid muscle

24. THE DIAPHRAGM IS INNERVATED BY WHICH OF THE FOLLOWING NERVES?
    1. Vagus
    2. Phrenic
    3. Upper intercostals
    4. Thoracoabdominal

25. WHICH STATEMENT(S) ABOUT THE FISSURES AND LOBES OF THE LUNGS IS(ARE) FALSE?
    1. The right oblique fissure originates at the head of the fourth or fifth rib
    2. The horizontal fissure ends anteriorly at the level of the fourth costal cartilage
    3. The right middle lobe is triangular in shape
    4. The oblique fissure follows the line of the fifth rib

26. WHICH OF THE FOLLOWING STATEMENTS ABOUT THE PLEURA IS(ARE) FALSE?
    1. Is a mucous membrane
    2. Lines the mediastinum
    3. Parietal pleura is insensitive to pain
    4. Encloses a potential space

27. THE EXTERNAL BRANCH OF THE SUPERIOR LARYNGEAL NERVE SUPPLIES WHICH OF THE FOLLOWING MUSCLES?
    1. Arytenoid
    2. Vocalis
    3. Posterior cricoarytenoid
    4. Cricothyroid

28. WHICH OF THE FOLLOWING NERVES SUPPLY(IES) GENERAL SENSORY INNERVATION TO THE NASAL CAVITY?
    1. Anterior ethmoidal
    2. Posterior superior nasal
    3. Greater palatine
    4. Nasopalatine

29. THE INFERIOR LOBE OF THE LEFT LUNG CONSISTS OF THE FOLLOWING BRONCHOPULMONARY SEGMENTS:
    1. Superior
    2. Anterior basal
    3. Posterior basal
    4. Lateral basal

30. WHICH STATEMENT(S) ABOUT THE LARYNGOPHARYNX IS(ARE) CORRECT?
    1. Contains the piriform recess
    2. Continuous with the esophagus
    3. Receives sensory innervation from the glossopharyngeal nerve
    4. Related posteriorly to the fourth to sixth cervical vertebrae

FIVE-CHOICE ASSOCIATION QUESTIONS

DIRECTIONS: Each group of questions below consists of a numbered list of descriptive words or phrases accompanied by a diagram with certain parts indicated by letters, or by a list of lettered headings. For each numbered word or phrase, SELECT THE LETTERED PART OR HEADING that matches it correctly. Then on the appropriate line of the answer sheet, blacken the space under the letter of that part or heading. Sometimes more than one numbered word or phrase may be correctly matched to the same lettered part or heading.

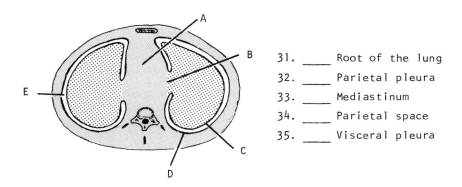

31. ____ Root of the lung
32. ____ Parietal pleura
33. ____ Mediastinum
34. ____ Parietal space
35. ____ Visceral pleura

Interpretation of Your Score

Number of Correct Responses	Level of Performance
32-35	Excellent - Exceptional
28-31	Superior - Very Superior
24-27	Average - Above Average
19-23	Poor - Marginal
18 or less	Very Poor - Failure

13. REVIEW EXAMINATION ON THE URINARY SYSTEM

INTRODUCTORY NOTE. The following 25 multiple-choice questions are based on the instructional objectives pertaining to the urinary system in Chapters Five and Six of this Study Guide & Review Manual. You should be able to answer these questions in 17-20 minutes. Before beginning, tear out an answer sheet from the back of the book and read the directions on how to use it. The key to the correct responses is on page 352.

FIVE-CHOICE COMPLETION QUESTIONS

DIRECTIONS: Each of the following questions or incomplete statements is followed by five suggested answers or completions. SELECT THE ONE BEST ANSWER in each case and then blacken the appropriate space on the answer sheet.

1. WHICH OF THE FOLLOWING SURFACES OF THE FEMALE URINARY BLADDER IS(ARE) COVERED BY PERITONEUM?
   A. Superior
   B. Posterior
   C. Lateral
   D. Anterior
   E. All of the above

2. WHICH STATEMENT REGARDING THE FEMALE URETER IS FALSE?
   A. Enters the true pelvis anterior to the sacroiliac joint
   B. Is related to the posterior border of the ovarian fossa
   C. Lies medial to the obturator vessels and nerve
   D. Crossed posteriorly by the uterine arteries
   E. Is approximately 25 cm long

3. THE NARROWEST PART OF THE MALE URETHRA IS THE:
   A. Membranous part
   B. External urethral orifice
   C. Internal urethral orifice
   D. Prostatic part
   E. None of the above

4. THE URINARY BLADDER IS INNERVATED BY:
   A. Afferent
   B. Somatic efferent
   C. Autonomic
   D. All of the above
   E. None of the above

5. ALL THE FOLLOWING STRUCTURES ARE LOCATED ANTERIOR TO THE RIGHT KIDNEY, EXCEPT THE:
   A. Suprarenal gland
   B. Duodenum
   C. Jejunum
   D. Colon
   E. Liver

6. THE FEMALE URETER IS SUPPLIED BY WHICH OF THE FOLLOWING ARTERIES?
   A. Uterine
   B. Internal iliac
   C. Common iliac
   D. Vesical
   E. All of the above

SELECT THE ONE BEST ANSWER

7. WHICH STATEMENT CONCERNING THE KIDNEYS IS FALSE?
   A. Both kidneys are located retroperitoneally
   B. Left kidney lies slightly lower than the right
   C. Has three capsules
   D. Left renal vein passes anterior to the aorta
   E. Lymph vessels drain directly into the para-aortic nodes

8. THE FEMALE URETHRA IS _____ CM LONG.
   A. 8-10
   B. 6-8
   C. 5-6
   D. 3-4
   E. 1-2

MULTI-COMPLETION QUESTIONS

DIRECTIONS: In each of the following questions or incomplete statements, ONE OR MORE of the completions given is correct. On the answer sheet blacken the space under A if 1, 2, and 3 are correct; B if 1 and 3 are correct; C if 2 and 4 are correct; D if only 4 is correct; and E if all are correct.

9. WHICH STRUCTURE(S) LIE(S) ANTERIOR TO THE LEFT KIDNEY?
   1. Pancreas
   2. Duodenum
   3. Jejunum
   4. Diaphragm

10. THE URINARY BLADDER:
    1. Has a capacity of approximately 500 ml
    2. Is lined by transitional epithelium
    3. Is pyramidal in shape when empty
    4. Has a trigone at its apex

11. THE MALE URINARY BLADDER IS:
    1. Separated from the pubic symphysis by a definite space
    2. Related to the rectum posteriorly
    3. Relatively immobile
    4. Not covered by peritoneum anteriorly

12. WHICH OF THE FOLLOWING ARTERIES SUPPLY(IES) THE FEMALE URETHRA?
    1. Common iliac
    2. Internal iliac
    3. Uterine
    4. Vesical

13. PERITONEUM COVERS WHICH OF THE FOLLOWING SURFACES OF THE FEMALE URINARY BLADDER?
    1. Anterior
    2. Posterior
    3. Lateral
    4. Superior

14. CONCERNING THE URETER:
    1. Abdominal part is twice as long as the pelvic part
    2. Less than 30 cm long
    3. Crosses the gonadal vessels anteriorly
    4. Intramural part is the narrowest

15. THE SPHINCTER URETHRAE MUSCLE IN THE FEMALE:
    1. Arises from the pubic bone
    2. Is inserted into the vaginal wall
    3. Forms a sphincter around the urethra
    4. Is supplied by the perineal nerve

A	B	C	D	E
1,2,3	1,3	2,4	only 4	all correct

16. WHICH OF THE FOLLOWING ARTERIES SUPPLY(IES) THE FEMALE URINARY BLADDER?
    1. Uterine
    2. Superior vesical
    3. Vaginal
    4. Inferior vesical

17. WHICH NERVE(S) IS(ARE) LOCATED POSTERIOR TO THE KIDNEYS?
    1. Subcostal
    2. Ilioinguinal
    3. Iliohypogastric
    4. Genitofemoral

18. THE PROSTATE GLAND IS:
    1. Traversed by the urethra
    2. Lobated
    3. Supplied by the inferior vesical artery
    4. Drained by lumbar lymph nodes

19. WHICH OF THE FOLLOWING STRUCTURES IS(ARE) PRESENT IN THE HILUM OF THE KIDNEY?
    1. Renal arteries
    2. Symphathetic fibers
    3. Ureter
    4. Lymph vessels

20. CONCERNING THE MALE URETHRA:
    1. Approximately 20 cm long
    2. Receives the ejaculatory ducts
    3. Prostatic part is the widest
    4. Passes below the symphysis pubis

FIVE-CHOICE ASSOCIATION QUESTIONS

DIRECTIONS: Each group of questions below consists of a numbered list of descriptive words or phrases accompanied by a diagram with certain parts indicated by letters, or by a list of lettered headings. For each numbered word or phrase, SELECT THE LETTERED PART OR HEADING that matches it correctly. Then on the appropriate line of the answer sheet, blacken the space under the letter of that part or heading. Sometimes more than one numbered word or phrase may be correctly matched to the same lettered part or heading.

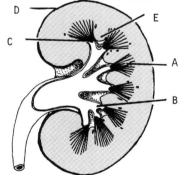

21. _____ Minor calyx
22. _____ Fibromuscular capsule
23. _____ Renal column
24. _____ Pyramid
25. _____ Papilla

Interpretation of Your Score

Number of Correct Responses	Level of Performance
23-25	Excellent - Exceptional
20-22	Superior - Very Superior
17-19	Average - Above Average
14-16	Poor - Marginal
13 or less	Very Poor - Failure

14. REVIEW EXAMINATION ON THE NERVOUS SYSTEM

INTRODUCTORY NOTE. The following 35 multiple-choice questions are based on the instructional objectives pertaining to the nervous system in Chapters One to Seven of this Study Guide & Review Manual. You should be able to answer these questions in 25-28 minutes. Before beginning, tear out an answer sheet from the back of the book and read the directions on how to use it. The key to the correct responses is on page 352.

FIVE-CHOICE COMPLETION QUESTIONS

DIRECTIONS: Each of the following questions or incomplete statements is followed by five suggested answers or completions. SELECT THE ONE BEST ANSWER in each case and then blacken the appropriate space on the answer sheet.

1. WHICH MUSCLE IS INNERVATED BY BOTH THE ULNAR AND MEDIAN NERVES?
   A. Pronator teres
   B. Pronator quadratus
   C. Flexor digitorum profundus
   D. Flexor digitorum superficialis
   E. None of the above

2. THE MANDIBULAR DIVISION OF THE TRIGEMINAL NERVE PASSES THROUGH WHICH FORAMEN?
   A. Ovale
   B. Spinosum
   C. Rotundum
   D. Lacerum
   E. Stylomastoid

3. THE GREATER SPLANCHNIC NERVE IS USUALLY DERIVED FROM WHICH OF THE FOLLOWING THORACIC GANGLIA?
   A. 1st - 5th
   B. 5th - 9th
   C. 9th - 10th
   D. Below the 10th
   E. All of the above

4. WHICH STATEMENT(S) ABOUT THE SPINAL CORD IS (ARE) CORRECT?
   A. Has a cervical enlargement
   B. Is flattened and cylindrical
   C. Averages 45 cm in length
   D. Ends between L1 and 2
   E. All of the above

5. THE MAJOR MOTOR SIGN IN PARALYSIS OF THE QUADRICEPS FEMORIS MUSCLE IS LOSS OF _____.
   A. Lateral rotation of the leg
   B. Extension of the thigh
   C. Adduction of the thigh
   D. Flexion of the leg
   E. Extension of the leg

6. PREGANGLIONIC SYMPATHETIC FIBERS TO THE CERVICAL SYMPATHETIC GANGLIA ARE USUALLY DERIVED FROM WHICH SEGMENTS OF THE SPINAL CORD?
   A. T1-T4
   B. C8-T3
   C. T1-L3
   D. C1-T5
   E. C2-C6

SELECT THE ONE BEST ANSWER

7. THE LONG THORACIC NERVE INNERVATES THE _____ MUSCLE.
   A. Trapezius
   B. Rhomboid major
   C. Serratus anterior
   D. Pectoralis major
   E. Latissimus dorsi

8. THE RIGHT VAGUS NERVE CONTRIBUTES SIGNIFICANTLY TO WHICH OF THE FOLLOWING PLEXUSES?
   A. Esophageal
   B. Cardiac
   C. Pulmonary
   D. All of the above
   E. None of the above

9. THE SUPERIOR GLUTEAL NERVE INNERVATES ALL THE FOLLOWING EXCEPT THE:
   A. Gluteus medius muscle
   B. Gluteus maximus muscle
   C. Gluteus minimus muscle
   D. Hip joint
   E. Tensor fasciae latae muscle

10. THE FACIAL NERVE INNERVATES ALL THE FOLLOWING MUSCLES EXCEPT:
    A. Orbicularis oris
    B. Platysma
    C. Buccinator
    D. Levator labii superioris
    E. Lateral pterygoid

11. THE LUNGS ARE INNERVATED BY ALL OF THE FOLLOWING EXCEPT THE:
    A. Anterior pulmonary plexus
    B. Posterior pulmonary plexus
    C. Thoracic sympathetic
    D. Vagus nerve
    E. Phrenic nerve

12. AN INJURY TO THE UPPER LATERAL MARGIN OF THE POPLITEAL FOSSA IS MOST LIKELY TO DAMAGE THE _____ NERVE.
    A. Genicular branch of the obturator
    B. Posterior femoral cutaneous
    C. Tibial
    D. Common peroneal
    E. Sciatic

MULTI-COMPLETION QUESTIONS

DIRECTIONS: In each of the following questions or incomplete statements, ONE OR MORE of the completions given is correct. On the answer sheet blacken the space under A if 1, 2, and 3 are correct; B if 1 and 3 are correct; C if 2 and 4 are correct; D if only 4 is correct; and E if all are correct.

13. THE AXILLARY NERVE INNERVATES THE FOLLOWING MUSCLE(S):
    1. Teres major
    2. Deltoid
    3. Serratus anterior
    4. Teres minor

14. WHICH OF THE FOLLOWING MUSCLES IS(ARE) INNERVATED BY THE ANSA CERVICALIS?
    1. Sternohyoid
    2. Sternothyroid
    3. Thyrohyoid
    4. Geniohyoid

15. THE DIAPHRAGM IS INNERVATED BY WHICH OF THE FOLLOWING NERVES?
    1. Vagus
    2. Phrenic
    3. Upper 3 intercostals
    4. Thoracoabdominal

A	B	C	D	E
1,2,3	1,3	2,4	only 4	all correct

16. THE OBTURATOR NERVE SUPPLIES THE:
    1. Adductor brevis muscle
    2. Quadriceps femoris
    3. Obturator externus muscle
    4. Skin of the anterior thigh

17. THE LUMBAR PLEXUS OF NERVES:
    1. Is located within the psoas major muscle
    2. Is formed by the anterior primary rami of L2-L4
    3. Gives rise to the femoral and obturator nerves
    4. Contributes to the lumbosacral trunk

18. THE STRUCTURE(S) CROSSING THE WRIST AND ENTERING THE PALM SUPERFICIAL TO THE FLEXOR RETINACULUM IS(ARE) THE:
    1. Ulnar nerve
    2. Flexor pollicis longus muscle
    3. Superficial palmar branch of the radial artery
    4. Median nerve

19. A PATIENT, DIAGNOSED TO HAVE NEURITIS OF THE MEDIAN NERVE AT THE WRIST, WOULD PRESENT THE FOLLOWING SYMPTOM(S):
    1. Paresthesia of the little finger
    2. Pain in the index finger
    3. Weakness in opposition of the little finger
    4. Weakness in abduction of the thumb

20. THE SKIN OF THE HEAD, FACE, AND NECK IS INNERVATED BY BRANCHES OF THE:
    1. Trigeminal nerve
    2. Dorsal rami of cervical nerves
    3. Cervical plexus
    4. Facial nerve

21. WHICH STATEMENT(S) ABOUT THE PUDENDAL NERVE IS(ARE) CORRECT?
    1. Has anterior scrotal or labial branches
    2. Lies in the deep perineal space
    3. Enters the perineum through the greater sciatic foramen
    4. Arises from S2, 3, and 4

22. WHICH STATEMENT(S) ABOUT THE SCIATIC NERVE IS(ARE) TRUE?
    1. Divides to form tibial and peroneal nerves
    2. Enters the gluteal region above the piriformis muscle
    3. Is the largest nerve in the body
    4. Supplies the gluteus maximus muscle

23. WHICH OF THE FOLLOWING MUSCLES IS(ARE) NOT INNERVATED BY THE MEDIAN NERVE?
    1. Abductor pollicis brevis
    2. Flexor pollicis brevis
    3. Opponens pollicis
    4. Adductor pollicis

A	B	C	D	E
1,2,3	1,3	2,4	only 4	all correct

24. THE LUMBAR SYMPATHETIC TRUNK:
    1. Runs along the anterior border of the psoas major muscle
    2. Is covered by the inferior vena cava on the right
    3. Begins beneath the medial arcuate ligament
    4. Has five ganglia

25. CONCERNING NERVES OF THE STOMACH:
    1. The right vagus forms a major part of the posterior gastric nerve
    2. The left vagus forms a major part of the anterior gastric nerve
    3. Division of the vagus nerves reduces gastric acidity
    4. Contain only parasympathetic fibers

26. THE LATISSIMUS DORSI MUSCLE IS SUPPLIED BY WHICH OF THE FOLLOWING NERVES?
    1. Lower subscapular
    2. Suprascapular
    3. Upper subscapular
    4. Thoracodorsal

27. MUSCLES INNERVATED BY MOTOR FIBERS OF THE FACIAL NERVE ARE THE:
    1. Anterior belly of digastric
    2. Orbicularis oculi
    3. Mylohyoid
    4. Buccinator

28. WHICH MUSCLE(S) IS(ARE) INNERVATED BY THE DEEP PERONEAL NERVE?
    1. Tibialis anterior
    2. Extensor digitorum longus
    3. Extensor hallucis longus
    4. Peroneus tertius

29. THE CELIAC GANGLIA ARE:
    1. Situated on the psoas major muscle
    2. Two in number
    3. Supplied by preganglionic fibers from the third and fourth lumbar sympathetic ganglia
    4. Located in relation to the celiac artery

30. THE LEFT LUMBAR SYMPATHETIC TRUNK:
    1. May be operated on to relieve vasospastic disease of the left lower limb
    2. Lies posterior to the abdominal aorta
    3. Runs along the medial border of the left psoas major muscle
    4. Is related posteriorly to the left renal vessels

FIVE-CHOICE ASSOCIATION QUESTIONS

DIRECTIONS: Each group of questions below consists of a numbered list of descriptive words or phrases accompanied by a diagram with certain parts indicated by letters, or by a list of lettered headings. For each numbered word or phrase, SELECT THE LETTERED PART OR HEADING that matches it correctly. Then on the appropriate line of the answer sheet, blacken the space under the letter of that part or heading. Sometimes more than one numbered word or phrase may be correctly matched to the same part or heading.

ASSOCIATION QUESTIONS

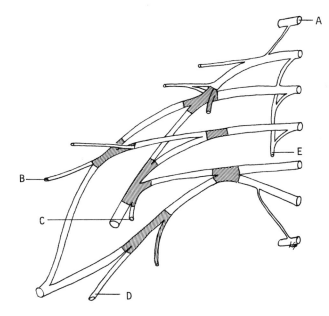

31. ___ A nerve when paralysed results in winging of the scapula

32. ___ An anterior primary ramus that contributes primarily to innervation of the diaphragm

33. ___ A motor nerve innervating muscles of the front of the arm

34. ___ Injury to this nerve results in claw-hand

35. ___ A lesion of this nerve results in loss of abduction at the shoulder joint

Interpretation of Your Score

Number of Correct Responses	Level of Performance
32-35	Excellent - Exceptional
28-31	Superior - Very Superior
24-27	Average - Above Average
19-23	Poor - Marginal
18 or less	Very Poor - Failure

PART THREE

REVIEW EXAMINATIONS ON THE REGIONS

15. REVIEW EXAMINATION ON THE UPPER LIMB

INTRODUCTORY NOTE. The following 50 multiple-choice questions are based on the instructional objectives for the upper limb listed in Chapter One of this Study Guide & Review Manual. You should be able to answer these questions in 35-40 minutes. Before beginning, tear out an answer sheet from the back of the book and read the directions on how to use it. The key to the correct responses is on page 352.

FIVE-CHOICE COMPLETION QUESTIONS

DIRECTIONS: Each of the following questions or incomplete statements is followed by five suggested answers or completions. SELECT THE ONE BEST ANSWER in each case and then blacken the appropriate space on the answer sheet.

1. THE FIRST AND SECOND LUMBRICAL MUSCLES ARE INNERVATED BY THE _____ NERVE.
   A. Deep ulnar
   B. Deep radial
   C. Median
   D. Anterior Interosseous
   E. Superficial radial

2. WHERE IS THE ULNAR NERVE MOST COMMONLY DAMAGED?
   A. At the elbow
   B. At the wrist
   C. In the axilla
   D. In the forearm
   E. In the midarm

3. WHICH MUSCLE(S) CONTRIBUTE(S) TO THE STABILITY OF THE SHOULDER JOINT?
   A. Infraspinatus
   B. Supraspinatus
   C. Subscapularis
   D. Teres minor
   E. All of the above

4. WHICH STRUCTURE CROSSES SUPERFICIAL TO THE FLEXOR RETINACULUM?
   A. Palmaris longus
   B. Flexor pollicis longus
   C. Median nerve
   D. Flexor digitorum profundus
   E. Flexor digitorum superficialis

5. THE MUSCLE INNERVATED BY THE LONG THORACIC NERVE IS THE:
   A. Trapezius
   B. Rhomboid major
   C. Serratus anterior
   D. Pectoralis major
   E. Latissimus dorsi

6. UPON EXAMINATION OF A CHILD WITH A DEEP CUT ON HER PALM, YOU FIND THAT SHE IS UNABLE TO ADDUCT HER THUMB. WHICH NERVE HAS MOST LIKELY BEEN SEVERED?
   A. Recurrent motor branch of the median
   B. Superficial ulnar
   C. Superficial radial
   D. Deep ulnar
   E. Deep radial

SELECT THE ONE BEST ANSWER

7. A RADIOGRAPH OF A YOUNG CHILD SHOWS A FRACTURE OF THE CLAVICLE AT THE JUNCTION OF THE LATERAL AND MIDDLE THIRDS. THE MEDIAL AND LATERAL FRAGMENTS ARE TILTED UPWARD AND DOWNWARD, RESPECTIVELY. UPWARD DISPLACEMENT OF THE MEDIAL HALF IS CAUSED BY WHICH OF THE FOLLOWING MUSCLES?
   A. Deltoid
   B. Trapezius
   C. Subclavius
   D. Sternocleidomastoid
   E. Pectoralis major

8. THE DISTAL END OF THE RADIUS USUALLY ARTICULATES WITH THE _____ BONES.
   A. Trapezium and trapezoid
   B. Scaphoid and lunate
   C. Capitate and hamate
   D. Trapezoid and scaphoid
   E. Lunate and trapezoid

9. THE BICEPS BRACHII MUSCLE IS A STRONG _____ OF THE FOREARM.
   A. Extensor
   B. Adductor
   C. Pronator
   D. Supinator
   E. Circumductor

10. THE PRONATOR TERES MUSCLE IS INNERVATED BY A BRANCH OF THE _____ NERVE.
    A. Musculocutaneous
    B. Ulnar
    C. Axillary
    D. Radial
    E. Median

11. WHICH MUSCLE IS INNERVATED BY BOTH THE ULNAR AND MEDIAN NERVES?
    A. Flexor digitorum superficialis
    B. Pronator teres
    C. Pronator quadratus
    D. Flexor digitorum profundus
    E. None of the above

12. THE STRUCTURE INDICATED BY THE ARROW IS THE:
    A. Radial notch
    B. Radial fossa
    C. Trochlea
    D. Coronoid fossa
    E. Trochlear notch

13. THE CORDS OF THE BRACHIAL PLEXUS OF NERVES ARE NAMED ACCORDING TO THEIR RELATIONSHIP TO:
    A. The first part of the axillary artery
    B. The second part of the axillary artery
    C. The third part of the axillary artery
    D. The subclavian artery

14. THE AXILLARY ARTERY TERMINATES AT THE:
    A. Lower border of pectoralis major muscle
    B. Lower border of teres major muscle
    C. Upper border of pectoralis minor muscle
    D. Outer border of the second rib
    E. Upper border of subscapularis muscle

SELECT THE ONE BEST ANSWER

15. WHICH MUSCLE DOES NOT FORM A BOUNDARY OF THE AXILLA?
    A. Biceps brachii
    B. Pectoralis major
    C. Intercostal
    D. Latissimus dorsi
    E. Brachialis

16. ABDUCTION OF THE THUMB IS WHEN IT IS MOVED:
    A. Laterally from the palm
    B. At right angles to the palmar plane
    C. Along side the index finger
    D. Straight across the palm
    E. None of the above

17. THE NERVE OF THE BRACHIAL PLEXUS DERIVED FROM THE VENTRAL PRIMARY RAMI OF SPINAL NERVES C7, 8 and T1 IS THE:
    A. Musculocutaneous
    B. Axillary
    C. Radial
    D. Median
    E. Ulnar

18. THE LATISSIMUS DORSI MUSCLE _____ _____ THE HUMERUS.
    A. Flexes
    B. Laterally rotates
    C. Abducts
    D. Adducts
    E. None of the above

19. WHICH OF THE FOLLOWING STATEMENTS IS NOT TRUE FOR THE TRICEPS BRACHII MUSCLE?
    A. Originates from the infraglenoid tubercle
    B. Extends the forearm
    C. Innervated by the radial nerve
    D. Inserts on the olecranon process
    E. Originates from the scapular spine

20. WHICH OF THE FOLLOWING STATEMENTS IS TRUE FOR THE PECTORALIS MAJOR MUSCLE?
    A. Arises from the humerus
    B. Inserts into the ribs
    C. Abducts the arm
    D. Innervated by the median nerve
    E. None of the above

MULTI-COMPLETION QUESTIONS

DIRECTIONS: In each of the following questions or incomplete statements, ONE OR MORE of the completions given is correct. On the answer sheet, blacken the space under A if 1, 2 and 3 are correct; B if 1 and 3 are correct; C if 2 and 4 are correct; D if only 4 is correct; and E if all are correct.

21. WHICH STATEMENT(S) ABOUT THE DELTOID MUSCLE IS(ARE) CORRECT?
    1. A lateral rotator of the arm
    2. An important stabilizer of the shoulder joint
    3. A medial rotator of the arm
    4. Innervated by C7 and 8 fibers

22. A BOY FELL ON HIS ELBOW WITH HIS ARM ABDUCTED, PRODUCING A FRACTURE OF THE SURGICAL NECK OF THE HUMERUS. THIS RESULTED IN ELEVATION AND ADDUCTION OF THE DISTAL FRAGMENT. THE ADDUCTION IS CAUSED BY WHICH MUSCLE(S)?
    1. Pectoralis major
    2. Teres major
    3. Latissimus dorsi
    4. Coracobrachialis

23. WHICH NERVE(S) ARISE(S) FROM THE POSTERIOR CORD OF THE BRANCHIAL PLEXUS?
    1. Thoracodorsal
    2. Dorsal scapular
    3. Subscapular
    4. Suprascapular

A	B	C	D	E
1,2,3	1,3	2,4	only 4	all correct

24. THE CONTENTS OF THE CUBITAL FOSSA ARE THE:
    1. Medial nerve
    2. Radial artery
    3. Radial nerve
    4. Ulnar nerve

25. WHICH MUSCLES ARE ATTACHED TO THE CLAVICLE?
    1. Subclavius
    2. Deltoid
    3. Pectoralis major
    4. Trapezius

26. WHICH MUSCLE(S) IS(ARE) INNERVATED BY THE MEDIAN NERVE?
    1. Opponens pollicis
    2. Abductor pollicis brevis
    3. Flexor pollicis brevis
    4. Adductor pollicis

27. WHICH STRUCTURE(S) IS(ARE) ASSOCIATED WITH THE "ANATOMICAL SNUFF-BOX"?
    1. Scaphoid bone
    2. Median nerve
    3. Radial artery
    4. Lunate bone

28. THE ATTACHMENT OF THE TRAPEZIUS MUSCLE IS TO THE FOLLOWING:
    1. Posterior aspect of the lateral third of the clavicle
    2. Superior lip of the spine of the scapula
    3. The triangular surface at the medial end of the spine of the scapula
    4. Lateral border of the acromial process and inferior lip of the spine of the scapula

29. THE HEAD OF THE RADIUS ARTICULATES WITH THE:
    1. Radial notch of the ulna
    2. Scaphoid
    3. Humerus
    4. Lunate

30. WHICH OF THE FOLLOWING NERVES IS(ARE) LOCATED WITHIN THE CUBITAL FOSSA?
    1. Musculocutaneous
    2. Axillary
    3. Ulnar
    4. Median

31. THE FOLLOWING MUSCLE(S) IS(ARE) INNERVATED BY THE AXILLARY NERVE:
    1. Teres major
    2. Deltoid
    3. Serratus anterior
    4. Teres minor

32. THE PECTORALIS MAJOR MUSCLE ORIGINATES FROM THE:
    1. Anterior aspect of the sternum
    2. Aponeurosis of the external oblique muscle
    3. Anterior aspect of the medial portion of the clavicle
    4. Upper six bony ribs

33. THE BRACHIAL ARTERY:
    1. Terminates as the radial and ulnar arteries
    2. Bifurcates opposite the neck of the humerus
    3. Lies medial to the humerus proximally
    4. In the cubital fossa is lateral to the biceps tendon

34. THE MUSCLE(S) FORMING THE POSTERIOR WALL OF THE AXILLA IS(ARE) THE:
    1. Subscapularis
    2. Teres minor
    3. Latissimus dorsi
    4. Pectoralis major

35. WHICH OF THE FOLLOWING TENDONS FORM(S) A BOUNDARY OF THE "ANATOMICAL SNUFF-BOX"?
    1. Extensor carpi radialis longus
    2. Extensor pollicis longus
    3. Extensor indicis
    4. Extensor pollicis brevis

A	B	C	D	E
1,2,3	1,3	2,4	only 4	all correct

36. BRACHIORADIALIS MUSCLE:
    1. Inserts into the base of the styloid process of the radius
    2. Acts only at the elbow joint
    3. Is innervated by the radial nerve
    4. Originates from the lower one-third of the lateral supracondylar ridge

37. THE FLEXOR RETINACULUM IS ATTACHED LATERALLY TO WHICH OF THE FOLLOWING BONE(S)?
    1. Pisiform
    2. Scaphoid
    3. Triquetrum
    4. Trapezium

38. NEURITIS OF THE MEDIAN NERVE AT THE WRIST WOULD PRESENT THE FOLLOWING SYMPTOM(S):
    1. Weakness in opposition of the little finger
    2. Weakness in abduction of the thumb
    3. Paresthesia of the little finger
    4. Pain in the index finger

39. WHICH OF THE FOLLOWING STATEMENTS ABOUT THE CLAVICLE IS(ARE) TRUE?
    1. Develops in membrane
    2. Its fulcrum of movement is at the costoclavicular ligament
    3. Commonly fractured bone
    4. Last bone to begin ossification in the embryo

40. WHICH BONE(S) IS(ARE) IN THE PROXIMAL ROW OF CARPAL BONES?
    1. Hamate
    2. Scaphoid
    3. Trapezium
    4. Pisiform

FIVE-CHOICE ASSOCIATION QUESTIONS

DIRECTIONS: Each group of questions below consists of a numbered list of descriptive words or phrases accompanied by a diagram with certain parts indicated by letters, or by a list of lettered headings. For each numbered word or phrase, SELECT THE LETTERED PART OR HEADING that matches it correctly. Then blacken the appropriate space on the answer sheet. Sometimes more than one numbered word or phrase may be correctly matched to the same lettered part or heading.

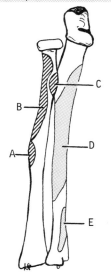

41. ___ Suprascapular artery
42. ___ Subscapular artery
43. ___ Circumflex scapular artery
44. ___ Dorsal scapular artery
45. ___ Posterior circumflex humeral artery

ASSOCIATION QUESTIONS

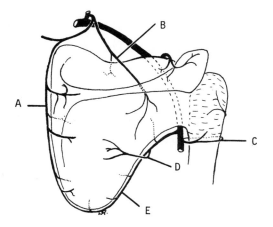

46. ____ Origin of a muscle which flexes the distal phalanx of a finger
47. ____ Insertion of a forearm pronator
48. ____ Insertion of a muscle innervated by the musculocutaneous nerve
49. ____ Origin of a pronator of the forearm
50. ____ Insertion of a supinator which originates from the ulna

Interpretation of Your Score

Number of Correct Responses	Level of Performance
43-50	Excellent - Exceptional
38-42	Superior - Very Superior
32-37	Average - Above Average
28-31	Poor - Marginal
27 or less	Very Poor - Failure

16. REVIEW EXAMINATION ON THE LOWER LIMB

INTRODUCTORY NOTE. The following 50 multiple-choice questions are based on the instructional objectives for the lower limb listed in Chapter Two of this Study Guide & Review Manual. You should be able to answer these questions in 35-40 minutes. Before beginning, tear out an answer sheet from the back of the book and read the directions on how to use it. The key to the correct responses is on page 353.

FIVE-CHOICE COMPLETION QUESTIONS

DIRECTIONS: Each of the following questions or incomplete statements is followed by five suggested answers or completions. SELECT THE ONE BEST ANSWER in each case and then blacken the appropriate space on the answer sheet.

1. WHICH OF THE FOLLOWING MUSCLES IS NOT A DORSIFLEXOR OF THE FOOT?
   A. Extensor digitorum longus
   B. Extensor digitorum brevis
   C. Extensor hallucis longus
   D. Tibialis anterior
   E. Peroneus tertius

2. WHICH MUSCLE(S) IS(ARE) INVOLVED IN INVERSION OF THE FOOT?
   A. Peroneus longus
   B. Extensor hallucis longus
   C. Flexor hallucis longus
   D. Tibialis posterior
   E. All of the above

3. THE GRACILIS MUSCLE IS A(AN) _____ OF THE THIGH AT THE HIP JOINT.
   A. Lateral rotator
   B. Medial rotator
   C. Adductor
   D. Abductor
   E. Flexor

4. WHICH MUSCLE IS NOT INVOLVED IN FLEXION OF THE THIGH?
   A. Iliacus
   B. Sartorius
   C. Psoas major
   D. Rectus femoris
   E. Obturator externus

5. THE HEEL OF THE FOOT RECEIVES ITS SENSORY INNERVATION FROM THE _____ NERVE.
   A. Lateral plantar
   B. Medial plantar
   C. Saphenous
   D. Tibial
   E. Sural

6. THE _____ MUSCLE MAINTAINS THE TRANSVERSE ARCH.
   A. Abductor hallucis longus
   B. Flexor digitorum brevis
   C. Abductor digiti minimi
   D. Adductor hallucis
   E. Peroneus brevis

SELECT THE ONE BEST ANSWER

7. THE STRUCTURE ATTACHED TO THE INTERCONDYLAR FOSSA OF THE FEMUR IS THE:
   A. Anterior cruciate ligament
   B. Posterior cruciate ligament
   C. Popliteus tendon
   D. Transverse ligament
   E. None of the above

8. WHICH STRUCTURE IS NOT A STABILIZER OF THE KNEE JOINT ON ITS LATERAL ASPECT?
   A. Iliotibial tract
   B. Popliteus tendon
   C. Biceps tendon
   D. Transverse ligament
   E. Fibular collateral ligament

9. THE MEDIAL ARCH IS SPANNED BY THE _____ MUSCLE.
   A. Peroneus longus
   B. Tibialis posterior
   C. Abductor hallucis
   D. Peroneus brevis
   E. Flexor hallucis longus

10. THE BASE OF THE FEMORAL TRIANGLE IS BOUNDED BY THE:
    A. Sartorius muscle
    B. Inguinal ligament
    C. Adductor longus muscle
    D. Femoral sheath
    E. None of the above

11. THE SUPERIOR GLUTEAL NERVE INNERVATES ALL STRUCTURES EXCEPT THE:
    A. Gluteus minimus muscle
    B. Gluteus maximus muscle
    C. Gluteus medius muscle
    D. Tensor fasciae latae muscle
    E. Hip joint

12. THE LIGAMENT OF THE HIP JOINT RESISTING EXTENSION IS THE:
    A. Ligament of the head
    B. Ischiofemoral
    C. Pubofemoral
    D. Transverse
    E. Iliofemoral

13. THE SACRAL PLEXUS OF NERVES IS DERIVED FROM THE ANTERIOR PRIMARY RAMI OF SPINAL NERVES:
    A. L2-5, S1,2
    B. L3,4, S1,2
    C. L3,4, S1-3
    D. L4,5, S1-4
    E. None of the above

14. THE MEDIAL COMPARTMENT OF THE FEMORAL SHEATH CONTAINS THE:
    A. Femoral artery
    B. Femoral canal
    C. Femoral vein
    D. Femoral nerve
    E. None of the above

15. A MUSCLE OF THE POSTERIOR COMPARTMENT OF THE LEG INSERTING INTO THE TENDO-CALCANEUS IS THE:
    A. Soleus
    B. Popliteus
    C. Tibialis posterior
    D. Flexor hallucis longus
    E. Flexor digitorum longus

16. WHICH OF THESE ARTERIES IS NOT A BRANCH OF THE FEMORAL?
    A. Superficial circumflex iliac
    B. Superficial epigastric
    C. Inferior epigastric
    D. Profunda femoris
    E. None of the above

17. THE MAJOR BURSA COMMUNICATING WITH THE SYNOVIAL JOINT SPACE OF THE KNEE IS THE:
    A. Infrapatellar
    B. Suprapatellar
    C. Prepatellar
    D. Semimembranosus
    E. None of the above

SELECT THE ONE BEST ANSWER

18. THE PERONEAL ARTERY ARISES FROM THE _____ ARTERY:
    A. Femoral
    B. Popliteal
    C. Posterior tibial
    D. Anterior tibial
    E. None of the above

19. THE FOLLOWING MUSCLES ARE EXTENSORS OF THE THIGH EXCEPT THE:
    A. Adductor magnus
    B. Biceps femoris
    C. Semitendinosus
    D. Adductor longus
    E. Gluteus maximus

20. AN INJURY TO THE UPPER LATERAL MARGIN OF THE POPLITEAL FOSSA IS MOST LIKELY TO DAMAGE THE _____ NERVE.
    A. Sciatic
    B. Tibial
    C. Common peroneal
    D. Posterior femoral cutaneous
    E. Genicular branch of the obturator

MULTI-COMPLETION QUESTIONS

DIRECTIONS: In each of the following questions or incomplete statements, ONE OR MORE of the completions given is correct. On the answer sheet blacken the space under A if 1, 2, and 3 are correct; B if 1 and 3 are correct; C if 2 and 4 are correct; D if only 4 is correct; and E if all are correct.

21. THE FLOOR OF THE FEMORAL TRIANGLE IS FORMED BY THE _____ MUSCLE(S).
    1. Iliopsoas
    2. Pectineus
    3. Adductor longus
    4. Rectus femoris

22. THE FEMORAL SHEATH IS FORMED BY THE:
    1. Fascia transversalis
    2. Cribriform fascia
    3. Fascia iliaca
    4. Fascia lata

23. THE SCIATIC NERVE:
    1. Lies deep to the gluteus maximus muscle
    2. Receives most of its fibers from L4, 5, S,2 and 3
    3. Leaves the pelvis via the greater sciatic notch
    4. Enters the thigh deep to the adductor magnus muscle

24. THE HEAD OF THE FEMUR IS SUPPLIED BY WHICH OF THE FOLLOWING ARTERIES?
    1. Superior gluteal
    2. Inferior gluteal
    3. Medial femoral circumflex
    4. Lateral femoral circumflex

25. WHICH STATEMENT(S) IS(ARE) CORRECT CONCERNING THE LATERAL ROTATORS OF THE HIP?
    1. Each is innervated by a branch of the sacral plexus
    2. Lie posterior to the hip joint
    3. Are inserted into the lesser trochanter of the femur
    4. Are covered by the gluteus maximus muscle

26. THE LATERAL LONGITUDINAL ARCH OF THE FOOT IS FORMED BY THE:
    1. Calcaneus
    2. Talus
    3. Cuboid
    4. Navicular

A	B	C	D	E
1,2,3	1,3	2,4	only 4	all correct

27. THE POPLITEUS MUSCLE:
    1. Inserts into the fibula
    2. Is located posterior to the knee joint
    3. Is innervated by the femoral nerve
    4. Unlocks the knee joint at the beginning of flexion

28. THE FOLLOWING MOVEMENTS OCCUR AT THE ANKLE JOINT:
    1. Plantar flexion
    2. Eversion
    3. Dorsiflexion
    4. Inversion

29. THE QUADRICEPS FEMORIS MUSCLE CONSISTS OF WHICH OF THE FOLLOWING MUSCLES?
    1. Rectus femoris
    2. Iliacus
    3. Vastus lateralis
    4. Biceps femoris

30. THE CALCANEUS OF THE FOOT ARTICULATES WITH THE _____ BONE(S).
    1. Fibula
    2. Cuboid
    3. First cuneiform
    4. Talus

31. WHICH STATEMENT(S) IS(ARE) CORRECT FOR THE ADDUCTOR CANAL?
    1. Contains the saphenous vein
    2. The vastus medialis muscle lies lateral to it
    3. Contains the femoral nerve
    4. Located on the medial aspect of the thigh

32. WHICH STATEMENT(S) IS(ARE) CORRECT FOR THE LUMBAR PLEXUS OF NERVES?
    1. Forms within the psoas major muscle
    2. Femoral nerve lies medial to the psoas major muscle
    3. A major branch is formed from contributions of L2,3, and 4
    4. Has a medial femoral cutaneous branch

33. THE LATERAL ROTATOR(S) OF THE THIGH IS (ARE) THE _____ MUSCLE(S).
    1. Adductor longus
    2. Adductor brevis
    3. Adductor magnus
    4. Pectineus

34. WHICH STATEMENT(S) ABOUT A FEMORAL HERNIA IS(ARE) CORRECT?
    1. The neck of the hernia lies deep to the inguinal ligament
    2. Lies below and lateral to the pubic tubercle
    3. Passes through the femoral canal
    4. Is more common in men than in women

35. LIGAMENTS ATTACHING THE TIBIA TO THE FIBULA ARE:
    1. Anterior tibiofibular
    2. Posterior tibiofibular
    3. Interosseous
    4. Deltoid

36. A MAN SEVERED HIS TIBIAL NERVE ABOVE THE ANKLE ON THE POSTERIOR ASPECT OF THE TIBIA. WHICH OF THE FOLLOWING SIGNS WOULD BE PRESENT?
    1. Sensory loss on the sole of the foot
    2. Sensory loss on the dorsum of the foot
    3. Clawing of the toes
    4. Foot drop

37. THE MUSCLE(S) MEDIAL TO THE KNEE JOINT IS(ARE) THE:
    1. Semimembranosus
    2. Semitendinosus
    3. Sartorius
    4. Gracilis

A	B	C	D	E
1,2,3	1,3	2,4	only 4	all correct

38. WHICH STRUCTURE(S) FORM(S) THE FLOOR OF THE POPLITEAL FOSSA?
    1. Oblique popliteal ligament
    2. Popliteus muscle
    3. Femur
    4. Plantaris muscle

39. THE INGUINAL GROUP OF LYMPH NODES DRAIN WHICH OF THE FOLLOWING AREAS?
    1. External genitalia
    2. Lower limb
    3. Gluteal region
    4. Abdominal wall

40. THE POPLITEAL FOSSA IS BOUNDED SUPERIORLY BY WHICH OF THE FOLLOWING MUSCLES?
    1. Semitendinosus
    2. Semimembranosus
    3. Biceps femoris
    4. Gracilis

41. THE ANTERIOR CRUCIATE LIGAMENT OF THE KNEE:
    1. Prevents backward dislocation of the femur
    2. Attaches to the lateral surface of the medial condyle of the femur
    3. Acquires its name from its tibial attachment
    4. Is partly extracapsular

42. THE HAMSTRING MUSCLES ARE:
    1. Attached to the ischial tuberosity
    2. Extensors of the thigh
    3. Flexors of the leg
    4. Extensors of the trunk

43. INVERTOR(S) OF THE FOOT IS(ARE) THE FOLLOWING MUSCLE(S):
    1. Tibialis posterior
    2. Peroneus tertius
    3. Tibialis anterior
    4. Plantaris

44. WHICH STATEMENT(S) IS(ARE) TRUE FOR THE GLUTEUS MEDIUS MUSCLE?
    1. Is innervated by the inferior gluteal nerve
    2. Inserts into the iliotibial tract
    3. Originates from the posterior iliac crest
    4. Paralysis of it leads to a waddling gait

45. WHICH MUSCLE(S) IS(ARE) DORSIFLEXORS OF THE FOOT?
    1. Peroneus tertius
    2. Extensor digitorum brevis
    3. Tibialis anterior
    4. Peroneus brevis

46. WHICH MUSCLE(S) IS(ARE) INNERVATED BY THE DEEP PERONEAL NERVE?
    1. Tibialis anterior
    2. Peroneus tertius
    3. Extensor digitorum longus
    4. Peroneus brevis

FIVE-CHOICE ASSOCIATION QUESTIONS

DIRECTIONS: Each group of questions below consists of a numbered list of descriptive words or phrases accompanied by a diagram with certain parts indicated by letters, or by a list of lettered headings. For each numbered word or phrase, SELECT THE LETTERED PART OR HEADING that matches it correctly. Then on the appropriate line of the answer sheet, blacken the space under the letter of that part or heading. Sometimes more than one numbered word or phrase may be correctly matched to the same lettered part or heading.

ASSOCIATION QUESTIONS

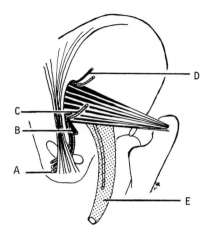

47. _____ Nerve to the gluteus maximus
48. _____ Inferior gluteal artery
49. _____ Nerve to the obturator internus
50. _____ Nerve to superior gemellus muscle

Interpretation of Your Score

Number of Correct Responses	Level of Performance
43-50	Excellent - Exceptional
38-42	Superior - Very Superior
32-37	Average - Above Average
28-31	Poor - Marginal
27 or less	Very Poor - Failure

17. REVIEW EXAMINATION ON THE HEAD AND NECK

INTRODUCTORY NOTE. The following 50 multiple-choice questions are based on the instructional objectives for the head and neck listed in Chapter Three of this Study Guide & Review Manual. You should be able to answer these questions in 35-40 minutes. Before beginning, tear out an answer sheet from the back of the book and read the directions on how to use it. The key to the correct responses is on page 353.

FIVE-CHOICE COMPLETION QUESTIONS

DIRECTIONS: Each of the following questions or incomplete statements is followed by five suggested answers or completions. SELECT THE ONE BEST ANSWER in each case and then blacken the appropriate space on the answer sheet.

1. WHICH VESSEL IS NOT A BRANCH OF THE MAXILLARY ARTERY?
   A. Buccal
   B. Deep temporal
   C. Sphenopalatine
   D. Middle meningeal
   E. Superficial temporal

2. WHICH STATEMENT ABOUT THE PHRENIC NERVE IS INCORRECT?
   A. Passes into the mediastinum behind the subclavian vein
   B. Consists mainly of the anterior primary ramus of C4
   C. Runs vertically over the middle scalene muscle
   D. Lies deep to the prevertebral fascia
   E. None of the above

3. THE BRANCH OF THE EXTERNAL CAROTID ARTERY ORIGINATING JUST BELOW THE LEVEL OF THE GREATER CORNU OF THE HYOID BONE IS THE _____ ARTERY.
   A. Posterior auricular
   B. Superior thyroid
   C. Maxillary
   D. Occipital
   E. Facial

4. SENSORY LOSS IN THE SKIN OVERLYING THE PAROTID GLAND INDICATES DAMAGE TO THE _____ NERVE.
   A. Zygomaticotemporal
   B. Transverse cervical
   C. Great auricular
   D. Great occipital
   E. None of the above

5. WHICH OF THE FOLLOWING MUSCLES IS SUPPLIED BY THE ABDUCENS NERVE?
   A. Superior oblique
   B. Lateral rectus
   C. Inferior oblique
   D. Inferior rectus
   E. Medial rectus

6. WHICH STATEMENT ABOUT THE TONSILS IS INCORRECT?
   A. The adenoids are enlarged pharyngeal tonsils
   B. The tubal tonsil is located in the nasopharynx
   C. The palatine tonsils are situated in the oral cavity
   D. The base of the tongue contains the lingual tonsil
   E. The pharyngeal tonsils are embedded in the wall of the nasopharynx

SELECT THE ONE BEST ANSWER

7. WHICH STATEMENT ABOUT THE TRAPEZIUS MUSCLE IS FALSE?
   A. Innervated by motor fibers of the accessory nerve
   B. Forms a boundary of the posterior triangle of the neck
   C. Inserts into the scapula and the length of the clavicle
   D. Receives its blood supply from the transverse cervical artery
   E. Weakness of this muscle results in drooping of the shoulder

8. THE MAXILLARY DIVISION OF THE TRIGEMINAL NERVE PASSES THROUGH WHICH OF THE FOLLOWING FORAMINA?
   A. Stylomastoid
   B. Rotundum
   C. Lacerum
   D. Spinosum
   E. Ovale

9. WHICH OF THE FOLLOWING NERVES IS NOT LOCATED WITHIN THE ORBIT?
   A. Infraorbital
   B. Trochlear
   C. Oculomotor
   D. Abducens
   E. Maxillary

10. WHICH OF THE FOLLOWING IS NOT LOCATED IN THE POSTERIOR TRIANGLE OF THE NECK?
    A. Accessory nerve
    B. Occipital lymph node
    C. Posterior auricular artery
    D. Brachial plexus of nerves
    E. Transverse cervical artery

11. WHICH STRUCTURE IS LOCATED OUTSIDE THE CAROTID SHEATH?
    A. Cervical sympathetic trunk
    B. Internal carotid artery
    C. Internal jugular vein
    D. Vagus nerve
    E. None of the above

12. THE FLOOR OF THE TYMPANIC CAVITY IS SUPERIOR TO THE:
    A. Internal acoustic meatus
    B. Facial canal
    C. Auditory tube
    D. Jugular fossa
    E. Carotid canal

13. THE ANTERIOR BELLY OF THE DIGASTRIC MUSCLE IS SUPPLIED BY THE:
    A. Facial nerve
    B. Ansa cervicalis
    C. Mylohyoid nerve
    D. Cervical plexus
    E. Accessory nerve

14. THE FACIAL NERVE INNERVATES ALL THESE MUSCLES EXCEPT THE:
    A. Levator labii superioris
    B. Orbicularis oris
    C. Buccinator
    D. Temporalis
    E. Platysma

15. THE HYPOPHYSEAL FOSSA IS LOCATED IN THE _____ BONE.
    A. Sphenoid
    B. Ethmoid
    C. Frontal
    D. Palatine
    E. Maxilla

16. WHICH STATEMENT ABOUT THE HARD PALATE IS INCORRECT?
    A. Has rugae that aid in gripping food
    B. Has mucous glands innervated by the facial nerve
    C. Formed by the maxilla and palatine bones and a lining mucous membrane
    D. The greater palatine vessels and the nasopalatine nerve pass through the greater palatine foramen
    E. Incisive papilla indicates the site of the incisive foramen

SELECT THE ONE BEST ANSWER

17. MOST SYMPATHETIC PREGANGLIONIC FIBERS TO THE HEAD REGION SYNAPSE IN THE:
    A. Inferior cervical ganglion
    B. Superior cervical ganglion
    C. Cervicothoracic ganglion
    D. Middle cervical ganglion
    E. Intermediolateral cell column

18. CEREBROSPINAL FLUID RETURNS TO THE BLOODSTREAM THROUGH THE:
    A. Veins in the subarachnoid space
    B. Interventricular foramen
    C. Lymphatic vessels
    D. Choroid plexuses
    E. Arachnoid villi

MULTI-COMPLETION QUESTIONS

DIRECTIONS: In each of the following questions or incomplete statements, ONE OR MORE of the completions given is correct. On the answer sheet blacken the space under A if 1, 2, and 3 are correct; B if 1 and 3 are correct; C if 2 and 4 are correct; D if only 4 is correct; and E if all are correct.

19. WHICH MUSCLE(S) IS(ARE) SUPPLIED BY THE HYPOGLOSSAL NERVE?
    1. Styloglossus
    2. Genioglossus
    3. Longitudinal lingual
    4. Palatoglossus

20. THE PHRENIC NERVE:
    1. Runs lateral to the carotid sheath
    2. Arises from cervical nerves 4, 5 and 6
    3. Runs anterior to the prevertebral fascia
    4. Runs downward in front of the scalenus anterior muscle

21. WHICH ARTERY(IES) SUPPLY(IES) THE PAROTID GLAND?
    1. Superficial temporal
    2. Transverse facial
    3. External carotid
    4. Maxillary

22. WHICH OF THE FOLLOWING ARTERIES CONTRIBUTE(S) TO THE FORMATION OF THE CIRCULUS ARTERIOSUS CEREBRI (CIRCLE OF WILLIS)?
    1. Anterior cerebral
    2. Posterior cerebral
    3. Posterior communicating
    4. Middle cerebral

23. THE FACIAL NERVE SUPPLIES:
    1. The mylohyoid muscle
    2. All muscles of facial expression
    3. Taste fibers to the root of the tongue
    4. Muscle in the embryonic hyoid arch

24. WHICH NERVE COMPONENT(S) IS(ARE) CONTAINED IN THE FACIAL NERVE?
    1. Special sensory
    2. General sensory
    3. Parasympathetic
    4. Somatic motor

25. THE RIGHT RECURRENT LARYNGEAL NERVE USUALLY LOOPS AROUND WHICH STRUCTURE(S)?
    1. Ligamentum arteriosum
    2. Axillary artery
    3. Aortic arch
    4. Subclavian artery

26. THE TRAPEZIUS MUSCLE ORIGINATES FROM THE:
    1. Spines of thoracic vertebrae
    2. Ligamentum nuchae
    3. Superior nuchal line
    4. Spine of the scapula

A	B	C	D	E
1,2,3	1,3	2,4	only 4	all correct

27. WHICH STATEMENT(S) ABOUT THE MIDDLE MENINGEAL ARTERY IS(ARE) CORRECT?
    1. Supplies the parietal lobe of the cerebral cortex
    2. Is accompanied by a sympathetic plexus of nerves
    3. Enters the skull through the foramen ovale
    4. Is extradural in position

28. THE MAXILLARY NERVE SUPPLIES THE:
    1. Skin over the upper jaw
    2. Anterior two-thirds of the tongue
    3. Mucous membrane of the soft palate
    4. Mandibular process of the first branchial arch

29. WHICH STATEMENT(S) ABOUT THE FACIAL NERVE IS(ARE) CORRECT?
    1. Motor to the posterior belly of the digastric muscle
    2. Nerve of the hyoid or second branchial arch
    3. Contains taste and secretomotor fibers
    4. Motor to the muscles of mastication

30. WHICH MUSCLE(S) IS(ARE) INNERVATED BY THE ANSA CERVICALIS?
    1. Sternohyoid   3. Sternothyroid
    2. Thyrohyoid   4. Omohyoid

31. PARALYSIS OF THE SYMPATHETIC NERVE SUPPLY TO THE EYE RESULTS IN:
    1. Slight drooping of the upper eyelid
    2. Constriction of the pupil
    3. Retraction of the eyeball
    4. Vasoconstriction

32. THE MIDDLE EAR CAVITY COMMUNICATES WITH THE:
    1. Internal acoustic meatus   3. External acoustic meatus
    2. Nasopharynx   4. Mastoid air cells

33. WHICH MUSCLE(S) IS(ARE) INNERVATED BY MOTOR FIBERS OF THE FACIAL NERVE?
    1. Buccinator   3. Orbicularis oculi
    2. Mylohyoid   4. Anterior belly of digastric

34. WHICH PARANASAL SINUS(ES) OPEN(S) INTO THE MIDDLE NASAL MEATUS?
    1. Maxillary   3. Middle ethmoidal
    2. Anterior ethmoidal   4. Posterior ethmoidal

35. THE BRACHIAL PLEXUS OF NERVES IS RELATED TO WHICH MUSCLE(S) IN THE NECK REGION?
    1. Superior belly of omohyoid 3. Sternohyoid
    2. Anterior scalene   4. Middle scalene

36. WHICH STRUCTURE(S) IS(ARE) LOCATED WITHIN THE POSTERIOR TRIANGLE OF THE NECK?
    1. Levator scapulae muscle   3. Accessory nerve
    2. Scalenus medius muscle   4. Trapezius muscle

37. THE SUPERIOR THYROID ARTERY:
    1. Is the first branch of the external carotid artery
    2. Runs parallel to the internal laryngeal nerve
    3. Has a branch that pierces the thyrohyoid membrane
    4. Is the only source of blood supply to the thyroid gland

A	B	C	D	E
1,2,3,	1,3	2,4	only 4	all correct

38. WHICH STRUCTURE(S) PASS(ES) THROUGH THE FORAMEN OVALE OF THE CRANIUM?
    1. Greater petrosal nerve
    2. Accessory meningeal artery
    3. Maxillary nerve
    4. Mandibular nerve

FIVE-CHOICE ASSOCIATION QUESTIONS

DIRECTIONS: Each group of questions below consists of a numbered list of descriptive words or phrases accompanied by a diagram with certain parts indicated by letters, or by a list of lettered headings. For each numbered word or phrase, SELECT THE LETTERED PART OR HEADING that matches it correctly. Then on the appropriate line of the answer sheet, blacken the space under the letter of that part or heading. Sometimes more than one numbered word or phrase may be correctly matched to the same lettered part or heading.

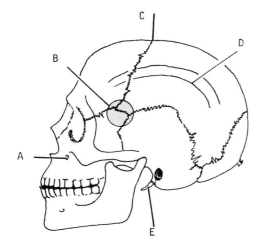

39. _____ Bregma

40. _____ Origin of the temporalis muscle

41. _____ Site of attachment for the stylopharyngeus muscle

42. _____ Site of exit for the infraorbital nerve

43. _____ Pterion

44. _____ Develops from the cartilage of the second branchial arch

ASSOCIATION QUESTIONS

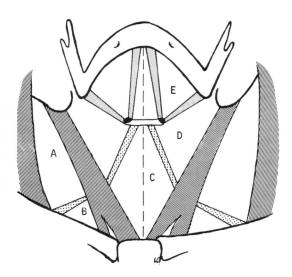

45. ____ Triangle containing the largest segments of the external carotid arteries

46. ____ Triangle containing the third part of the subclavian artery

47. ____ Triangle containing the thyroid cartilage

48. ____ Triangle in which the facial artery branches

49. ____ Triangle containing the branchial plexus

50. ____ Triangle containing the accessory nerve

Interpretation of Your Score

Number of Correct Responses	Level of Performance
43-50	Excellent - Exceptional
38-42	Superior - Very Superior
32-37	Average - Above Average
28-31	Poor - Marginal
27 or less	Very Poor - Failure

18. REVIEW EXAMINATION ON THE THORAX

INTRODUCTORY NOTE. The following 50 multiple-choice questions are based on the instructional objectives for the thorax listed in Chapter Four of this Study Guide & Review Manual. You should be able to answer these questions in 35-40 minutes. Before beginning, tear out an answer sheet from the back of the book and read the directions on how to use it. The key to the correct responses is on page 353.

FIVE-CHOICE COMPLETION QUESTIONS

DIRECTIONS: Each of the following questions or incomplete statements is followed by five suggested answers or completions. SELECT THE ONE BEST ANSWER in each case and then blacken the appropriate space on the answer sheet.

1. THE RIGHT VAGUS NERVE CONTRIBUTES SIGNIFICANTLY TO WHICH PLEXUS(ES)?
   A. Esophageal
   B. Pulmonary
   C. Cardiac
   D. All of the above
   E. None of the above

2. WHICH OF THE FOLLOWING DIAMETERS OF THE THORAX IS(ARE) INCREASED DURING INSPIRATION?
   A. Vertical
   B. Transverse
   C. Anteroposterior
   D. All of the above
   E. None of the above

3. THE COSTOMEDIASTINAL RECESS IS INDICATED BY _____ IN THIS HORIZONTAL SECTION OF THE UPPER PART OF THE RIGHT LUNG.

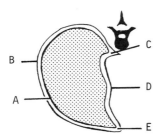

4. FROM WHICH STRUCTURE(S) IS THE CORONARY SINUS DERIVED?
   A. Right anterior cardinal and right common cardinal veins
   B. Right horn of the sinus venosus
   C. Left horn of the sinus venosus
   D. Sinus venosus
   E. None of the above

SELECT THE ONE BEST ANSWER

5. IN THE POSTERIOR MEDIASTINUM, THE ANTERIOR RELATION OF THE ESOPHAGUS IS THE:
   A. Azygos vein
   B. Pericardium
   C. Phrenic nerve
   D. Descending aorta
   E. Superior vena cava

6. WHICH STATEMENT ABOUT THE LEFT LUNG IS FALSE?
   A. Has a groove for the azygos vein on its mediastinal surface
   B. Is subdivided into eight bronchopulmonary segments
   C. The cardiac notch is in its upper lobe
   D. Is supplied by bronchial arteries
   E. Has only two lobes

7. WHICH STRUCTURE(S) PASS(ES) THROUGH THE CENTRAL TENDON OF THE DIAPHRAGM?
   A. Inferior vena cava
   B. Thoracic duct
   C. Azygos vein
   D. Aorta
   E. All of the above

8. THE BLOOD SUPPLY TO THE INTERCOSTAL SPACES IS FROM THE:
   A. Musculophrenic artery
   B. Internal thoracic artery
   C. Descending thoracic aorta
   D. Costocervical trunk
   E. All of the above

9. THE DIAPHRAGM IS SUPPLIED BY:
   A. Branches of the internal thoracic artery
   B. Phrenic nerves (C3, 4 and 5)
   C. Branches from the aorta
   D. Lower intercostal nerves
   E. All of the above

10. THE NEUROVASCULAR BUNDLE LIES:
    A. Between the middle and internal layers of intercostal muscles
    B. Between the external and middle layers of intercostal muscles
    C. Deep to the internal layer of intercostal muscles
    D. Superficial to the external intercostal muscles
    E. None of the above

11. WHICH STATEMENT ABOUT THE RIGHT VENTRICLE IS FALSE?
    A. Communicates with the right atrium via the mitral valve
    B. Has papillary muscles attached to its internal wall
    C. Its infundibulum has a smooth wall
    D. Contains the moderator band
    E. Is triangular in shape

12. THE INTERNAL THORACIC ARTERY BIFURCATES INTO WHICH TWO ARTERIES?
    A. Intercostal and musculophrenic
    B. Anterior and posterior intercostal
    C. Superior epigastric and musculophrenic
    D. Superior epigastric and anterior intercostal
    E. Pericardiacophrenic and musculophrenic

13. THE HEART RECEIVES POSTGANGLIONIC SYMPATHETIC FIBERS VIA WHICH NERVE(S)?
    A. Thoracic cardiac
    B. Cervicothoracic
    C. Cervical
    D. All of the above
    E. None of the above

SELECT THE ONE BEST ANSWER

14. THE AORTIC HIATUS IS LOCATED OPPOSITE WHICH OF THE FOLLOWING THORACIC VERTEBRAE?
    A. 8
    B. 9
    C. 10
    D. 11
    E. 12

15. ALL OF THE FOLLOWING VESSELS ENTER THE RIGHT ATRIUM EXCEPT THE:
    A. Small coronary veins
    B. Pulmonary veins
    C. Superior vena cava
    D. Inferior vena cava
    E. Coronary sinus

16. THE LYMPHATIC DRAINAGE OF THE LUNGS IS TO WHICH GROUP(S) OF NODES?
    A. Bronchopulmonary
    B. Tracheobronchial
    C. Paratracheal
    D. Pulmonary
    E. All of the above

17. WHICH STATEMENT(S) ABOUT THE DIAPHRAGM IS(ARE) TRUE?
    A. Intrapleural pressure is decreased following contraction
    B. Contraction promotes venous return
    C. Is bilaterally innervated
    D. Involved in respiration
    E. All of the above

18. THE STERNAL ANGLE IS LOCATED AT THE ARTICULATION BETWEEN THE:
    A. Body of the sternum and the xiphoid process
    B. Manubrium and the body of the sternum
    C. Manubrium and xiphoid process
    D. Sternum and clavicle
    E. None of the above

19. THE MITRAL VALVE IS MOST AUDIBLE OVER THE:
    A. Right junction of the sternal body and the costal margin
    B. Left junction of the sternal body and the costal margin
    C. Left second intercostal space
    D. Right second intercostal space
    E. Left fifth intercostal space

20. THE MAJOR PART OF THE THYMUS GLAND IS LOCATED IN THE:
    A. Anterior mediastinum
    B. Middle mediastinum
    C. Superior mediastinum
    D. Posterior mediastinum
    E. Suprasternal region

MULTI-COMPLETION QUESTIONS

DIRECTIONS: In each of the following questions or incomplete statements, ONE OR MORE of the completions given is correct. On the answer sheet, blacken the space under A if 1, 2 and 3 are correct; B if 1 and 3 are correct; C if 2 and 4 are correct; D if only 4 is correct; and E if all are correct.

21. THE DEEP CARDIAC PLEXUS OF NERVES IS FORMED BY BRANCHES FROM THE:
    1. Vagus
    2. Thoracic sympathetic ganglia
    3. Recurrent laryngeal
    4. Cervical sympathetic ganglia

A	B	C	D	E
1,2,3	1,3	2,4	only 4	all correct

22. NORMALLY THE RIGHT PRIMARY BRONCHUS IS:
    1. More vertical than the left
    2. Superior to the azygos vein
    3. Shorter than the left
    4. Smaller in diameter than the left

23. THE RIGHT CORONARY ARTERY:
    1. Anastomoses with the left coronary artery
    2. Passes between the right auricle and the pulmonary trunk
    3. Is located in the coronary sulcus
    4. Arises from the anterior aortic sinus

24. WHICH NERVE(S) INNERVATE(S) THE DIAPHRAGM?
    1. Thoracoabdominal
    2. Upper intercostals
    3. Phrenic
    4. Vagus

25. THE INFERIOR LOBE OF THE LEFT LUNG IS SUBDIVIDED INTO WHICH OF THE FOLLOWING NAMED BRONCHOPULMONARY SEGMENTS?
    1. Superior
    2. Anterior basal
    3. Posterior basal
    4. Lateral basal

26. WHICH STRUCTURE(S) IS(ARE) NOT IN THE ANTERIOR MEDIASTINUM?
    1. Thymus gland
    2. Phrenic nerve
    3. Lymph nodes
    4. Thoracic duct

27. THE MAJOR NAMED BRANCH(ES) OF THE LEFT CORONARY ARTERY IS(ARE) THE:
    1. Posterior interventricular
    2. Circumflex
    3. Marginal
    4. Anterior interventricular

28. THE THORACIC DUCT:
    1. Passes through the aortic hiatus of the diaphragm
    2. Lies between the aorta and the azygos vein
    3. Begins in the abdomen
    4. Enters the external jugular vein

29. WHICH STATEMENT(S) ABOUT THE HEART IS(ARE) CORRECT?
    1. The floor of the fossa ovalis represents the persistent portion of the septum primum of the embryonic heart
    2. The interior of the right atrium is divided into two parts by the sulcus terminalis
    3. The infundibulum of the right ventricle is derived from the bulbus cordis of the embryonic heart
    4. Structural defects of the ventricular septum are rare

30. THE LEFT CORONARY ARTERY:
    1. Is located in the atrioventricular groove
    2. Arises from the anterior aortic sinus
    3. Lies between the left auricle and the pulmonary trunk
    4. Is smaller than the right coronary artery

31. WHICH STRUCTURE(S) IS(ARE) LOCATED IN THE WALLS OF THE VENTRICLES OF THE HEART?
    1. Musculi pectinati
    2. Chordae tendineae
    3. Crista terminalis
    4. Trabeculae carneae

A	B	C	D	E
1,2,3	1,3	2,4	only 4	all correct

32. THE PARIETAL PLEURA RECEIVES ITS BLOOD SUPPLY FROM THE FOLLOWING ARTERY(IES)?
    1. Posterior intercostal
    2. Internal thoracic
    3. Superior phrenic
    4. Bronchial

33. WHICH STRUCTURE(S) LIE(S) ANTERIOR TO THE ESOPHAGUS IN THE THORAX?
    1. Thoracic duct
    2. Left bronchus
    3. Right vagus nerve
    4. Pericardium

34. WHICH STRUCTURE(S) PASS(ES) THROUGH THE ESOPHAGEAL HIATUS?
    1. Vagus nerve
    2. Azygos vein
    3. Esophagus
    4. Thoracic duct

35. THE CORONARY ARTERIES:
    1. Lie superficial to the myocardium
    2. Flow of blood through them occurs during diastole
    3. Their branches are end arteries
    4. The second branches of the aorta

36. WHICH STATEMENT(S) ABOUT THE HEART IS(ARE) FALSE?
    1. The atria form its base
    2. The long axis of the heart is commonly transverse in late pregnancy
    3. Blood flow from the atria to the ventricles is horizontal and forward
    4. Diaphragmatic movements have little to do with heart positioning.

37. THE AREA LABELLED "A" IN THE DIAGRAM IS:

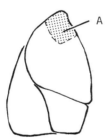

    1. The apical bronchopulmonary segment
    2. An uncommon site of pulmonary tuberculosis
    3. Auscultated over the supraclavicular region
    4. Supplied by the left apical segmental bronchus

38. THE THORACIC AORTA:
    1. Passes through the diaphragm at the level of the 10th thoracic vertebra
    2. Gives rise to all posterior intercostal arteries
    3. Descends on the right side of the vertebral column
    4. Lies behind the esophagus

39. THE ESOPHAGUS IS DRAINED BY WHICH VEIN(S)?
    1. Inferior thyroid
    2. Hemiazygos
    3. Azygos
    4. Gastric

40. THE MUSCULAR FIBERS OF THE DIAPHRAGM ARISE FROM THE:
    1. Xiphoid process of the sternum
    2. Lumbocostal arches and crura
    3. Lower costal cartilages
    4. Lower ribs

FIVE-CHOICE ASSOCIATION QUESTIONS

DIRECTIONS: Each group of questions below consists of a numbered list of descriptive words or phrases accompanied by a diagram with certain parts indicated by letters, or by a list of lettered headings. For each numbered word or phrase, SELECT THE LETTERED PART OR HEADING that matches it correctly and then blacken the appropriate space on the answer sheet. Sometimes more than one numbered word or phrase may be correctly matched to the same lettered part or heading.

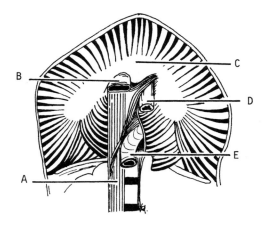

41. ___ Crural attachment of the diaphragm
42. ___ Structure passing through the central tendon of the diaphragm
43. ___ The tendinous part of the diaphragm
44. ___ Pierces the diaphragm at the level of the 12th thoracic vertebra
45. ___ Passes behind the median arcuate ligament

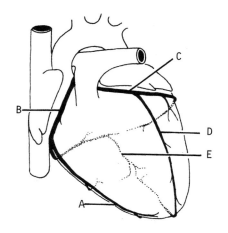

46. ___ Marginal artery
47. ___ Larger of the coronary arteries
48. ___ Give branches to the ventricles and most of the interventricular septum
49. ___ Anterior interventricular branch
50. ___ Usually supplies the atrioventricular node

Interpretation of Your Score

Number of Correct Responses	Level of Performance
43-50	Excellent - Exceptional
38-42	Superior - Very Superior
32-37	Average - Above Average
28-31	Poor - Marginal
27 or less	Very Poor - Failure

19. REVIEW EXAMINATION ON THE ABDOMEN

INTRODUCTORY NOTE. The following 50 multiple-choice questions are based on the instructional objectives for the abdomen listed in Chapter Five of this Study Guide & Review Manual. You should be able to answer these questions in 35-40 minutes. Before beginning, tear out an answer sheet from the back of the book and read the directions on how to use it. The key to the correct responses is on page 353.

FIVE-CHOICE COMPLETION QUESTIONS

DIRECTIONS: Each of the following questions or incomplete statements is followed by five suggested answers or completions. SELECT THE ONE BEST ANSWER in each case and then blacken the appropriate space on the answer sheet.

1. WHICH STATEMENT(S) ABOUT THE INTERTUBERCULAR PLANE IS(ARE) TRUE?
   A. Is one of the planes used in dividing the abdomen into zones
   B. Is used as a guide when performing a lumbar puncture
   C. Corresponds to the highest point of the iliac crest
   D. Is at the level of the fifth lumbar vertebra
   E. All of the above

2. WHICH STRUCTURE(S) PASS(ES) THROUGH THE INGUINAL CANAL IN A MALE FETUS?
   A. Cremaster muscle
   B. Testicular artery
   C. Ductus deferens
   D. Processus vaginalis
   E. All of the above

3. WHICH STATEMENT(S) ABOUT THE VENOUS DRAINAGE OF THE STOMACH IS(ARE) TRUE?
   A. The right gastric vein drains into the portal vein
   B. The left gastroepiploic vein drains into the splenic vein
   C. The left gastric veins communicate with the esophageal veins
   D. The left gastric vein drains the upper half of the lesser curvature
   E. All of the above

4. THE HILUM OF THE RIGHT KIDNEY IS COVERED BY THE:
   A. Right colic vessels
   B. Quadrate lobe of the liver
   C. Second part of the duodenum
   D. Right colic flexure
   E. Gall bladder

5. THE INFERIOR VENA CAVA IS CROSSED ANTERIORLY BY ALL OF THE FOLLOWING STRUCTURES EXCEPT THE_____.
   A. Third part of the duodenum
   B. Superior mesenteric artery
   C. Right suprarenal gland
   D. Root of the mesentery
   E. Right gonadal artery

6. WHICH STATEMENT ABOUT THE MESENTERY IS FALSE?
   A. Suspends the small gut to the posterior abdominal wall
   B. Its attached border runs downwards and to the right
   C. Has an extensive free border enclosing the gut
   D. Its attached border is about 6 cm in length
   E. Has a short attached border

SELECT THE ONE BEST ANSWER

7. WHICH STATEMENT ABOUT THE STOMACH IS FALSE?
   A. Is a common site of ulcers
   B. Receives its blood supply from the celiac artery
   C. Has the greater omentum attached to its greater curvature
   D. Is divided into cardiac and pyloric portions
   E. Its fundic portion receives its blood supply from the left gastroepiploic artery

8. ALL STATEMENTS ABOUT THE VERMIFORM APPENDIX ARE TRUE EXCEPT:
   A. Opens into the apex of the cecum in fetuses
   B. Subject to considerable variation in position
   C. Retroileal appendices are the most common type
   D. Opens about two cm below the ileocecal junction in adults
   E. When the cecum is full, the free appendix usually hangs in the pelvis

9. THE ABDOMINAL AORTA BIFURCATES AT THE LEVEL OF THE _____ LUMBAR VERTEBRA.
   A. First
   B. Second
   C. Third
   D. Fourth
   E. Fifth

10. WHICH STATEMENT ABOUT AN INGUINAL HERNIA IS FALSE?
    A. May be present at birth
    B. More common in females
    C. Are of two types
    D. Is associated with the processus vaginalis
    E. May be associated with weak abdominal musculature

11. WHICH STATEMENT REGARDING THE SUBCOSTAL PLANE IS FALSE?
    A. Passes horizontally through the trunk
    B. Is at the level of the third lumbar vertebra
    C. Passes through the lowest part of the inferior costal margin
    D. Indicates the level of origin of the inferior mesenteric artery
    E. None of the above

12. THE VISCERAL SURFACE OF THE SPLEEN IS NOT IN CONTACT WITH THE:
    A. Tail of the pancreas
    B. Left adrenal gland
    C. Left colic flexure
    D. Left kidney
    E. Stomach

13. WHICH STATEMENT(S) ABOUT THE FALCIFORM LIGAMENT IS(ARE) TRUE?
    A. A sickle-shaped peritoneal fold
    B. Attached to the anterior abdominal wall
    C. Contains the ligamentum teres in its free border
    D. Attached to the anterior and superior aspects of the liver
    E. All of the above

14. WHICH STATEMENT ABOUT THE QUADRATE LOBE OF THE LIVER IS FALSE?
    A. Located medial to the gall bladder
    B. Related to the pylorus of the stomach
    C. In contact with the right colic flexure
    D. Found on the inferior surface of the liver
    E. Supplied by the left branch of the portal vein

15. WHICH OF THE FOLLOWING IS NOT A BRANCH OF THE SUPERIOR MESENTERIC ARTERY?
    A. Jejunal
    B. Ileocolic
    C. Left colic
    D. Right colic
    E. Middle colic

SELECT THE ONE BEST ANSWER

16. WHICH STATEMENT ABOUT THE RIGHT VERTICAL PLANE IS INCORRECT?
    A. Used for dividing the abdominal cavity into zones
    B. Passes through the fundus of the gall bladder
    C. Drawn upwards from the right pubic tubercle
    D. Meets the ninth right costal cartilage
    E. Meets the transpyloric plane

17. WHICH STATEMENT ABOUT THE SECOND PART OF THE DUODENUM IS FALSE?
    A. Related superiorly to the uncinate process of the pancreas
    B. Related laterally to the hepatic flexure of the colon
    C. Related posteriorly to the hilum of the right kidney
    D. Related medially to the head of the pancreas
    E. Related anteriorly to the transverse colon

18. THE PART OF THE ABDOMINAL CAVITY DELINEATED BY THE XIPHOID PROCESS AND THE ADJACENT COSTAL MARGINS, THE TWO VERTICAL PLANES, AND THE SUBCOSTAL PLANE IS KNOWN AS THE _____ REGION:
    A. Hypochondriac
    B. Epigastric
    C. Hypogastric
    D. Suprapubic
    E. Iliac

19. WHICH STATEMENT(S) ABOUT THE GALL BLADDER IS(ARE) TRUE?
    A. Consists of a fundus, body, and neck
    B. Is supplied only by the cystic artery
    C. Forms part of the extrahepatic biliary system
    D. Inflammation of it is common in older obese persons
    E. All of the above

20. WHICH STATEMENT ABOUT THE BILE DUCT, AFTER IT LEAVES THE LESSER OMENTUM, IS FALSE?
    A. Traverses the medial part of the body of the pancreas
    B. Related to the superior pancreaticoduodenal artery
    C. Related medially to the gastroduodenal artery
    D. Lies behind the first part of the duodenum
    E. Related posteriorly to the portal vein

MULTI-COMPLETION QUESTIONS

DIRECTIONS: In each of the following questions or incomplete statements, ONE OR MORE of the completions given is correct. On the answer sheet blacken the space under A if 1, 2, and 3 are correct; B if 1 and 3 are correct; C if 2 and 4 are correct; D if only 4 is correct; and E if all are correct.

21. THE FEATURE(S) OF THE LUMBAR PLEXUS OF NERVES IS(ARE):
    1. Located within the substance of the psoas major muscle
    2. Formed by the anterior primary rami of L1-L5
    3. Gives rise to the femoral and obturator nerves
    4. Contributes to the lumbosacral trunk

22. THE BARE AREA OF THE LIVER IS:
    1. An area not covered by peritoneum
    2. Located principally on the superior surface of the liver
    3. Limited medially by the groove for the inferior vena cava
    4. Limited laterally by the left triangular ligament

A	B	C	D	E
1,2,3	1,3	2,4	only 4	all correct

23. THE GALL BLADDER:
    1. Is related to the inferior surface of the liver
    2. Is completely enclosed by peritoneum
    3. Has an average capacity of 30 ml
    4. Opens into the bile duct

24. THE UMBILICAL REGION SEPARATES THE TERRITORIES:
    1. Innervated by the 9th and 11th thoracic segments
    2. Of lymphatic drainage
    3. Of venous drainage
    4. Of arterial supply

25. IN DISEASES OF THE RIGHT KIDNEY, PAIN IS USUALLY:
    1. Felt at the tip of the left shoulder
    2. Felt in the region of the right renal angle
    3. Felt in the right hypochondrium
    4. Radiated to the external genitalia

26. THE RECTUS ABDOMINIS MUSCLE:
    1. Has a segmental innervation
    2. Originates from the pubic crest
    3. Inserts into the 5th, 6th and 7th costal cartilages
    4. Is covered anteriorly in its entire length by the aponeurosis of the three abdominal muscles

27. THE RIGHT SUPRARENAL GLAND IS:
    1. Semilunar in shape
    2. Related posteriorly to the diaphragm
    3. Drained by a vein that joins the right renal vein
    4. Related anteriorly to the inferior vena cava

28. THE RECTUS SHEATH CONTAINS THE:
    1. Musculophrenic artery
    2. Rectus abdominis
    3. Intercostal nerves T3-T6
    4. Inferior epigastric artery

29. THE PSOAS MAJOR MUSCLE:
    1. Inserts into the greater trochanter of the femur
    2. Is a powerful flexor of the thigh
    3. Is innervated by the femoral nerve
    4. Lies posterior to the kidney

30. THE INTERNAL OBLIQUE MUSCLE:
    1. Takes origin from the lateral half of the inguinal ligament
    2. Inserts into the linea alba and the lower ribs
    3. Originates from the intermediate part of the anterior two-thirds of the iliac crest
    4. Forms the conjoint tendon supporting the lateral half of the inguinal canal

31. THE QUADRATUS LUMBORUM MUSCLE:
    1. Has the subcostal, ilioinguinal and iliohypogastric nerves on its anterior aspect
    2. Forms a posterior relation of the kidney
    3. Is attached inferiorly to the iliac crest
    4. Is a lateral flexor of the trunk

A	B	C	D	E
1,2,3	1,3	2,4	only 4	all correct

32. THE ARTERIAL SUPPLY OF THE ANTERIOR ABDOMINAL WALL IS FROM THE:
    1. Inferior epigastric
    2. Lateral thoracic
    3. Superior epigastric
    4. Superficial external pudendal

33. THE COMMON BILE DUCT IS:
    1. Formed by the union of the common hepatic and cystic ducts
    2. Located in the lesser omentum
    3. 4-6 cm long
    4. Enters third part of the duodenum

34. CONCERNING MECKEL'S DIVERTICULUM:
    1. May produce gross bleeding from the bowel during infancy
    2. Common congenital malformation of the digestive tract
    3. Occurs in about two per cent of persons
    4. Represents a remnant of the yolk stalk

35. THE ANTERIOR RELATIONS OF THE STOMACH INCLUDE THE:
    1. Diaphragm
    2. Left lobe of the liver
    3. Anterior abdominal wall
    4. Gall bladder

36. CONCERNING THE DUODENUM:
    1. Derived from the foregut and midgut of the embryo
    2. The beginning of its first part is called the free part
    3. Is relatively fixed except for its first part
    4. Supplied by the celiac trunk and the superior mesenteric artery

37. THE POSTERIOR SURFACE OF THE STOMACH IS RELATED TO THE:
    1. Splenic artery
    2. Left kidney
    3. Left suprarenal
    4. Head of the pancreas

38. THE PYLORUS OF THE STOMACH:
    1. Contains the pyloric antrum
    2. Has the prepyloric vein on its anterior surface
    3. Is related posteriorly to the hepatic artery
    4. Is the junction between the stomach and duodenum

39. THE FIRST PART OF THE DUODENUM:
    1. Is the most mobile part
    2. Is a common site for ulcers
    3. Commences in the transpyloric plane
    4. Terminates at the neck of the gall bladder

40. THE INFERIOR VENA CAVA IS:
    1. Formed by union of the common iliac veins
    2. Anterior to the right psoas major muscle
    3. Related medially to the abdominal aorta
    4. Located on the posterior abdominal wall

41. THE ANTERIOR WALL OF THE OMENTAL BURSA (LESSER SAC) IS COMPOSED OF:
    1. The posterior layer of the lesser omentum
    2. Peritoneum on the posterior surface of the liver
    3. Peritoneum covering the posterior surface of the stomach
    4. The posterior peritoneal layer extending downwards from the greater curvature

A	B	C	D	E
1,2,3	1,3	2,4	only 4	all correct

42. THE LESSER OMENTUM:
    1. Has a free right margin
    2. Is attached superiorly to the porta hepatis of the liver
    3. Is attached inferiorly to the lesser curvature of the stomach
    4. Encloses the inferior vena cava in its free margin

43. THE FIRST INCH OF THE DUODENUM:
    1. Forms the inferior boundary of the epiploic foramen
    2. Is referred to as the "duodenal bulb or cap"
    3. Is covered by peritoneum
    4. Is in contact with the neck of the gall bladder

44. THE DUODENUM IS SUPPLIED BY WHICH OF THE FOLLOWING ARTERIES?
    1. Gastroduodenal
    2. Superior mesenteric
    3. Pancreaticoduodenal
    4. Common hepatic

45. THE TRANSVERSE COLON:
    1. Begins anterior to the right kidney
    2. Terminates immediately below the spleen
    3. Is supplied mainly by the middle colic artery
    4. Is the shortest segment of the colon

FIVE-CHOICE ASSOCIATION QUESTIONS

DIRECTIONS: Each group of questions below consists of a numbered list of descriptive words or phrases accompanied by a diagram with certain parts indicated by letters, or by a list of lettered headings. For each numbered word or phrase, SELECT THE LETTERED PART OR HEADING that matches it correctly. Then on the appropriate line of the answer sheet, blacken the space under the letter of that part or heading. Sometimes more than one numbered word or phrase may be correctly matched to the same lettered part or heading.

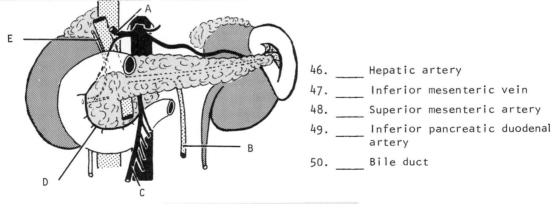

46. _____ Hepatic artery
47. _____ Inferior mesenteric vein
48. _____ Superior mesenteric artery
49. _____ Inferior pancreatic duodenal artery
50. _____ Bile duct

Interpretation of Your Score

Number of Correct Responses	Level of Performance
43-50	Excellent - Exceptional
38-42	Superior - Very Superior
32-37	Average - Above Average
28-31	Poor - Marginal
27 or less	Very Poor - Failure

20. REVIEW EXAMINATION ON THE PELVIS AND PERINEUM

INTRODUCTORY NOTE. The following 50 multiple-choice questions are based on the instructional objectives for the pelvis and perineum listed in Chapter Six of this Study Guide & Review Manual. You should be able to answer these questions in 35-40 minutes. Before beginning, tear out an answer sheet from the back of the book and read the directions on how to use it. The key to the correct responses is on page 354.

FIVE-CHOICE COMPLETION QUESTIONS

DIRECTIONS: Each of the following questions or incomplete statements is followed by five suggested answers or completions. SELECT THE ONE BEST ANSWER in each case and then blacken the appropriate space on the answer sheet.

1. WHICH JOINT OF THE PELVIS IS SYNOVIAL?
   A. Sacroiliac
   B. Lumbosacral
   C. Pubic symphysis
   D. Sacrococcygeal
   E. None of the above

2. THE BROAD LIGAMENT CONTAINS ALL THE FOLLOWING STRUCTURES EXCEPT THE:
   A. Ovary
   B. Uterine tube
   C. Uterine artery
   D. Ovarian ligament
   E. Part of the ureter

3. WHICH OF THE FOLLOWING NERVES SUPPLY(IES) THE FEMALE EXTERNAL GENITALIA?
   A. Pudendal
   B. Ilioinguinal
   C. Uterovaginal plexus
   D. All of the above
   E. None of the above

4. THE TRUE CONJUGATE OF THE PELVIS IS USUALLY ESTIMATED FROM WHICH OF THE FOLLOWING?
   A. Obstetrical conjugate
   B. Transverse conjugate
   C. Oblique diameter
   D. Diagonal conjugate
   E. Pelvic outlet

5. WHICH OF THE FOLLOWING ARTERIES SUPPLY(IES) THE FEMALE URETER?
   A. Uterine
   B. Vesical
   C. Common iliac
   D. Internal iliac
   E. All of the above

6. WHICH OF THE FOLLOWING IS NOT IN THE DEEP PERINEAL POUCH?
   A. Deep transverse perineal muscle
   B. Bulbospongiosus muscle
   C. Bulbourethral gland
   D. Sphincter urethrae
   E. Perineal nerve

SELECT THE ONE BEST ANSWER

7. THE RECTUM IS DRAINED BY WHICH OF THE FOLLOWING GROUPS OF LYMPH NODES?
   A. Sacral
   B. Internal iliac
   C. Common iliac
   D. Inferior mesenteric
   E. All of the above

8. WHICH SURFACE(S) OF THE FEMALE URINARY BLADDER IS(ARE) COVERED BY PERITONEUM?
   A. Anterior
   B. Superior
   C. Posterior
   D. Lateral
   E. All of the above

9. THE CREMASTERIC FASCIA OF THE SPERMATIC CORD IS A CONTINUATION OF THE _____ OF THE ANTEROLATERAL ABDOMINAL WALL.
   A. Transversus abdominus muscle
   B. Internal oblique muscle
   C. Transversalis fascia
   D. Peritoneum
   E. None of the above

10. WHICH PELVIC MEASUREMENT IS THE SHORTEST?
    A. True conjugate
    B. Oblique diameter
    C. Diagonal conjugate
    D. Transverse diameter
    E. Obstetrical conjugate

11. WHICH STATEMENT ABOUT THE EXTERNAL ANAL SPHINCTER IS FALSE?
    A. Is under voluntary control
    B. Consists of striated muscle
    C. Is the axis of the birth canal
    D. Lies between the iliac fossae
    E. Subcutaneous part inserts in the central tendon of the perineum

12. WHICH STATEMENT REGARDING THE GREATER PELVIC CAVITY IS FALSE?
    A. Located above the level of the arcuate lines
    B. Is part of the abdominal cavity
    C. Is the axis of the birth canal
    D. Lies between the iliac fossae
    E. None of the above

13. WHICH STATEMENT ABOUT THE OVARY IS FALSE?
    A. The ovarian ligament contains the ovarian vessels and nerves
    B. Innervated by the abdominal aortic plexus of nerves
    C. The left ovarian vein drains into the renal vein
    D. Supplied by the ovarian artery
    E. Supplied by the uterine artery

14. WHICH MUSCLE IS NOT ATTACHED TO THE PERINEAL BODY?
    A. Superficial transverse perineal
    B. Deep transverse perineal
    C. Ischiocavernosus
    D. Bulbospongiosus
    E. Levator ani

15. WHICH STATEMENT CONCERNING THE VENOUS DRAINAGE OF THE ANAL CANAL IS FALSE?
    A. Venous channels draining the upper and lower parts communicate
    B. Internal hemorrhoids involve the inferior rectal vessels
    C. Inferior rectal veins drain below the pectinate line
    D. Upper half is drained only by the superior rectal veins
    E. Internal are more common than external hemorrhoids

SELECT THE ONE BEST ANSWER

16. WHICH STATEMENT ABOUT THE OBTURATOR INTERNUS MUSCLE IS FALSE?
    A. Its tendon leaves the pelvis via the obturator canal
    B. Inserts on the greater trochanter of the femur
    C. Originates from the obturator membrane
    D. Forms the lateral wall of the pelvis
    E. None of the above

17. LYMPH FROM THE UTERUS DRAINS INTO WHICH GROUP(S) OF LYMPH NODES?
    A. Inguinal
    B. Lumbar
    C. Sacral
    D. Iliac
    E. All of the above

18. TUMORS OF THE TESTIS TEND TO METASTASIZE TO THE _____ LYMPHATIC NODES.
    A. Lumbar
    B. Sacral
    C. Common iliac
    D. External iliac
    E. Internal iliac

19. WHICH STATEMENT CONCERNING THE VAGINA IS FALSE?
    A. Its axis is directed downwards and forwards
    B. The cervix projects into its posterior wall
    C. The posterior wall is longer than the anterior
    D. Its posterior fornix is deeper than the anterior
    E. The lower half is located below the pelvic diaphragm

MULTI-COMPLETION QUESTIONS

DIRECTIONS: In each of the following questions or incomplete statements, ONE OR MORE of the completions given is correct. On the answer sheet blacken the space under A if 1, 2, and 3 are correct; B if 1 and 3 are correct; C if 2 and 4 are correct; D if only 4 is correct; and E if all are correct.

20. THE LEVATOR ANI MUSCLE:
    1. Consists of several parts
    2. Has free anteromedial borders
    3. Is innervated by the sacral plexus
    4. Separates the pelvis from the ischiorectal fossa

21. THE FEMALE URINARY BLADDER IS SUPPLIED BY WHICH OF THE FOLLOWING ARTERIES?
    1. Vaginal
    2. Uterine
    3. Superior vesical
    4. Inferior vesical

22. THE ANAL CANAL:
    1. Is 2-4 cm long
    2. Begins just above the pelvic diaphragm
    3. Is surrounded by the levator ani muscle
    4. Has a voluntary internal sphincter

23. THE ISCHIORECTAL FOSSA IS:
    1. Located behind an imaginary line joining the ischial tuberosities
    2. Frequently a site of infection
    3. Is of clinical importance
    4. A space containing fat

A	B	C	D	E
1,2,3	1,3	2,4	only 4	all correct

24. THE COCCYGEUS MUSCLE:
    1. Arises from the ischial spine
    2. Is innervated by only sacral nerves
    3. Inserts into both the sacrum and coccyx
    4. Is attached to the sacrospinous ligament

25. CONCERNING THE MALE URINARY BLADDER:
    1. The neck is relatively immobile
    2. Its posterior surface is related to the rectum
    3. Separated from the pubic symphysis by a definite space
    4. Its anterior surface is covered by peritoneum

26. WHICH JOINT(S) IS(ARE) LOCATED IN THE PELVIS?
    1. Pubic symphysis
    2. Sacrococcygeal
    3. Lumbosacral
    4. Sacroiliac

27. WHICH GROUP(S) OF LYMPH NODES DRAIN(S) THE ANAL CANAL?
    1. External iliac
    2. Internal iliac
    3. Inferior mesenteric
    4. Superficial inguinal

28. THE PIRIFORMIS MUSCLE:
    1. Leaves the pelvis via the lesser sciatic foramen
    2. Inserts on the lesser trochanter of the femur
    3. Originates from the obturator membrane
    4. Is a lateral rotator of the thigh

29. THE PELVIC DIAPHRAGM CONSISTS OF WHICH MUSCLE(S)?
    1. Piriformis
    2. Levator ani
    3. Obturator internus
    4. Coccygeus

30. THE OVARY IS:
    1. Attached to the uterine tube by the ovarian ligament
    2. Enclosed in the broad ligament
    3. Partly covered by peritoneum
    4. Located in the ovarian fossa

31. THE ISCHIORECTAL FOSSA HAS A:
    1. Medial wall formed by the rectum
    2. Floor formed by the skin over the anal triangle
    3. Lateral wall formed by the ischium
    4. Posterior wall contributed by the gluteus maximus muscle

32. THE UTERUS IS SUPPORTED BY WHICH LIGAMENT(S)?
    1. Transverse cervical
    2. Sacrocervical
    3. Pubocervical
    4. Broad

33. WHICH STATEMENT(S) IS(ARE) TRUE FOR THE PELVIC DIAPHRAGM?
    1. Resists intra-abdominal pressure
    2. Lies deep to the ischiorectal fossa
    3. Formed in part by the puborectus muscle
    4. Lies anterior and superficial to the urogenital diaphragm

34. WHICH OF THE FOLLOWING ARTERIES SUPPLY(IES) THE UTERUS?
    1. Ovarian
    2. Umbilical
    3. Uterine
    4. Internal pudendal

A	B	C	D	E
1,2,3,	1,3	2,4	only 4	all correct

35. THE RECTUM:
    1. Is 20-35 cm long
    2. Has several flexures
    3. Is narrowest at the ampulla
    4. Begins about the level of the 3rd sacral vertebra

36. THE UTERINE TUBE IS:
    1. Supplied by the uterine artery
    2. Supplied by the ovarian artery
    3. Drained by the lumbar lymph nodes
    4. Innervated by the hypogastric plexus of nerves

37. THE BONY PELVIS IS FORMED BY THE:
    1. Ilium
    2. Pubis
    3. Sacrum
    4. Ischium

38. PARASYMPATHETIC FIBERS ARE CONVEYED TO THE PELVIC VISCERA BY WHICH SPLANCHNIC NERVE(S)?
    1. Sacral
    2. Greater
    3. Lumbar
    4. Pelvic

39. THE DEEP TRANSVERSE PERINEAL MUSCLE IN THE MALE:
    1. Originates from the ischium
    2. Inserts into the perineal body
    3. Is innervated by the perineal nerve
    4. Assists in fixing the central tendon of the perineum

40. THE VAGINA IS:
    1. Supplied by the vaginal artery
    2. Supplied by the uterine artery
    3. Innervated by the pelvic plexus of nerves
    4. Partly drained by the superficial inguinal lymph nodes

41. THE TRUE PELVIC CAVITY:
    1. Is funnel-shaped
    2. Forms part of the birth canal
    3. Is longer at the back than in front
    4. Lower aperture lies at the plane of the terminal lines

42. WHICH STATEMENT(S) ABOUT THE PUDENDAL NERVE IS(ARE) CORRECT?
    1. Arises from L2, 3 and 4
    2. Lies in the deep perineal space
    3. Has anterior scrotal or labial branches
    4. Enters the perineum through the lesser sciatic foramen

43. THE MUSCLE(S) MAKING UP THE UROGENITAL DIAPHRAGM IS(ARE) THE:
    1. Bulbospongiosus
    2. Deep transverse perineal
    3. Ischiocavernosus
    4. Sphincter urethrae

44. WHICH OF THE FOLLOWING CONTRIBUTE(S) TO FORMATION OF THE PERINEUM?
    1. Sacrospinous ligament
    2. Tip of the coccyx
    3. Symphysis pubis
    4. Arcuate pubic ligament

45. THE PERINEAL BODY IS:
    1. Located anterior to the anus
    2. A fibromuscular mass
    3. A midline structure
    4. Vulnerable during parturition

FIVE-CHOICE ASSOCIATION QUESTIONS

DIRECTIONS: Each group of questions below consists of a numbered list of descriptive words or phrases accompanied by a diagram with certain parts indicated by letters, or by a list of lettered headings. For each numbered word or phrase, SELECT THE LETTERED PART OR HEADING that matches it correctly. Then on the appropriate line of the answer sheet, blacken the space under the letter of that part or heading. Sometimes more than one numbered word or phrase may be correctly matched to the same lettered part or heading.

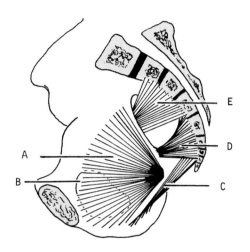

46. ____ Obturator internus muscle
47. ____ Sacrospinous ligament
48. ____ Transmits the obturator nerve
49. ____ Gives rise to the falciform process
50. ____ Piriformis muscle

Interpretation of Your Score

Number of Correct Responses	Level of Performance
43-50	Excellent - Exceptional
38-42	Superior - Very Superior
32-37	Average - Above Average
28-31	Poor - Marginal
27 or less	Very Poor - Failure

21. REVIEW EXAMINATION ON THE BACK

INTRODUCTORY NOTE. The following 25 multiple-choice questions are based on the instructional objectives for the back listed in Chapter Seven of this Study Guide & Review Manual. You should be able to answer these questions in 17-20 minutes. Before beginning, tear out an answer sheet from the back of the book and read the directions on how to use it. The key to the correct responses is on page 354.

FIVE-CHOICE COMPLETION QUESTIONS

DIRECTIONS: Each of the following questions or incomplete statements is followed by five suggested answers or completions. SELECT THE ONE BEST ANSWER in each case and then blacken the appropriate space on the answer sheet.

1.

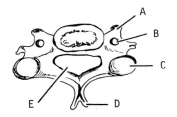

The anterior tubercle of this vertebra is indicated by ____.

2. WHICH OF THE FOLLOWING IS NOT TRUE FOR THE LUMBAR VERTEBRAE?
   A. Bodies heavy
   B. Five in number
   C. Free rotation
   D. Sagittal articulating plane
   E. Transverse processes project upward

3. THE FILUM TERMINALE IS A CONTINUATION OF THE:
   A. Denticulate ligament
   B. Coccygeal ligament
   C. Arachnoid
   D. Dura mater
   E. Pia mater

4. WHICH OF THE FOLLOWING BELONGS TO THE ERECTOR SPINAE GROUP OF MUSCLES?
   A. Sternocleidomastoid
   B. Longissimus thoracis
   C. Levator scapulae
   D. Rhomboid major
   E. Trapezius

SELECT THE ONE BEST ANSWER

5. A STRUCTURE NOT SUPPORTING THE DENS OF THE AXIS IS THE:
   A. Alar ligament
   B. Membrana tectoria
   C. Transverse ligament
   D. Apical ligament
   E. Ligamentum flavum

6. KYPHOSIS OF THE VERTEBRAL COLUMN IS A(AN):
   A. Accentuated primary curve
   B. Accentuated secondary curvature
   C. Deviation at the lumbosacral angle
   D. Lateral deviation
   E. None of the above

7. WHICH ARTERY DOES NOT SUPPLY STRUCTURES OF THE BACK?
   A. External iliac
   B. Vertebral
   C. Lumbar
   D. Internal iliac
   E. Lateral sacral

8. THE FIRST CERVICAL VERTEBRA HAS NO:
   A. Superior articular facet
   B. Transverse foramen
   C. Posterior tubercles
   D. Transverse processes
   E. Body

9. A DISTINCTIVE FEATURE OF THE SECOND CERVICAL VERTEBRA IS THE:
   A. Absence of a spinous process
   B. Presence of cylindrical pedicles
   C. Absence of a foramen transversarium
   D. Presence of an odontoid process
   E. None of the above

10. THE INTERNAL VERTEBRAL VENOUS PLEXUSES ARE FOUND IN THE _____.
    A. Subarachnoid space
    B. Subdural space
    C. Epidural space
    D. Dural sac
    E. None of the above

MULTI-COMPLETION QUESTIONS

DIRECTIONS: In each of the following questions or incomplete statements, ONE OR MORE of the completions given is correct. On the answer sheet blacken the space under A if 1, 2, and 3 are correct; B if 1 and 3 are correct; C if 2 and 4 are correct; D if only 4 is correct; and E if all are correct.

11. WHICH STATEMENT(S) IS(ARE) CORRECT FOR THE SPINAL NERVES?
    1. The cervical roots are the thickest
    2. Length of the roots increase from above downwards
    3. Cervical spinal ganglia lie within the intervertebral foramina
    4. Emerge below their sites of origin in the thoracic region

12. THE SPINAL DURA MATER:
    1. Attaches to the walls of the vertebral canal
    2. Is the outermost protective layer
    3. Adheres to the arachnoid
    4. Ends as a blind sac

A	B	C	D	E
1,2,3	1,3	2,4	only 4	all correct

13. THE SUBOCCIPITAL NERVE:
    1. Is derived from the dorsal ramus of C2
    2. Emerges above the posterior arch of the axis
    3. Supplies the levator scapulae muscle
    4. Lies inferior to the vertebral artery in the suboccipital triangle

14. THE SPLENIUS MUSCLE:
    1. Runs obliquely across the back of the neck
    2. Is considered one of the erector spinae group
    3. Rotates the head
    4. Consists of three parts

15. CONCERNING THE LUMBROSACRAL JOINT:
    1. Sharpest change in direction of the vertebral column
    2. Articular processes at this junction allow anterior displacement
    3. Separated by a disc thicker in front than behind
    4. Upper surface of the sacrum forms a 22-33° angle with the horizontal plane

16. WHICH STATEMENT(S) IS(ARE) CORRECT FOR THE ATLANTOAXIAL JOINTS?
    1. Their major movement is rotation
    2. Are synovial
    3. Supported by the accessory atlantoaxial ligament
    4. Are four in number

17. WHICH STATEMENT(S) IS(ARE) CORRECT FOR THE ERECTOR SPINAE GROUP OF MUSCLES?
    1. Lie in the "gutters" of the vertebral column
    2. Extend from the sacrum to the skull
    3. Form a longitudinal group
    4. Supplied by ventral rami of spinal nerves

18. WHICH STATEMENT(S) IS(ARE) CORRECT FOR THE ADULT SPINAL CORD?
    1. Averages 45 cm in length
    2. Has a cervical enlargement
    3. Is flattened and cylindrical
    4. Ends at the disc between L1 and 2

19. WHICH STATEMENT(S) IS(ARE) CORRECT FOR THE LUMBAR VERTEBRAE?
    1. Have short, narrow spinous processes
    2. Distinguished by their thin transverse processes
    3. Superior vertebral notches are deep
    4. Bodies are wider transversely than anteroposteriorly

20. WHICH STATEMENT(S) IS(ARE) CORRECT FOR THE BLOOD VESSELS SUPPLYING THE SPINAL CORD?
    1. Consist of four longitudinal arteries
    2. Do not form an adequate anastomosis
    3. Located in the subdural space
    4. Receive branches from segmental arteries

21. THE ERECTOR SPINAE MUSCLES FUNCTION AS_____OF THE VERTEBRAL COLUMN.
    1. Lateral flexors   3. Extensors
    2. Rotators   4. Flexors

FIVE-CHOICE ASSOCIATION QUESTIONS

DIRECTIONS: Each group of questions below consists of a numbered list of descriptive words or phrases accompanied by a diagram with certain parts indicated by letters, or by a list of lettered headings. For each numbered word or phrase, SELECT THE LETTERED PART OR HEADING that matches it correctly. Then on the appropriate line of the answer sheet, blacken the space under the letter of that part or heading. Sometimes more than one numbered word or phrase may be correctly matched to the same lettered part or heading.

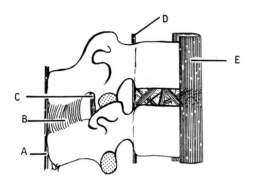

22. ____ A ligament continuing superiorly with the ligamentum nuchae

23. ____ Ligament within the epidural space

24. ____ Ligamentum flavum

25. ____ Ligament checking dorsiflexion of the vertebral column

Interpretation of Your Score

Number of Correct Responses	Level of Performance
23-25	Excellent - Exceptional
20-22	Superior - Very Superior
17-19	Average - Above Average
14-16	Poor - Marginal
13 or less	Very Poor - Failure

KEY TO CORRECT RESPONSES

REVIEW EXAMINATIONS ON THE SYSTEMS

8. THE ARTICULAR, MUSCULAR, AND SKELETAL SYSTEMS

1. C	9. A	17. A	25. C	33. C	41. A	49. A
2. C	10. B	18. E	26. E	34. E	42. C	50. B
3. D	11. D	19. E	27. A	35. E	43. A	
4. D	12. D	20. B	28. A	36. E	44. B	
5. D	13. D	21. E	29. A	37. B	45. D	
6. A	14. D	22. B	30. A	38. E	46. E	
7. B	15. B	23. B	31. B	39. C	47. E	
8. B	16. C	24. A	32. D	40. E	48. C	

9. THE CARDIOVASCULAR SYSTEM

1. C	9. E	17. C	25. E	33. B	41. A	49. B
2. C	10. A	18. E	26. C	34. E	42. E	50. E
3. E	11. E	19. D	27. E	35. C	43. A	
4. C	12. C	20. D	28. E	36. B	44. D	
5. C	13. E	21. D	29. C	37. C	45. B	
6. C	14. D	22. A	30. E	38. E	46. C	
7. E	15. D	23. C	31. B	39. A	47. E	
8. B	16. C	24. E	32. E	40. E	48. A	

10. THE DIGESTIVE SYSTEM

1. C	9. E	17. E	25. E	33. B	41. A	49. E
2. E	10. D	18. C	26. A	34. A	42. A	50. D
3. B	11. A	19. A	27. A	35. A	43. B	
4. A	12. E	20. E	28. C	36. B	44. C	
5. B	13. E	21. D	29. C	37. C	45. A	
6. A	14. D	22. B	30. A	38. E	46. D	
7. C	15. E	23. C	31. E	39. A	47. B	
8. E	16. A	24. A	32. A	40. A	48. C	

11. THE REPRODUCTIVE SYSTEM

1. E	6. C	11. C	16. A	21. E	26. E	31. B
2. D	7. A	12. B	17. A	22. A	27. B	32. C
3. E	8. B	13. B	18. E	23. A	28. A	33. E
4. B	9. A	14. C	19. E	24. C	29. E	34. D
5. A	10. D	15. B	20. A	25. C	30. B	35. A

12. THE RESPIRATORY SYSTEM

1. D	6. E	11. B	16. E	21. D	26. B	31. B
2. B	7. E	12. A	17. A	22. B	27. D	32. D
3. B	8. E	13. E	18. E	23. C	28. E	33. A
4. C	9. E	14. A	19. A	24. C	29. E	34. E
5. C	10. D	15. A	20. D	25. D	30. E	35. C

13. THE URINARY SYSTEM

1. A	5. C	9. B	13. D	17. A	21. B	25. C
2. D	6. E	10. A	14. C	18. A	22. D	
3. B	7. B	11. E	15. E	19. E	23. E	
4. D	8. D	12. E	16. E	20. E	24. A	

14. THE NERVOUS SYSTEM

1. C	6. A	11. E	16. B	21. D	26. D	31. E
2. A	7. C	12. D	17. A	22. B	27. C	32. A
3. B	8. D	13. C	18. B	23. D	28. E	33. B
4. E	9. B	14. A	19. C	24. A	29. C	34. D
5. E	10. E	15. C	20. B	25. A	30. B	35. C

REVIEW EXAMINATIONS ON THE REGIONS

15. THE UPPER LIMB

1. C	9. D	17. E	25. E	33. B	41. B	49. E
2. A	10. E	18. E	26. A	34. B	42. E	50. B
3. E	11. D	19. E	27. B	35. C	43. D	
4. A	12. D	20. E	28. A	36. A	44. A	
5. C	13. B	21. A	29. B	37. C	45. C	
6. D	14. B	22. A	30. D	38. C	46. D	
7. D	15. E	23. B	31. C	39. A	47. A	
8. B	16. B	24. A	32. A	40. C	48. C	

16. THE LOWER LIMB

1. B	9. C	17. B	25. C	33. A	41. B	49. A
2. D	10. B	18. C	26. B	34. A	42. E	50. A
3. C	11. B	19. D	27. C	35. A	43. B	
4. E	12. E	20. C	28. B	36. B	44. D	
5. D	13. D	21. A	29. B	37. E	45. B	
6. D	14. B	22. B	30. C	38. A	46. A	
7. E	15. A	23. E	31. C	39. E	47. C	
8. D	16. C	24. B	32. B	40. A	48. B	

17. THE HEAD AND NECK

1. E	9. E	17. B	25. D	33. B	41. E	49. A
2. C	10. C	18. E	26. A	34. A	42. A	50. A
3. B	11. A	19. B	27. C	35. C	43. B	
4. C	12. D	20. D	28. B	36. A	44. E	
5. B	13. C	21. E	29. A	37. B	45. D	
6. C	14. D	22. A	30. E	38. C	46. B	
7. C	15. A	23. C	31. A	39. C	47. C	
8. B	16. D	24. E	32. C	40. D	48. E	

18. THE THORAX

1. D	9. E	17. E	25. E	33. C	41. A	49. D
2. D	10. A	18. B	26. C	34. B	42. B	50. E
3. C	11. A	19. E	27. C	35. A	43. C	
4. C	12. C	20. C	28. A	36. D	44. E	
5. B	13. D	21. E	29. B	37. B	45. E	
6. A	14. E	22. B	30. B	38. D	46. A	
7. A	15. B	23. E	31. C	39. E	47. C	
8. E	16. E	24. B	32. A	40. E	48. D	

19. THE ABDOMEN

1. E	9. D	17. A	25. C	33. A	41. E	49. D
2. E	10. B	18. B	26. A	34. E	42. A	50. E
3. E	11. E	19. E	27. C	35. A	43. A	
4. C	12. B	20. A	28. C	36. E	44. E	
5. C	13. E	21. B	29. C	37. A	45. A	
6. D	14. C	22. B	30. A	38. C	46. A	
7. E	15. C	23. B	31. E	39. E	47. B	
8. C	16. C	24. E	32. B	40. E	48. C	

20. THE PELVIS AND PERINEUM

1. A	9. B	17. E	25. A	33. D	41. A	49. C
2. A	10. E	18. A	26. E	34. B	42. D	50. E
3. D	11. E	19. B	27. C	35. C	43. C	
4. D	12. C	20. E	28. D	36. E	44. C	
5. E	13. A	21. E	29. C	37. E	45. E	
6. B	14. C	22. A	30. D	38. D	46. A	
7. E	15. B	23. E	31. C	39. E	47. D	
8. B	16. A	24. E	32. E	40. E	48. B	

21. THE BACK

1. A	5. E	9. D	13. D	17. A	21. B	25. E
2. C	6. A	10. C	14. A	18. E	22. A	
3. E	7. A	11. C	15. B	19. C	23. D	
4. B	8. E	12. C	16. A	20. D	24. C	

ANSWER SHEET

DIRECTIONS: Indicate your answer by blackening between the guidelines as shown in the sample.

Sample
A==== B==== C==== D==== E====

1 A=== B=== C=== D=== E===
2 A=== B=== C=== D=== E===
3 A=== B=== C=== D=== E===
4 A=== B=== C=== D=== E===
5 A=== B=== C=== D=== E===
6 A=== B=== C=== D=== E===
7 A=== B=== C=== D=== E===
8 A=== B=== C=== D=== E===
9 A=== B=== C=== D=== E===
10 A=== B=== C=== D=== E===
11 A=== B=== C=== D=== E===
12 A=== B=== C=== D=== E===
13 A=== B=== C=== D=== E===
14 A=== B=== C=== D=== E===
15 A=== B=== C=== D=== E===
16 A=== B=== C=== D=== E===
17 A=== B=== C=== D=== E===
18 A=== B=== C=== D=== E===
19 A=== B=== C=== D=== E===
20 A=== B=== C=== D=== E===
21 A=== B=== C=== D=== E===
22 A=== B=== C=== D=== E===
23 A=== B=== C=== D=== E===
24 A=== B=== C=== D=== E===
25 A=== B=== C=== D=== E===
26 A=== B=== C=== D=== E===
27 A=== B=== C=== D=== E===
28 A=== B=== C=== D=== E===
29 A=== B=== C=== D=== E===
30 A=== B=== C=== D=== E===
31 A=== B=== C=== D=== E===
32 A=== B=== C=== D=== E===
33 A=== B=== C=== D=== E===
34 A=== B=== C=== D=== E===
35 A=== B=== C=== D=== E===
36 A=== B=== C=== D=== E===
37 A=== B=== C=== D=== E===
38 A=== B=== C=== D=== E===
39 A=== B=== C=== D=== E===
40 A=== B=== C=== D=== E===
41 A=== B=== C=== D=== E===
42 A=== B=== C=== D=== E===
43 A=== B=== C=== D=== E===
44 A=== B=== C=== D=== E===
45 A=== B=== C=== D=== E===
46 A=== B=== C=== D=== E===
47 A=== B=== C=== D=== E===
48 A=== B=== C=== D=== E===
49 A=== B=== C=== D=== E===
50 A=== B=== C=== D=== E===

Cut here

ANSWER SHEET

DIRECTIONS: Indicate your answer by blackening between the guidelines as shown in the sample.

Sample
A==== B==== C==== D==== E====

1 A=== B=== C=== D=== E===   2 A=== B=== C=== D=== E===   3 A=== B=== C=== D=== E===
4 A=== B=== C=== D=== E===   5 A=== B=== C=== D=== E===   6 A=== B=== C=== D=== E===
7 A=== B=== C=== D=== E===   8 A=== B=== C=== D=== E===   9 A=== B=== C=== D=== E===
10 A=== B=== C=== D=== E===   11 A=== B=== C=== D=== E===   12 A=== B=== C=== D=== E===
13 A=== B=== C=== D=== E===   14 A=== B=== C=== D=== E===   15 A=== B=== C=== D=== E===
16 A=== B=== C=== D=== E===   17 A=== B=== C=== D=== E===   18 A=== B=== C=== D=== E===
19 A=== B=== C=== D=== E===   20 A=== B=== C=== D=== E===   21 A=== B=== C=== D=== E===
22 A=== B=== C=== D=== E===   23 A=== B=== C=== D=== E===   24 A=== B=== C=== D=== E===
25 A=== B=== C=== D=== E===   26 A=== B=== C=== D=== E===   27 A=== B=== C=== D=== E===
28 A=== B=== C=== D=== E===   29 A=== B=== C=== D=== E===   30 A=== B=== C=== D=== E===
31 A=== B=== C=== D=== E===   32 A=== B=== C=== D=== E===   33 A=== B=== C=== D=== E===
34 A=== B=== C=== D=== E===   35 A=== B=== C=== D=== E===   36 A=== B=== C=== D=== E===
37 A=== B=== C=== D=== E===   38 A=== B=== C=== D=== E===   39 A=== B=== C=== D=== E===
40 A=== B=== C=== D=== E===   41 A=== B=== C=== D=== E===   42 A=== B=== C=== D=== E===
43 A=== B=== C=== D=== E===   44 A=== B=== C=== D=== E===   45 A=== B=== C=== D=== E===
46 A=== B=== C=== D=== E===   47 A=== B=== C=== D=== E===   48 A=== B=== C=== D=== E===
49 A=== B=== C=== D=== E===   50 A=== B=== C=== D=== E===

Cut here

ANSWER SHEET

DIRECTIONS: Indicate your answer by blackening between the guidelines as shown in the sample.

Sample
A==== B==== C==== D==== E====

1 A=== B=== C=== D=== E===   2 A=== B=== C=== D=== E===   3 A=== B=== C=== D=== E===
4 A=== B=== C=== D=== E===   5 A=== B=== C=== D=== E===   6 A=== B=== C=== D=== E===
7 A=== B=== C=== D=== E===   8 A=== B=== C=== D=== E===   9 A=== B=== C=== D=== E===
10 A=== B=== C=== D=== E===   11 A=== B=== C=== D=== E===   12 A=== B=== C=== D=== E===
13 A=== B=== C=== D=== E===   14 A=== B=== C=== D=== E===   15 A=== B=== C=== D=== E===
16 A=== B=== C=== D=== E===   17 A=== B=== C=== D=== E===   18 A=== B=== C=== D=== E===
19 A=== B=== C=== D=== E===   20 A=== B=== C=== D=== E===   21 A=== B=== C=== D=== E===
22 A=== B=== C=== D=== E===   23 A=== B=== C=== D=== E===   24 A=== B=== C=== D=== E===
25 A=== B=== C=== D=== E===   26 A=== B=== C=== D=== E===   27 A=== B=== C=== D=== E===
28 A=== B=== C=== D=== E===   29 A=== B=== C=== D=== E===   30 A=== B=== C=== D=== E===
31 A=== B=== C=== D=== E===   32 A=== B=== C=== D=== E===   33 A=== B=== C=== D=== E===
34 A=== B=== C=== D=== E===   35 A=== B=== C=== D=== E===   36 A=== B=== C=== D=== E===
37 A=== B=== C=== D=== E===   38 A=== B=== C=== D=== E===   39 A=== B=== C=== D=== E===
40 A=== B=== C=== D=== E===   41 A=== B=== C=== D=== E===   42 A=== B=== C=== D=== E===
43 A=== B=== C=== D=== E===   44 A=== B=== C=== D=== E===   45 A=== B=== C=== D=== E===
46 A=== B=== C=== D=== E===   47 A=== B=== C=== D=== E===   48 A=== B=== C=== D=== E===
49 A=== B=== C=== D=== E===   50 A=== B=== C=== D=== E===

Cut here

ANSWER SHEET

DIRECTIONS: Indicate your answer by blackening between the guidelines as shown in the sample.

Sample
A==== B==== C==== D==== E====

1 A=== B=== C=== D=== E===
2 A=== B=== C=== D=== E===
3 A=== B=== C=== D=== E===
4 A=== B=== C=== D=== E===
5 A=== B=== C=== D=== E===
6 A=== B=== C=== D=== E===
7 A=== B=== C=== D=== E===
8 A=== B=== C=== D=== E===
9 A=== B=== C=== D=== E===
10 A=== B=== C=== D=== E===
11 A=== B=== C=== D=== E===
12 A=== B=== C=== D=== E===
13 A=== B=== C=== D=== E===
14 A=== B=== C=== D=== E===
15 A=== B=== C=== D=== E===
16 A=== B=== C=== D=== E===
17 A=== B=== C=== D=== E===
18 A=== B=== C=== D=== E===
19 A=== B=== C=== D=== E===
20 A=== B=== C=== D=== E===
21 A=== B=== C=== D=== E===
22 A=== B=== C=== D=== E===
23 A=== B=== C=== D=== E===
24 A=== B=== C=== D=== E===
25 A=== B=== C=== D=== E===
26 A=== B=== C=== D=== E===
27 A=== B=== C=== D=== E===
28 A=== B=== C=== D=== E===
29 A=== B=== C=== D=== E===
30 A=== B=== C=== D=== E===
31 A=== B=== C=== D=== E===
32 A=== B=== C=== D=== E===
33 A=== B=== C=== D=== E===
34 A=== B=== C=== D=== E===
35 A=== B=== C=== D=== E===
36 A=== B=== C=== D=== E===
37 A=== B=== C=== D=== E===
38 A=== B=== C=== D=== E===
39 A=== B=== C=== D=== E===
40 A=== B=== C=== D=== E===
41 A=== B=== C=== D=== E===
42 A=== B=== C=== D=== E===
43 A=== B=== C=== D=== E===
44 A=== B=== C=== D=== E===
45 A=== B=== C=== D=== E===
46 A=== B=== C=== D=== E===
47 A=== B=== C=== D=== E===
48 A=== B=== C=== D=== E===
49 A=== B=== C=== D=== E===
50 A=== B=== C=== D=== E===

Cut here

ANSWER SHEET

DIRECTIONS: Indicate your answer by blackening between the guidelines as shown in the sample.

Sample
A==== B==== C==== D==== E====

1 A=== B=== C=== D=== E===   2 A=== B=== C=== D=== E===   3 A=== B=== C=== D=== E===
4 A=== B=== C=== D=== E===   5 A=== B=== C=== D=== E===   6 A=== B=== C=== D=== E===
7 A=== B=== C=== D=== E===   8 A=== B=== C=== D=== E===   9 A=== B=== C=== D=== E===
10 A=== B=== C=== D=== E=== 11 A=== B=== C=== D=== E=== 12 A=== B=== C=== D=== E===
13 A=== B=== C=== D=== E=== 14 A=== B=== C=== D=== E=== 15 A=== B=== C=== D=== E===
16 A=== B=== C=== D=== E=== 17 A=== B=== C=== D=== E=== 18 A=== B=== C=== D=== E===
19 A=== B=== C=== D=== E=== 20 A=== B=== C=== D=== E=== 21 A=== B=== C=== D=== E===
22 A=== B=== C=== D=== E=== 23 A=== B=== C=== D=== E=== 24 A=== B=== C=== D=== E===
25 A=== B=== C=== D=== E=== 26 A=== B=== C=== D=== E=== 27 A=== B=== C=== D=== E===
28 A=== B=== C=== D=== E=== 29 A=== B=== C=== D=== E=== 30 A=== B=== C=== D=== E===
31 A=== B=== C=== D=== E=== 32 A=== B=== C=== D=== E=== 33 A=== B=== C=== D=== E===
34 A=== B=== C=== D=== E=== 35 A=== B=== C=== D=== E=== 36 A=== B=== C=== D=== E===
37 A=== B=== C=== D=== E=== 38 A=== B=== C=== D=== E=== 39 A=== B=== C=== D=== E===
40 A=== B=== C=== D=== E=== 41 A=== B=== C=== D=== E=== 42 A=== B=== C=== D=== E===
43 A=== B=== C=== D=== E=== 44 A=== B=== C=== D=== E=== 45 A=== B=== C=== D=== E===
46 A=== B=== C=== D=== E=== 47 A=== B=== C=== D=== E=== 48 A=== B=== C=== D=== E===
49 A=== B=== C=== D=== E=== 50 A=== B=== C=== D=== E===

Cut here

ANSWER SHEET

DIRECTIONS: Indicate your answer by blackening between the guidelines as shown in the sample.

Sample
A==== B==== C==== D==== E====

1 A=== B=== C=== D=== E===
2 A=== B=== C=== D=== E===
3 A=== B=== C=== D=== E===
4 A=== B=== C=== D=== E===
5 A=== B=== C=== D=== E===
6 A=== B=== C=== D=== E===
7 A=== B=== C=== D=== E===
8 A=== B=== C=== D=== E===
9 A=== B=== C=== D=== E===
10 A=== B=== C=== D=== E===
11 A=== B=== C=== D=== E===
12 A=== B=== C=== D=== E===
13 A=== B=== C=== D=== E===
14 A=== B=== C=== D=== E===
15 A=== B=== C=== D=== E===
16 A=== B=== C=== D=== E===
17 A=== B=== C=== D=== E===
18 A=== B=== C=== D=== E===
19 A=== B=== C=== D=== E===
20 A=== B=== C=== D=== E===
21 A=== B=== C=== D=== E===
22 A=== B=== C=== D=== E===
23 A=== B=== C=== D=== E===
24 A=== B=== C=== D=== E===
25 A=== B=== C=== D=== E===
26 A=== B=== C=== D=== E===
27 A=== B=== C=== D=== E===
28 A=== B=== C=== D=== E===
29 A=== B=== C=== D=== E===
30 A=== B=== C=== D=== E===
31 A=== B=== C=== D=== E===
32 A=== B=== C=== D=== E===
33 A=== B=== C=== D=== E===
34 A=== B=== C=== D=== E===
35 A=== B=== C=== D=== E===
36 A=== B=== C=== D=== E===
37 A=== B=== C=== D=== E===
38 A=== B=== C=== D=== E===
39 A=== B=== C=== D=== E===
40 A=== B=== C=== D=== E===
41 A=== B=== C=== D=== E===
42 A=== B=== C=== D=== E===
43 A=== B=== C=== D=== E===
44 A=== B=== C=== D=== E===
45 A=== B=== C=== D=== E===
46 A=== B=== C=== D=== E===
47 A=== B=== C=== D=== E===
48 A=== B=== C=== D=== E===
49 A=== B=== C=== D=== E===
50 A=== B=== C=== D=== E===

Cut here

ANSWER SHEET

DIRECTIONS: Indicate your answer by blackening between the guidelines as shown in the sample.

Sample
A==== B==== C==== D==== E====

1 A=== B=== C=== D=== E===   2 A=== B=== C=== D=== E===   3 A=== B=== C=== D=== E===
4 A=== B=== C=== D=== E===   5 A=== B=== C=== D=== E===   6 A=== B=== C=== D=== E===
7 A=== B=== C=== D=== E===   8 A=== B=== C=== D=== E===   9 A=== B=== C=== D=== E===
10 A=== B=== C=== D=== E===   11 A=== B=== C=== D=== E===   12 A=== B=== C=== D=== E===
13 A=== B=== C=== D=== E===   14 A=== B=== C=== D=== E===   15 A=== B=== C=== D=== E===
16 A=== B=== C=== D=== E===   17 A=== B=== C=== D=== E===   18 A=== B=== C=== D=== E===
19 A=== B=== C=== D=== E===   20 A=== B=== C=== D=== E===   21 A=== B=== C=== D=== E===
22 A=== B=== C=== D=== E===   23 A=== B=== C=== D=== E===   24 A=== B=== C=== D=== E===
25 A=== B=== C=== D=== E===   26 A=== B=== C=== D=== E===   27 A=== B=== C=== D=== E===
28 A=== B=== C=== D=== E===   29 A=== B=== C=== D=== E===   30 A=== B=== C=== D=== E===
31 A=== B=== C=== D=== E===   32 A=== B=== C=== D=== E===   33 A=== B=== C=== D=== E===
34 A=== B=== C=== D=== E===   35 A=== B=== C=== D=== E===   36 A=== B=== C=== D=== E===
37 A=== B=== C=== D=== E===   38 A=== B=== C=== D=== E===   39 A=== B=== C=== D=== E===
40 A=== B=== C=== D=== E===   41 A=== B=== C=== D=== E===   42 A=== B=== C=== D=== E===
43 A=== B=== C=== D=== E===   44 A=== B=== C=== D=== E===   45 A=== B=== C=== D=== E===
46 A=== B=== C=== D=== E===   47 A=== B=== C=== D=== E===   48 A=== B=== C=== D=== E===
49 A=== B=== C=== D=== E===   50 A=== B=== C=== D=== E===

Cut here

ANSWER SHEET

DIRECTIONS: Indicate your answer by blackening between the guidelines as shown in the sample.

Sample
A==== B==== C==== D==== E====

1 A=== B=== C=== D=== E===   2 A=== B=== C=== D=== E===   3 A=== B=== C=== D=== E===
4 A=== B=== C=== D=== E===   5 A=== B=== C=== D=== E===   6 A=== B=== C=== D=== E===
7 A=== B=== C=== D=== E===   8 A=== B=== C=== D=== E===   9 A=== B=== C=== D=== E===
10 A=== B=== C=== D=== E=== 11 A=== B=== C=== D=== E=== 12 A=== B=== C=== D=== E===
13 A=== B=== C=== D=== E=== 14 A=== B=== C=== D=== E=== 15 A=== B=== C=== D=== E===
16 A=== B=== C=== D=== E=== 17 A=== B=== C=== D=== E=== 18 A=== B=== C=== D=== E===
19 A=== B=== C=== D=== E=== 20 A=== B=== C=== D=== E=== 21 A=== B=== C=== D=== E===
22 A=== B=== C=== D=== E=== 23 A=== B=== C=== D=== E=== 24 A=== B=== C=== D=== E===
25 A=== B=== C=== D=== E=== 26 A=== B=== C=== D=== E=== 27 A=== B=== C=== D=== E===
28 A=== B=== C=== D=== E=== 29 A=== B=== C=== D=== E=== 30 A=== B=== C=== D=== E===
31 A=== B=== C=== D=== E=== 32 A=== B=== C=== D=== E=== 33 A=== B=== C=== D=== E===
34 A=== B=== C=== D=== E=== 35 A=== B=== C=== D=== E=== 36 A=== B=== C=== D=== E===
37 A=== B=== C=== D=== E=== 38 A=== B=== C=== D=== E=== 39 A=== B=== C=== D=== E===
40 A=== B=== C=== D=== E=== 41 A=== B=== C=== D=== E=== 42 A=== B=== C=== D=== E===
43 A=== B=== C=== D=== E=== 44 A=== B=== C=== D=== E=== 45 A=== B=== C=== D=== E===
46 A=== B=== C=== D=== E=== 47 A=== B=== C=== D=== E=== 48 A=== B=== C=== D=== E===
49 A=== B=== C=== D=== E=== 50 A=== B=== C=== D=== E===

Cut here

ANSWER SHEET

DIRECTIONS: Indicate your answer by blackening between the guidelines as shown in the sample.

Sample
A B C D E

1 A B C D E	2 A B C D E	3 A B C D E
4 A B C D E	5 A B C D E	6 A B C D E
7 A B C D E	8 A B C D E	9 A B C D E
10 A B C D E	11 A B C D E	12 A B C D E
13 A B C D E	14 A B C D E	15 A B C D E
16 A B C D E	17 A B C D E	18 A B C D E
19 A B C D E	20 A B C D E	21 A B C D E
22 A B C D E	23 A B C D E	24 A B C D E
25 A B C D E	26 A B C D E	27 A B C D E
28 A B C D E	29 A B C D E	30 A B C D E
31 A B C D E	32 A B C D E	33 A B C D E
34 A B C D E	35 A B C D E	36 A B C D E
37 A B C D E	38 A B C D E	39 A B C D E
40 A B C D E	41 A B C D E	42 A B C D E
43 A B C D E	44 A B C D E	45 A B C D E
46 A B C D E	47 A B C D E	48 A B C D E
49 A B C D E	50 A B C D E	

Cut here

ANSWER SHEET

DIRECTIONS: Indicate your answer by blackening between the guidelines as shown in the sample.

Sample
A==== B==== C==== D==== E====

1 A=== B=== C=== D=== E=== 2 A=== B=== C=== D=== E=== 3 A=== B=== C=== D=== E===
4 A=== B=== C=== D=== E=== 5 A=== B=== C=== D=== E=== 6 A=== B=== C=== D=== E===
7 A=== B=== C=== D=== E=== 8 A=== B=== C=== D=== E=== 9 A=== B=== C=== D=== E===
10 A=== B=== C=== D=== E=== 11 A=== B=== C=== D=== E=== 12 A=== B=== C=== D=== E===
13 A=== B=== C=== D=== E=== 14 A=== B=== C=== D=== E=== 15 A=== B=== C=== D=== E===
16 A=== B=== C=== D=== E=== 17 A=== B=== C=== D=== E=== 18 A=== B=== C=== D=== E===
19 A=== B=== C=== D=== E=== 20 A=== B=== C=== D=== E=== 21 A=== B=== C=== D=== E===
22 A=== B=== C=== D=== E=== 23 A=== B=== C=== D=== E=== 24 A=== B=== C=== D=== E===
25 A=== B=== C=== D=== E=== 26 A=== B=== C=== D=== E=== 27 A=== B=== C=== D=== E===
28 A=== B=== C=== D=== E=== 29 A=== B=== C=== D=== E=== 30 A=== B=== C=== D=== E===
31 A=== B=== C=== D=== E=== 32 A=== B=== C=== D=== E=== 33 A=== B=== C=== D=== E===
34 A=== B=== C=== D=== E=== 35 A=== B=== C=== D=== E=== 36 A=== B=== C=== D=== E===
37 A=== B=== C=== D=== E=== 38 A=== B=== C=== D=== E=== 39 A=== B=== C=== D=== E===
40 A=== B=== C=== D=== E=== 41 A=== B=== C=== D=== E=== 42 A=== B=== C=== D=== E===
43 A=== B=== C=== D=== E=== 44 A=== B=== C=== D=== E=== 45 A=== B=== C=== D=== E===
46 A=== B=== C=== D=== E=== 47 A=== B=== C=== D=== E=== 48 A=== B=== C=== D=== E===
49 A=== B=== C=== D=== E=== 50 A=== B=== C=== D=== E===

Cut here

ANSWER SHEET

DIRECTIONS: Indicate your answer by blackening between the guidelines as shown in the sample.

Sample
A==== B==== C==== D==== E====

1 A=== B=== C=== D=== E===
2 A=== B=== C=== D=== E===
3 A=== B=== C=== D=== E===
4 A=== B=== C=== D=== E===
5 A=== B=== C=== D=== E===
6 A=== B=== C=== D=== E===
7 A=== B=== C=== D=== E===
8 A=== B=== C=== D=== E===
9 A=== B=== C=== D=== E===
10 A=== B=== C=== D=== E===
11 A=== B=== C=== D=== E===
12 A=== B=== C=== D=== E===
13 A=== B=== C=== D=== E===
14 A=== B=== C=== D=== E===
15 A=== B=== C=== D=== E===
16 A=== B=== C=== D=== E===
17 A=== B=== C=== D=== E===
18 A=== B=== C=== D=== E===
19 A=== B=== C=== D=== E===
20 A=== B=== C=== D=== E===
21 A=== B=== C=== D=== E===
22 A=== B=== C=== D=== E===
23 A=== B=== C=== D=== E===
24 A=== B=== C=== D=== E===
25 A=== B=== C=== D=== E===
26 A=== B=== C=== D=== E===
27 A=== B=== C=== D=== E===
28 A=== B=== C=== D=== E===
29 A=== B=== C=== D=== E===
30 A=== B=== C=== D=== E===
31 A=== B=== C=== D=== E===
32 A=== B=== C=== D=== E===
33 A=== B=== C=== D=== E===
34 A=== B=== C=== D=== E===
35 A=== B=== C=== D=== E===
36 A=== B=== C=== D=== E===
37 A=== B=== C=== D=== E===
38 A=== B=== C=== D=== E===
39 A=== B=== C=== D=== E===
40 A=== B=== C=== D=== E===
41 A=== B=== C=== D=== E===
42 A=== B=== C=== D=== E===
43 A=== B=== C=== D=== E===
44 A=== B=== C=== D=== E===
45 A=== B=== C=== D=== E===
46 A=== B=== C=== D=== E===
47 A=== B=== C=== D=== E===
48 A=== B=== C=== D=== E===
49 A=== B=== C=== D=== E===
50 A=== B=== C=== D=== E===

Cut here

ANSWER SHEET

DIRECTIONS: Indicate your answer by blackening between the guidelines as shown in the sample.

Sample
A==== B==== C==== D==== E====

1 A=== B=== C=== D=== E===	2 A=== B=== C=== D=== E===	3 A=== B=== C=== D=== E===
4 A=== B=== C=== D=== E===	5 A=== B=== C=== D=== E===	6 A=== B=== C=== D=== E===
7 A=== B=== C=== D=== E===	8 A=== B=== C=== D=== E===	9 A=== B=== C=== D=== E===
10 A=== B=== C=== D=== E===	11 A=== B=== C=== D=== E===	12 A=== B=== C=== D=== E===
13 A=== B=== C=== D=== E===	14 A=== B=== C=== D=== E===	15 A=== B=== C=== D=== E===
16 A=== B=== C=== D=== E===	17 A=== B=== C=== D=== E===	18 A=== B=== C=== D=== E===
19 A=== B=== C=== D=== E===	20 A=== B=== C=== D=== E===	21 A=== B=== C=== D=== E===
22 A=== B=== C=== D=== E===	23 A=== B=== C=== D=== E===	24 A=== B=== C=== D=== E===
25 A=== B=== C=== D=== E===	26 A=== B=== C=== D=== E===	27 A=== B=== C=== D=== E===
28 A=== B=== C=== D=== E===	29 A=== B=== C=== D=== E===	30 A=== B=== C=== D=== E===
31 A=== B=== C=== D=== E===	32 A=== B=== C=== D=== E===	33 A=== B=== C=== D=== E===
34 A=== B=== C=== D=== E===	35 A=== B=== C=== D=== E===	36 A=== B=== C=== D=== E===
37 A=== B=== C=== D=== E===	38 A=== B=== C=== D=== E===	39 A=== B=== C=== D=== E===
40 A=== B=== C=== D=== E===	41 A=== B=== C=== D=== E===	42 A=== B=== C=== D=== E===
43 A=== B=== C=== D=== E===	44 A=== B=== C=== D=== E===	45 A=== B=== C=== D=== E===
46 A=== B=== C=== D=== E===	47 A=== B=== C=== D=== E===	48 A=== B=== C=== D=== E===
49 A=== B=== C=== D=== E===	50 A=== B=== C=== D=== E===	

Cut here

ANSWER SHEET

DIRECTIONS: Indicate your answer by blackening between the guidelines as shown in the sample.

Sample
A ==== B ==== C ==== D ==== E ====

Cut here

1 A === B === C === D === E ===   2 A === B === C === D === E ===   3 A === B === C === D === E ===
4 A === B === C === D === E ===   5 A === B === C === D === E ===   6 A === B === C === D === E ===
7 A === B === C === D === E ===   8 A === B === C === D === E ===   9 A === B === C === D === E ===
10 A === B === C === D === E ===   11 A === B === C === D === E ===   12 A === B === C === D === E ===
13 A === B === C === D === E ===   14 A === B === C === D === E ===   15 A === B === C === D === E ===
16 A === B === C === D === E ===   17 A === B === C === D === E ===   18 A === B === C === D === E ===
19 A === B === C === D === E ===   20 A === B === C === D === E ===   21 A === B === C === D === E ===
22 A === B === C === D === E ===   23 A === B === C === D === E ===   24 A === B === C === D === E ===
25 A === B === C === D === E ===   26 A === B === C === D === E ===   27 A === B === C === D === E ===
28 A === B === C === D === E ===   29 A === B === C === D === E ===   30 A === B === C === D === E ===
31 A === B === C === D === E ===   32 A === B === C === D === E ===   33 A === B === C === D === E ===
34 A === B === C === D === E ===   35 A === B === C === D === E ===   36 A === B === C === D === E ===
37 A === B === C === D === E ===   38 A === B === C === D === E ===   39 A === B === C === D === E ===
40 A === B === C === D === E ===   41 A === B === C === D === E ===   42 A === B === C === D === E ===
43 A === B === C === D === E ===   44 A === B === C === D === E ===   45 A === B === C === D === E ===
46 A === B === C === D === E ===   47 A === B === C === D === E ===   48 A === B === C === D === E ===
49 A === B === C === D === E ===   50 A === B === C === D === E ===

ANSWER SHEET

DIRECTIONS: Indicate your answer by blackening between the guidelines as shown in the sample.

Sample
A==== B==== C==== D==== E====

1 A=== B=== C=== D=== E===	2 A=== B=== C=== D=== E===	3 A=== B=== C=== D=== E===
4 A=== B=== C=== D=== E===	5 A=== B=== C=== D=== E===	6 A=== B=== C=== D=== E===
7 A=== B=== C=== D=== E===	8 A=== B=== C=== D=== E===	9 A=== B=== C=== D=== E===
10 A=== B=== C=== D=== E===	11 A=== B=== C=== D=== E===	12 A=== B=== C=== D=== E===
13 A=== B=== C=== D=== E===	14 A=== B=== C=== D=== E===	15 A=== B=== C=== D=== E===
16 A=== B=== C=== D=== E===	17 A=== B=== C=== D=== E===	18 A=== B=== C=== D=== E===
19 A=== B=== C=== D=== E===	20 A=== B=== C=== D=== E===	21 A=== B=== C=== D=== E===
22 A=== B=== C=== D=== E===	23 A=== B=== C=== D=== E===	24 A=== B=== C=== D=== E===
25 A=== B=== C=== D=== E===	26 A=== B=== C=== D=== E===	27 A=== B=== C=== D=== E===
28 A=== B=== C=== D=== E===	29 A=== B=== C=== D=== E===	30 A=== B=== C=== D=== E===
31 A=== B=== C=== D=== E===	32 A=== B=== C=== D=== E===	33 A=== B=== C=== D=== E===
34 A=== B=== C=== D=== E===	35 A=== B=== C=== D=== E===	36 A=== B=== C=== D=== E===
37 A=== B=== C=== D=== E===	38 A=== B=== C=== D=== E===	39 A=== B=== C=== D=== E===
40 A=== B=== C=== D=== E===	41 A=== B=== C=== D=== E===	42 A=== B=== C=== D=== E===
43 A=== B=== C=== D=== E===	44 A=== B=== C=== D=== E===	45 A=== B=== C=== D=== E===
46 A=== B=== C=== D=== E===	47 A=== B=== C=== D=== E===	48 A=== B=== C=== D=== E===
49 A=== B=== C=== D=== E===	50 A=== B=== C=== D=== E===	

Cut here

ANSWER SHEET

DIRECTIONS: Indicate your answer by blackening between the guidelines as shown in the sample.

Sample
A=== B=== C=== D=== E===

Cut here

1 A=B=C=D=E=	2 A=B=C=D=E=	3 A=B=C=D=E=
4 A=B=C=D=E=	5 A=B=C=D=E=	6 A=B=C=D=E=
7 A=B=C=D=E=	8 A=B=C=D=E=	9 A=B=C=D=E=
10 A=B=C=D=E=	11 A=B=C=D=E=	12 A=B=C=D=E=
13 A=B=C=D=E=	14 A=B=C=D=E=	15 A=B=C=D=E=
16 A=B=C=D=E=	17 A=B=C=D=E=	18 A=B=C=D=E=
19 A=B=C=D=E=	20 A=B=C=D=E=	21 A=B=C=D=E=
22 A=B=C=D=E=	23 A=B=C=D=E=	24 A=B=C=D=E=
25 A=B=C=D=E=	26 A=B=C=D=E=	27 A=B=C=D=E=
28 A=B=C=D=E=	29 A=B=C=D=E=	30 A=B=C=D=E=
31 A=B=C=D=E=	32 A=B=C=D=E=	33 A=B=C=D=E=
34 A=B=C=D=E=	35 A=B=C=D=E=	36 A=B=C=D=E=
37 A=B=C=D=E=	38 A=B=C=D=E=	39 A=B=C=D=E=
40 A=B=C=D=E=	41 A=B=C=D=E=	42 A=B=C=D=E=
43 A=B=C=D=E=	44 A=B=C=D=E=	45 A=B=C=D=E=
46 A=B=C=D=E=	47 A=B=C=D=E=	48 A=B=C=D=E=
49 A=B=C=D=E=	50 A=B=C=D=E=	

ANSWER SHEET

DIRECTIONS: Indicate your answer by blackening between the guidelines as shown in the sample.

Sample
A==== B==== C==== D==== E====

1 A=== B=== C=== D=== E===   2 A=== B=== C=== D=== E===   3 A=== B=== C=== D=== E===
4 A=== B=== C=== D=== E===   5 A=== B=== C=== D=== E===   6 A=== B=== C=== D=== E===
7 A=== B=== C=== D=== E===   8 A=== B=== C=== D=== E===   9 A=== B=== C=== D=== E===
10 A=== B=== C=== D=== E===   11 A=== B=== C=== D=== E===   12 A=== B=== C=== D=== E===
13 A=== B=== C=== D=== E===   14 A=== B=== C=== D=== E===   15 A=== B=== C=== D=== E===
16 A=== B=== C=== D=== E===   17 A=== B=== C=== D=== E===   18 A=== B=== C=== D=== E===
19 A=== B=== C=== D=== E===   20 A=== B=== C=== D=== E===   21 A=== B=== C=== D=== E===
22 A=== B=== C=== D=== E===   23 A=== B=== C=== D=== E===   24 A=== B=== C=== D=== E===
25 A=== B=== C=== D=== E===   26 A=== B=== C=== D=== E===   27 A=== B=== C=== D=== E===
28 A=== B=== C=== D=== E===   29 A=== B=== C=== D=== E===   30 A=== B=== C=== D=== E===
31 A=== B=== C=== D=== E===   32 A=== B=== C=== D=== E===   33 A=== B=== C=== D=== E===
34 A=== B=== C=== D=== E===   35 A=== B=== C=== D=== E===   36 A=== B=== C=== D=== E===
37 A=== B=== C=== D=== E===   38 A=== B=== C=== D=== E===   39 A=== B=== C=== D=== E===
40 A=== B=== C=== D=== E===   41 A=== B=== C=== D=== E===   42 A=== B=== C=== D=== E===
43 A=== B=== C=== D=== E===   44 A=== B=== C=== D=== E===   45 A=== B=== C=== D=== E===
46 A=== B=== C=== D=== E===   47 A=== B=== C=== D=== E===   48 A=== B=== C=== D=== E===
49 A=== B=== C=== D=== E===   50 A=== B=== C=== D=== E===

Cut here

ANSWER SHEET

DIRECTIONS: Indicate your answer by blackening between the guidelines as shown in the sample.

Sample
A==== B==== C==== D==== E====

1 A=== B=== C=== D=== E===	2 A=== B=== C=== D=== E===	3 A=== B=== C=== D=== E===
4 A=== B=== C=== D=== E===	5 A=== B=== C=== D=== E===	6 A=== B=== C=== D=== E===
7 A=== B=== C=== D=== E===	8 A=== B=== C=== D=== E===	9 A=== B=== C=== D=== E===
10 A=== B=== C=== D=== E===	11 A=== B=== C=== D=== E===	12 A=== B=== C=== D=== E===
13 A=== B=== C=== D=== E===	14 A=== B=== C=== D=== E===	15 A=== B=== C=== D=== E===
16 A=== B=== C=== D=== E===	17 A=== B=== C=== D=== E===	18 A=== B=== C=== D=== E===
19 A=== B=== C=== D=== E===	20 A=== B=== C=== D=== E===	21 A=== B=== C=== D=== E===
22 A=== B=== C=== D=== E===	23 A=== B=== C=== D=== E===	24 A=== B=== C=== D=== E===
25 A=== B=== C=== D=== E===	26 A=== B=== C=== D=== E===	27 A=== B=== C=== D=== E===
28 A=== B=== C=== D=== E===	29 A=== B=== C=== D=== E===	30 A=== B=== C=== D=== E===
31 A=== B=== C=== D=== E===	32 A=== B=== C=== D=== E===	33 A=== B=== C=== D=== E===
34 A=== B=== C=== D=== E===	35 A=== B=== C=== D=== E===	36 A=== B=== C=== D=== E===
37 A=== B=== C=== D=== E===	38 A=== B=== C=== D=== E===	39 A=== B=== C=== D=== E===
40 A=== B=== C=== D=== E===	41 A=== B=== C=== D=== E===	42 A=== B=== C=== D=== E===
43 A=== B=== C=== D=== E===	44 A=== B=== C=== D=== E===	45 A=== B=== C=== D=== E===
46 A=== B=== C=== D=== E===	47 A=== B=== C=== D=== E===	48 A=== B=== C=== D=== E===
49 A=== B=== C=== D=== E===	50 A=== B=== C=== D=== E===	

Cut here

ANSWER SHEET

DIRECTIONS: Indicate your answer by blackening between the guidelines as shown in the sample.

Sample
A==== B==== C==== D==== E====

1 A=== B=== C=== D=== E===	2 A=== B=== C=== D=== E===	3 A=== B=== C=== D=== E===
4 A=== B=== C=== D=== E===	5 A=== B=== C=== D=== E===	6 A=== B=== C=== D=== E===
7 A=== B=== C=== D=== E===	8 A=== B=== C=== D=== E===	9 A=== B=== C=== D=== E===
10 A=== B=== C=== D=== E===	11 A=== B=== C=== D=== E===	12 A=== B=== C=== D=== E===
13 A=== B=== C=== D=== E===	14 A=== B=== C=== D=== E===	15 A=== B=== C=== D=== E===
16 A=== B=== C=== D=== E===	17 A=== B=== C=== D=== E===	18 A=== B=== C=== D=== E===
19 A=== B=== C=== D=== E===	20 A=== B=== C=== D=== E===	21 A=== B=== C=== D=== E===
22 A=== B=== C=== D=== E===	23 A=== B=== C=== D=== E===	24 A=== B=== C=== D=== E===
25 A=== B=== C=== D=== E===	26 A=== B=== C=== D=== E===	27 A=== B=== C=== D=== E===
28 A=== B=== C=== D=== E===	29 A=== B=== C=== D=== E===	30 A=== B=== C=== D=== E===
31 A=== B=== C=== D=== E===	32 A=== B=== C=== D=== E===	33 A=== B=== C=== D=== E===
34 A=== B=== C=== D=== E===	35 A=== B=== C=== D=== E===	36 A=== B=== C=== D=== E===
37 A=== B=== C=== D=== E===	38 A=== B=== C=== D=== E===	39 A=== B=== C=== D=== E===
40 A=== B=== C=== D=== E===	41 A=== B=== C=== D=== E===	42 A=== B=== C=== D=== E===
43 A=== B=== C=== D=== E===	44 A=== B=== C=== D=== E===	45 A=== B=== C=== D=== E===
46 A=== B=== C=== D=== E===	47 A=== B=== C=== D=== E===	48 A=== B=== C=== D=== E===
49 A=== B=== C=== D=== E===	50 A=== B=== C=== D=== E===	

Cut here

ANSWER SHEET

DIRECTIONS: Indicate your answer by blackening between the guidelines as shown in the sample.

Sample
A==== B==== C==== D==== E====

1 A=== B=== C=== D=== E===   2 A=== B=== C=== D=== E===   3 A=== B=== C=== D=== E===
4 A=== B=== C=== D=== E===   5 A=== B=== C=== D=== E===   6 A=== B=== C=== D=== E===
7 A=== B=== C=== D=== E===   8 A=== B=== C=== D=== E===   9 A=== B=== C=== D=== E===
10 A=== B=== C=== D=== E===   11 A=== B=== C=== D=== E===   12 A=== B=== C=== D=== E===
13 A=== B=== C=== D=== E===   14 A=== B=== C=== D=== E===   15 A=== B=== C=== D=== E===
16 A=== B=== C=== D=== E===   17 A=== B=== C=== D=== E===   18 A=== B=== C=== D=== E===
19 A=== B=== C=== D=== E===   20 A=== B=== C=== D=== E===   21 A=== B=== C=== D=== E===
22 A=== B=== C=== D=== E===   23 A=== B=== C=== D=== E===   24 A=== B=== C=== D=== E===
25 A=== B=== C=== D=== E===   26 A=== B=== C=== D=== E===   27 A=== B=== C=== D=== E===
28 A=== B=== C=== D=== E===   29 A=== B=== C=== D=== E===   30 A=== B=== C=== D=== E===
31 A=== B=== C=== D=== E===   32 A=== B=== C=== D=== E===   33 A=== B=== C=== D=== E===
34 A=== B=== C=== D=== E===   35 A=== B=== C=== D=== E===   36 A=== B=== C=== D=== E===
37 A=== B=== C=== D=== E===   38 A=== B=== C=== D=== E===   39 A=== B=== C=== D=== E===
40 A=== B=== C=== D=== E===   41 A=== B=== C=== D=== E===   42 A=== B=== C=== D=== E===
43 A=== B=== C=== D=== E===   44 A=== B=== C=== D=== E===   45 A=== B=== C=== D=== E===
46 A=== B=== C=== D=== E===   47 A=== B=== C=== D=== E===   48 A=== B=== C=== D=== E===
49 A=== B=== C=== D=== E===   50 A=== B=== C=== D=== E===

Cut here

ANSWER SHEET

DIRECTIONS: Indicate your answer by blackening between the guidelines as shown in the sample.

Sample
A==== B==== C==== D==== E====

1 A=== B=== C=== D=== E===   2 A=== B=== C=== D=== E===   3 A=== B=== C=== D=== E===
4 A=== B=== C=== D=== E===   5 A=== B=== C=== D=== E===   6 A=== B=== C=== D=== E===
7 A=== B=== C=== D=== E===   8 A=== B=== C=== D=== E===   9 A=== B=== C=== D=== E===
10 A=== B=== C=== D=== E===   11 A=== B=== C=== D=== E===   12 A=== B=== C=== D=== E===
13 A=== B=== C=== D=== E===   14 A=== B=== C=== D=== E===   15 A=== B=== C=== D=== E===
16 A=== B=== C=== D=== E===   17 A=== B=== C=== D=== E===   18 A=== B=== C=== D=== E===
19 A=== B=== C=== D=== E===   20 A=== B=== C=== D=== E===   21 A=== B=== C=== D=== E===
22 A=== B=== C=== D=== E===   23 A=== B=== C=== D=== E===   24 A=== B=== C=== D=== E===
25 A=== B=== C=== D=== E===   26 A=== B=== C=== D=== E===   27 A=== B=== C=== D=== E===
28 A=== B=== C=== D=== E===   29 A=== B=== C=== D=== E===   30 A=== B=== C=== D=== E===
31 A=== B=== C=== D=== E===   32 A=== B=== C=== D=== E===   33 A=== B=== C=== D=== E===
34 A=== B=== C=== D=== E===   35 A=== B=== C=== D=== E===   36 A=== B=== C=== D=== E===
37 A=== B=== C=== D=== E===   38 A=== B=== C=== D=== E===   39 A=== B=== C=== D=== E===
40 A=== B=== C=== D=== E===   41 A=== B=== C=== D=== E===   42 A=== B=== C=== D=== E===
43 A=== B=== C=== D=== E===   44 A=== B=== C=== D=== E===   45 A=== B=== C=== D=== E===
46 A=== B=== C=== D=== E===   47 A=== B=== C=== D=== E===   48 A=== B=== C=== D=== E===
49 A=== B=== C=== D=== E===   50 A=== B=== C=== D=== E===

Cut here

ANSWER SHEET

DIRECTIONS: Indicate your answer by blackening between the guidelines as shown in the sample.

Sample
A==== B==== C==== D==== E====

1 A=== B=== C=== D=== E===
2 A=== B=== C=== D=== E===
3 A=== B=== C=== D=== E===
4 A=== B=== C=== D=== E===
5 A=== B=== C=== D=== E===
6 A=== B=== C=== D=== E===
7 A=== B=== C=== D=== E===
8 A=== B=== C=== D=== E===
9 A=== B=== C=== D=== E===
10 A=== B=== C=== D=== E===
11 A=== B=== C=== D=== E===
12 A=== B=== C=== D=== E===
13 A=== B=== C=== D=== E===
14 A=== B=== C=== D=== E===
15 A=== B=== C=== D=== E===
16 A=== B=== C=== D=== E===
17 A=== B=== C=== D=== E===
18 A=== B=== C=== D=== E===
19 A=== B=== C=== D=== E===
20 A=== B=== C=== D=== E===
21 A=== B=== C=== D=== E===
22 A=== B=== C=== D=== E===
23 A=== B=== C=== D=== E===
24 A=== B=== C=== D=== E===
25 A=== B=== C=== D=== E===
26 A=== B=== C=== D=== E===
27 A=== B=== C=== D=== E===
28 A=== B=== C=== D=== E===
29 A=== B=== C=== D=== E===
30 A=== B=== C=== D=== E===
31 A=== B=== C=== D=== E===
32 A=== B=== C=== D=== E===
33 A=== B=== C=== D=== E===
34 A=== B=== C=== D=== E===
35 A=== B=== C=== D=== E===
36 A=== B=== C=== D=== E===
37 A=== B=== C=== D=== E===
38 A=== B=== C=== D=== E===
39 A=== B=== C=== D=== E===
40 A=== B=== C=== D=== E===
41 A=== B=== C=== D=== E===
42 A=== B=== C=== D=== E===
43 A=== B=== C=== D=== E===
44 A=== B=== C=== D=== E===
45 A=== B=== C=== D=== E===
46 A=== B=== C=== D=== E===
47 A=== B=== C=== D=== E===
48 A=== B=== C=== D=== E===
49 A=== B=== C=== D=== E===
50 A=== B=== C=== D=== E===

Cut here

ANSWER SHEET

DIRECTIONS: Indicate your answer by blackening between the guidelines as shown in the sample.

Sample
A==== B==== C==== D==== E====

1 A=== B=== C=== D=== E===	2 A=== B=== C=== D=== E===	3 A=== B=== C=== D=== E===
4 A=== B=== C=== D=== E===	5 A=== B=== C=== D=== E===	6 A=== B=== C=== D=== E===
7 A=== B=== C=== D=== E===	8 A=== B=== C=== D=== E===	9 A=== B=== C=== D=== E===
10 A=== B=== C=== D=== E===	11 A=== B=== C=== D=== E===	12 A=== B=== C=== D=== E===
13 A=== B=== C=== D=== E===	14 A=== B=== C=== D=== E===	15 A=== B=== C=== D=== E===
16 A=== B=== C=== D=== E===	17 A=== B=== C=== D=== E===	18 A=== B=== C=== D=== E===
19 A=== B=== C=== D=== E===	20 A=== B=== C=== D=== E===	21 A=== B=== C=== D=== E===
22 A=== B=== C=== D=== E===	23 A=== B=== C=== D=== E===	24 A=== B=== C=== D=== E===
25 A=== B=== C=== D=== E===	26 A=== B=== C=== D=== E===	27 A=== B=== C=== D=== E===
28 A=== B=== C=== D=== E===	29 A=== B=== C=== D=== E===	30 A=== B=== C=== D=== E===
31 A=== B=== C=== D=== E===	32 A=== B=== C=== D=== E===	33 A=== B=== C=== D=== E===
34 A=== B=== C=== D=== E===	35 A=== B=== C=== D=== E===	36 A=== B=== C=== D=== E===
37 A=== B=== C=== D=== E===	38 A=== B=== C=== D=== E===	39 A=== B=== C=== D=== E===
40 A=== B=== C=== D=== E===	41 A=== B=== C=== D=== E===	42 A=== B=== C=== D=== E===
43 A=== B=== C=== D=== E===	44 A=== B=== C=== D=== E===	45 A=== B=== C=== D=== E===
46 A=== B=== C=== D=== E===	47 A=== B=== C=== D=== E===	48 A=== B=== C=== D=== E===
49 A=== B=== C=== D=== E===	50 A=== B=== C=== D=== E===	

Cut here

ANSWER SHEET

DIRECTIONS: Indicate your answer by blackening between the guidelines as shown in the sample.

Sample
A==== B==== C==== D==== E====

1 A=== B=== C=== D=== E===	2 A=== B=== C=== D=== E===	3 A=== B=== C=== D=== E===
4 A=== B=== C=== D=== E===	5 A=== B=== C=== D=== E===	6 A=== B=== C=== D=== E===
7 A=== B=== C=== D=== E===	8 A=== B=== C=== D=== E===	9 A=== B=== C=== D=== E===
10 A=== B=== C=== D=== E===	11 A=== B=== C=== D=== E===	12 A=== B=== C=== D=== E===
13 A=== B=== C=== D=== E===	14 A=== B=== C=== D=== E===	15 A=== B=== C=== D=== E===
16 A=== B=== C=== D=== E===	17 A=== B=== C=== D=== E===	18 A=== B=== C=== D=== E===
19 A=== B=== C=== D=== E===	20 A=== B=== C=== D=== E===	21 A=== B=== C=== D=== E===
22 A=== B=== C=== D=== E===	23 A=== B=== C=== D=== E===	24 A=== B=== C=== D=== E===
25 A=== B=== C=== D=== E===	26 A=== B=== C=== D=== E===	27 A=== B=== C=== D=== E===
28 A=== B=== C=== D=== E===	29 A=== B=== C=== D=== E===	30 A=== B=== C=== D=== E===
31 A=== B=== C=== D=== E===	32 A=== B=== C=== D=== E===	33 A=== B=== C=== D=== E===
34 A=== B=== C=== D=== E===	35 A=== B=== C=== D=== E===	36 A=== B=== C=== D=== E===
37 A=== B=== C=== D=== E===	38 A=== B=== C=== D=== E===	39 A=== B=== C=== D=== E===
40 A=== B=== C=== D=== E===	41 A=== B=== C=== D=== E===	42 A=== B=== C=== D=== E===
43 A=== B=== C=== D=== E===	44 A=== B=== C=== D=== E===	45 A=== B=== C=== D=== E===
46 A=== B=== C=== D=== E===	47 A=== B=== C=== D=== E===	48 A=== B=== C=== D=== E===
49 A=== B=== C=== D=== E===	50 A=== B=== C=== D=== E===	

Cut here

ANSWER SHEET

DIRECTIONS: Indicate your answer by blackening between the guidelines as shown in the sample.

Sample
A==== B==== C==== D==== E====

1 A=== B=== C=== D=== E===   2 A=== B=== C=== D=== E===   3 A=== B=== C=== D=== E===
4 A=== B=== C=== D=== E===   5 A=== B=== C=== D=== E===   6 A=== B=== C=== D=== E===
7 A=== B=== C=== D=== E===   8 A=== B=== C=== D=== E===   9 A=== B=== C=== D=== E===
10 A=== B=== C=== D=== E===   11 A=== B=== C=== D=== E===   12 A=== B=== C=== D=== E===
13 A=== B=== C=== D=== E===   14 A=== B=== C=== D=== E===   15 A=== B=== C=== D=== E===
16 A=== B=== C=== D=== E===   17 A=== B=== C=== D=== E===   18 A=== B=== C=== D=== E===
19 A=== B=== C=== D=== E===   20 A=== B=== C=== D=== E===   21 A=== B=== C=== D=== E===
22 A=== B=== C=== D=== E===   23 A=== B=== C=== D=== E===   24 A=== B=== C=== D=== E===
25 A=== B=== C=== D=== E===   26 A=== B=== C=== D=== E===   27 A=== B=== C=== D=== E===
28 A=== B=== C=== D=== E===   29 A=== B=== C=== D=== E===   30 A=== B=== C=== D=== E===
31 A=== B=== C=== D=== E===   32 A=== B=== C=== D=== E===   33 A=== B=== C=== D=== E===
34 A=== B=== C=== D=== E===   35 A=== B=== C=== D=== E===   36 A=== B=== C=== D=== E===
37 A=== B=== C=== D=== E===   38 A=== B=== C=== D=== E===   39 A=== B=== C=== D=== E===
40 A=== B=== C=== D=== E===   41 A=== B=== C=== D=== E===   42 A=== B=== C=== D=== E===
43 A=== B=== C=== D=== E===   44 A=== B=== C=== D=== E===   45 A=== B=== C=== D=== E===
46 A=== B=== C=== D=== E===   47 A=== B=== C=== D=== E===   48 A=== B=== C=== D=== E===
49 A=== B=== C=== D=== E===   50 A=== B=== C=== D=== E===

Cut here

ANSWER SHEET

DIRECTIONS: Indicate your answer by blackening between the guidelines as shown in the sample.

Sample
A B C D E

1 A B C D E
2 A B C D E
3 A B C D E
4 A B C D E
5 A B C D E
6 A B C D E
7 A B C D E
8 A B C D E
9 A B C D E
10 A B C D E
11 A B C D E
12 A B C D E
13 A B C D E
14 A B C D E
15 A B C D E
16 A B C D E
17 A B C D E
18 A B C D E
19 A B C D E
20 A B C D E
21 A B C D E
22 A B C D E
23 A B C D E
24 A B C D E
25 A B C D E
26 A B C D E
27 A B C D E
28 A B C D E
29 A B C D E
30 A B C D E
31 A B C D E
32 A B C D E
33 A B C D E
34 A B C D E
35 A B C D E
36 A B C D E
37 A B C D E
38 A B C D E
39 A B C D E
40 A B C D E
41 A B C D E
42 A B C D E
43 A B C D E
44 A B C D E
45 A B C D E
46 A B C D E
47 A B C D E
48 A B C D E
49 A B C D E
50 A B C D E

Cut here

ANSWER SHEET

DIRECTIONS: Indicate your answer by blackening between the guidelines as shown in the sample.

Sample
A==== B==== C==== D==== E====

1 A B C D E	2 A B C D E	3 A B C D E
4 A B C D E	5 A B C D E	6 A B C D E
7 A B C D E	8 A B C D E	9 A B C D E
10 A B C D E	11 A B C D E	12 A B C D E
13 A B C D E	14 A B C D E	15 A B C D E
16 A B C D E	17 A B C D E	18 A B C D E
19 A B C D E	20 A B C D E	21 A B C D E
22 A B C D E	23 A B C D E	24 A B C D E
25 A B C D E	26 A B C D E	27 A B C D E
28 A B C D E	29 A B C D E	30 A B C D E
31 A B C D E	32 A B C D E	33 A B C D E
34 A B C D E	35 A B C D E	36 A B C D E
37 A B C D E	38 A B C D E	39 A B C D E
40 A B C D E	41 A B C D E	42 A B C D E
43 A B C D E	44 A B C D E	45 A B C D E
46 A B C D E	47 A B C D E	48 A B C D E
49 A B C D E	50 A B C D E	

Cut here

ANSWER SHEET

DIRECTIONS: Indicate your answer by blackening between the guidelines as shown in the sample.

Sample
A==== B==== C==== D==== E====

1 A=== B=== C=== D=== E===   2 A=== B=== C=== D=== E===   3 A=== B=== C=== D=== E===
4 A=== B=== C=== D=== E===   5 A=== B=== C=== D=== E===   6 A=== B=== C=== D=== E===
7 A=== B=== C=== D=== E===   8 A=== B=== C=== D=== E===   9 A=== B=== C=== D=== E===
10 A=== B=== C=== D=== E===   11 A=== B=== C=== D=== E===   12 A=== B=== C=== D=== E===
13 A=== B=== C=== D=== E===   14 A=== B=== C=== D=== E===   15 A=== B=== C=== D=== E===
16 A=== B=== C=== D=== E===   17 A=== B=== C=== D=== E===   18 A=== B=== C=== D=== E===
19 A=== B=== C=== D=== E===   20 A=== B=== C=== D=== E===   21 A=== B=== C=== D=== E===
22 A=== B=== C=== D=== E===   23 A=== B=== C=== D=== E===   24 A=== B=== C=== D=== E===
25 A=== B=== C=== D=== E===   26 A=== B=== C=== D=== E===   27 A=== B=== C=== D=== E===
28 A=== B=== C=== D=== E===   29 A=== B=== C=== D=== E===   30 A=== B=== C=== D=== E===
31 A=== B=== C=== D=== E===   32 A=== B=== C=== D=== E===   33 A=== B=== C=== D=== E===
34 A=== B=== C=== D=== E===   35 A=== B=== C=== D=== E===   36 A=== B=== C=== D=== E===
37 A=== B=== C=== D=== E===   38 A=== B=== C=== D=== E===   39 A=== B=== C=== D=== E===
40 A=== B=== C=== D=== E===   41 A=== B=== C=== D=== E===   42 A=== B=== C=== D=== E===
43 A=== B=== C=== D=== E===   44 A=== B=== C=== D=== E===   45 A=== B=== C=== D=== E===
46 A=== B=== C=== D=== E===   47 A=== B=== C=== D=== E===   48 A=== B=== C=== D=== E===
49 A=== B=== C=== D=== E===   50 A=== B=== C=== D=== E===

Cut here

ANSWER SHEET

DIRECTIONS: Indicate your answer by blackening between the guidelines as shown in the sample.

Sample
A==== B==== C==== D==== E====

1 A=== B=== C=== D=== E===
2 A=== B=== C=== D=== E===
3 A=== B=== C=== D=== E===
4 A=== B=== C=== D=== E===
5 A=== B=== C=== D=== E===
6 A=== B=== C=== D=== E===
7 A=== B=== C=== D=== E===
8 A=== B=== C=== D=== E===
9 A=== B=== C=== D=== E===
10 A=== B=== C=== D=== E===
11 A=== B=== C=== D=== E===
12 A=== B=== C=== D=== E===
13 A=== B=== C=== D=== E===
14 A=== B=== C=== D=== E===
15 A=== B=== C=== D=== E===
16 A=== B=== C=== D=== E===
17 A=== B=== C=== D=== E===
18 A=== B=== C=== D=== E===
19 A=== B=== C=== D=== E===
20 A=== B=== C=== D=== E===
21 A=== B=== C=== D=== E===
22 A=== B=== C=== D=== E===
23 A=== B=== C=== D=== E===
24 A=== B=== C=== D=== E===
25 A=== B=== C=== D=== E===
26 A=== B=== C=== D=== E===
27 A=== B=== C=== D=== E===
28 A=== B=== C=== D=== E===
29 A=== B=== C=== D=== E===
30 A=== B=== C=== D=== E===
31 A=== B=== C=== D=== E===
32 A=== B=== C=== D=== E===
33 A=== B=== C=== D=== E===
34 A=== B=== C=== D=== E===
35 A=== B=== C=== D=== E===
36 A=== B=== C=== D=== E===
37 A=== B=== C=== D=== E===
38 A=== B=== C=== D=== E===
39 A=== B=== C=== D=== E===
40 A=== B=== C=== D=== E===
41 A=== B=== C=== D=== E===
42 A=== B=== C=== D=== E===
43 A=== B=== C=== D=== E===
44 A=== B=== C=== D=== E===
45 A=== B=== C=== D=== E===
46 A=== B=== C=== D=== E===
47 A=== B=== C=== D=== E===
48 A=== B=== C=== D=== E===
49 A=== B=== C=== D=== E===
50 A=== B=== C=== D=== E===

Cut here